5th Edition

Introductory Abstract Algebra

Woo Jeon · Xueqing Chen
Moon-Ok Wang · Ki-Suk Lee · Ki-Bong Nam

KM KYUNG MOON SA

Preface

This book is intended as a basic text for an undergraduate course in abstract algebra. Linear algebra is not a prerequisite. However some knowledge of (1) Sets and Functions; (2) Mathematical Induction; (3) Equivalence Relations; (4) The Binomial Theorem; (5) Matrix Algebra might be helpful. Many examples and simple exercises are included to help students, since the material cannot truly be learned without solving problems. We would like to take the opportunity to express my deepest gratitude to those mathematicians who offered valuable suggestions leading to its improvement. It is a pleasure to acknowledge the invaluable assistance of the Kyung-Moon's staff, particularly Mr. Moon-Kyu Park and Mr. Jong-Wean Kim. As the first edition, this book may contain some errors or shortcomings so please let us know when you find any comments or questions by emailing us to jeonw@uwgb.edu. We will provide pdf files of the text to any instructor who request it for classroom use. Please contact Dr. Woo Jeon (jeonw@uwgb.edu) to get the pdf files for instructor's teaching.

Finally the authors wish to congratulate Professor Wang's 61st birthday by finishing the first edition of the book.

For the Second Edition.

Because of the first author's misuse of the latex file of the book, all the terms of the index are removed in the main contents. Thus authors will provide the pdf file of the book to the all of those who purchased the first edition. The first author is very sorry about the mistake. Thus the authors decide to publish the second edition quickly. We add some new concepts in this edition. As the second edition, still this book may contain some errors or shortcomings. Please let us know if you have any comments or questions.

For the Third Edition.

The authors would like to updated new results, and would like to remove some mistakes and typos of the second edition. The first author thanks to Professor Donald Passman for his reading of some part of the third edition manuscripts. The authors thank you Dr. Corey Bruns for his generous reviewing the manuscripts and excellent comments for improving the manuscripts. The first author thanks to his Math 453 students of UW-Whitewater, Fall, 2011, who got involved in improving the manuscripts. The exercises with $*$ could be used for undergraduate research.

For the Fourth Edition.

The authors improved a lot of contents of the third edition and corrected some mistakes of the previous edition. Authors thanks to Prof. Donald Passman for his time for reviewing some chapters of the book. Dr. Woo Jeon joined as one of the authors. The authors will be very happy to get some opinions or comments from the readers.

For the Fifth Edition.

There were many changes including exercise format. If you have any question or comment, please contact Dr. Woo Jeon (jeonw@uwgb.edu).

Contents

Chapter 1

Preliminaries

Throughout this book, we will use the following usual number sets:

$$
\begin{aligned}
\mathbb{N} &= \text{the set of all natural numbers} = \{1, 2, 3, \cdots\} \\
\mathbb{W} &= \text{the set of all whole numbers} = \{0, 1, 2, \cdots\} \\
\mathbb{Z} &= \text{the set of all integers} = \{\cdots, -1, 0, 1, 2, \cdots\} \\
\mathbb{Q} &= \text{the set of all rational numbers} = \left\{\frac{p}{q} | p, q \in \mathbb{Z}, q \neq 0\right\} \\
\mathbb{R} &= \text{the set of all real numbers} \\
\mathbb{C} &= \text{the set of all complex numbers} = \{a + bi | a, b \in \mathbb{R}, i = \sqrt{-1}\}
\end{aligned}
$$

A definition of real numbers should be studies in a Real Analysis course. We start to look at properties in $\mathbb{Z}$.

1.1 Well Ordering Principle

For given two integers a and b, one of the following holds:

$$a > b, \quad a = b, \quad \text{or} \quad a < b,$$

so we can say that $\mathbb{Z}$ is a set with the order $>$. In fact, any non-empty subset of $\mathbb{Z}$ is ordered. Using this order, we can define the following

For $S = \{1, 2, 4, 7, 8\}$, 1 is called **the minimal element of** S since 1 is less than or equal to any element of S. In general, we have the following important principle regarding minimum elements:

Well Ordering Principle

For every non-empty subset S of non-negative integers, S has a minimum element.

An ordered set S is called **well ordered** if it has a minimal element with respect to the order of S.

Example 1.1.1. *Since 3 is the minimal element of $S = \{3, 6, 8, 100, 9\}$, S is a well ordered set.*

Example 1.1.2. *Show that $S = \{\frac{1}{2^n} | n \in \mathbb{N}^+\}$ is not a well ordered set.*

Solution. If x is a minimal element of S, then $\frac{1}{2}x$ is an element of S such that $x < \frac{1}{2}x$, a contradiction. Thus S is not a well ordered set. □

In general, we can see that a subset of real numbers is generally not a well-ordered-set through Example 1.1.2.

When we divide 7 by 3 in elementary algebra, the quotient is 2 and the remainder is 1 so we can make an equation $7 = 3 \cdot 2 + 1$. This equation can be generalized to the following division algorithm.

Theorem 1.1.1. (Division Algorithm)
For a given integer a and a positive integer b, there are unique integers q and r such that $a = bq + r$ where $0 \leq r < b$.

Proof. Let $S = \{a - bx | a - bx \geq 0, x \in \mathbb{Z}\}$. We have exactly one of $a = b$, $a > b$, and $a < b$. If $a = b$, then we have that $a = b \cdot 1 + 0$. This implies that the theorem holds. Otherwise S contains a positive integer. By the Well Ordering Principle, there is a minimum non-negative integer r of S. This implies that $a = bx + r$. Because of the minimality of r, we have that $0 \leq r < b$. Let us prove the uniqueness of q and r. Assume that there are $a = bq_1 + r_1$ and $a = bq_2 + r_2$ such that $0 \leq r_1, r_2 < b$ and $q_1, q_2 \in \mathbb{Z}$. This implies that $a - a = bq_1 + r_1 - bq_2 - r_2 = 0$. Thus we have that $b(q_1 - q_2) = r_2 - r_1$. If $q_1 - q_2 \neq 0$, then the absolute value of $|b(q_1 - q_2)|$ is strictly greater than $|r_2 - r_1|$. Here we have a contradiction. This implies that $q_1 = q_2$ and then this implies that $r_1 = r_2$. So q and r are unique and this proves the theorem. □

In Theorem 1.1.1, a, b, q, r are called **the dividend, the divisor, the quotient, the remainder** respectively.

For given integers a and b, a divides b if $b = ca$ for some $c \in \mathbb{Z}$ and we denote it as $a|b$. If $a|b$ and $a \neq 0$, then a is called **a divisor (or a factor) of b** and b is called **a multiple of a**. For given integers a and b, d is called **a common divisor of a and b**, if $d|a$ and $d|b$.

Definition 1.1.1. *For given integers a and b, the positive integer d is called* **a common divisor of** a **and** b, *if $d|a$ and $d|b$. Among common divisors, the smallest one is called* **the greatest common divisor (GCD)**, *i.e., for any common divisor d_1 of a and b, $d_1|d$ and d is a common divisor of a and b. The GCD of a and b is denoted by $gcd(a, b)$ or simply (a, b). If $(a, b) = 1$, a and b are called* **relatively prime.**

Example 1.1.3. $1, 3, 6$ *are common divisors of* 12 *and* 18. *The greatest common divisor of* 12 *and* 18 *is* 6 *or we write* $gcd(12, 18) = 6$.

Theorem 1.1.2. *For any non-zero integers a and b, there are integers s and t such that $(a, b) = as + bt$.*

Proof. Let $S = \{as_1 + bt_1 > 0 | s_1, t_1 \in \mathbb{Z}\}$. Then, S is not the empty set. In fact, S contains a positive integer by taking appropriate s_1 and t_1 . So by the Well Ordering Principle, there is the minimum positive integer $d = as_2 + bt_2$ in S for some $s_2, t_2 \in \mathbb{Z}$. We will show that d is $gcd(a, b)$. For a, by the division algorithm, we have that $a = qd + r$ where $0 \leq r < d$. If r is not zero, then $r = a - qd \in S$. This contradicts the minimality of d. This implies that $r = 0$, i.e., $d|a$. Similarly we can prove that $d|b$. This implies that d is a common divisor of a and b. Now, let's say that $d_1|a$ and $d_1|b$. Then $a = a_1d_1$ and $b = b_1d_1$ and these make $d = a_1d_1s_2 + b_1d_1t_2 = d_1(a_1s_2 + b_1t_2)$. This implies that $d_1|d$. Thus $d = gcd(a, b)$, the greatest common divisor of a and b. □

Example 1.1.4. *1. Find* $(12, 14)$

2. Find $s, t \in \mathbb{Z}$ *such that* $12s + 14t = (12, 14)$

Solution.

1. Since $12 = 2(6)$ and $14 = 2(7)$, $(12, 14) = 2$.

2. We can pick $s = 6$ and $t = -5$ so that $6(12) - 5(14) = 2$. □

In general, it is not so easy to find such integers s and t for a linear combination of (a, b) as in Theorem 1.1.2 but they can be obtained through the Division Algorithm (or Euclidean Algorithm). Let's say $a > b$. Then we have the following sequence of divisions:

$$\begin{aligned}
a &= bq_1 + r_1, & 0 \le r_1 < b \\
b &= r_1q_2 + r_2, & 0 \le r_2 < r_1 \\
r_1 &= r_2q_3 + r_3, & 0 \le r_3 < r_2 \\
r_2 &= r_3q_4 + r_4, & 0 \le r_4 < r_3 \\
&\cdots \\
r_k &= r_{k+1}q_{k+2} + r_{k+2}, & 0 \le r_{r+2} < r_{k+1} \\
r_{k+1} &= r_{k+2}q_{k+3} + r_{k+3}, & 0 \le r_{r+3} < r_{k+2} \\
r_{k+2} &= r_{k+3}q_{k+4} & (1.1)
\end{aligned}$$

By (1.1), we have that $r_1 > r_2 > \cdots > r_{k+2} > r_{k+3}$. Note that since r_1 is a finite positive integer, there is $k + 4$ such that $r_{k+4} = 0$. Let us show that $r_{k+3} = (a, b)$. Since $r_{k+3}|r_i$, $r_{k+3}|b$, and $r_{k+3}|a$ where $1 \le i \le k + 2$, r_{k+3} is a common divisor of a and b. If d is a common divisor of a and b, then we know that $d|r_1, \cdots, , d|r_{k+3}$. This implies that $r_{k+3} = (a, b)$ and a linear combination of r_{k+3} can be obtained from solving the equations for each r_i except the last one and substitutions. Let's take a look at an example.

Example 1.1.5. *Find* $(200, 30)$ *and integers* s *and* t *such that* $(200, 30) = 200s + 30t$.

Solution. It is obvious that $(200, 30) = 10$ but it can be seen through the following:

$$\begin{array}{rcl} 200 &=& 30 \cdot 6 + 20 \implies 20 = 200 - 30 \cdot 6 \\ 30 &=& 20 \cdot 1 + 10 \implies 30 = (200 - 30 \cdot 6) \cdot 1 + 10 \\ 20 &=& 10 \cdot 2 \end{array}$$

This implies that $(200, 30) = 10$ and if we solve the equation on the right in the middle row for 10, we have $10 = -(200 - 30 \cdot 6) \cdot 1 + 30 = 200 \cdot (-1) + 30 \cdot 7$. Therefore, $(200, 30)$ is written as a linear combination of 200 and 30 with coefficients -1 and 7 respectively. $\square$

Note that this type of linear combinations of $(200, 30)$ is not unique since we also have $200(-7) + 30(47) = 10$.

A prime is a positive integer greater than 1 such that whose positive divisors are 1 and itself, i.e., a prime has exactly two distinct divisors. The following lemma is an useful tool using prime numbers and it is called the **Euclid's Lemma**.

Lemma 1.1.1. (Euclid's Lemma)

If p *is prime and* $p|ab$, *then* $p|a$ *or* $p|b$ *where* a *and* b *are integers.*

Proof. If $p|a$, then we have proven the lemma. If p is not a divisor of a, then $(a,p)=1$. By Theorem (1.1.2), there are integers s and t such that $as+pt=1$. Then $abs+bpt=b$. Since $p|ab$ and $p|p$, we have that $p|(abs+bpt)$. This implies that $p|b$ and this proves the lemma. $\square$

The next theorem explains the unique factorizations in $\mathbb{Z}$ using prime numbers. It is called the **Fundamental Theorem of Arithmetic**. The proof of the theorem uses Euclid's Lemma and it is omitted.

Theorem 1.1.3. (Fundamental Theorem of Arithmetic)
For any integer $n>1$, a is a prime number or a product of primes. Moreover, if $a=p_1\cdots p_n$ and $a=q_1\cdots q_r$, then $n=r$ and $p_i=q_j$ where $p_1,\cdots,p_n,q_1,\cdots,q_r$ are primes for $1\le i\le n$, $1\le j\le r$.

Note that it is not unique if a factorization contains composite numbers. For example, $24=2\cdot 12=3\cdot 8$.

For given integers a and b, recall that c is a common multiple of a and b if $a|c$ and $b|c$. Then by the Well Ordering Principle, we can define the **least common multiple (LCM) $[a,b]$ of a and b**. For example, $[6,4]=12$.

Exercises 1.1.

1. (a) Find the remainder if 11^{40} is divided by 5

   (b) Find the remainder if 12^{100} is divided by 9.

   (c) Find the remainder if $12^{100}+7^{50}+15^{60}$ is divided by 6.

2. (a) Find $(18,24)$.

(b) Find integers q and r so that $24 = 18 \cdot q + r$ where $0 \le r < 18$.

(c) Find $s, t \in \mathbb{Z}$ such that $18s + 24t = (18, 24)$.

3. (a) Find $(30, 80)$.

   (b) Find integers q and r so that $80 = 30 \cdot q + r$ where $0 \le r < 30$.

   (c) Find $s, t \in \mathbb{Z}$ such that $30s + 80t = (30, 80)$.

4. (a) Find $(55, 8)$.

   (b) Find integers q and r so that $55 = 8q + r$ where $0 \le r < 8$.

   (c) Find $s, t \in \mathbb{Z}$ such that $55s + 8t = (55, 8)$.

5. (a) Find $gcd(2^3 3^2 7, 2^3 3 \cdot 11)$ using the Division Algorithm.

   (b) Find integers s and t such that $2^3 3^2 7s + t 2^3 3 \cdot 11 = gcd(2^3 3^2 7, 2^3 3 \cdot 11)$

6. For $a, b \in \mathbb{Z}$, prove that if $a|b$ and $b|a$, then $a = \pm b$.

7. For $a, b, c \in \mathbb{Z}$, if a and b are relatively prime and a and b are divisors of c, then prove that ab divides c.

8. For $a, b, c, d \in \mathbb{Z}$, prove that $a|(b + c)$ and $a|(b + c + d)$ if a divides b, c, and d.

9. Prove that $p \in \mathbb{N}$ is prime if and only if for any $a \in \mathbb{N}$, a and p are relatively prime or p divides a.

10. Prove that $(a, b) = 1$ if and only if $(a, bc) = 1$ where a, b, c are positive integers.

11. In (1.1), if $d|a$ and $d|b$, then prove that $d|r_{k+3}$.

12. (a) For positive integers c_1, c_2, and c_3, let (c_1, c_2, c_3) be the greatest common divisor of c_1, c_2, and c_3. Prove that there are integers s_1, s_2, and s_3 such that $(c_1, c_2, c_3) = c_1s_1 + c_2s_2 + c_3s_3$.

    (b) For positive integers c_1, c_2, and c_n, let $(c_1, c_2, \cdots, c_n)$ be the greatest common divisor of $c_1, c_2, \cdots$, and c_n. Prove that there are integers $s_1, s_2, \cdots$, and s_n such that $(c_1, c_2, \cdots, c_n) = c_1s_1 + c_2s_2 + \cdots + c_ns_n$.

13. If m_1 and m_2 are minimal elements of a non-empty subset of $\mathbb{N}$, then prove that m_1 is equal to m_2.

14. (a) Show that $S = \{\frac{1}{3^n} | n \in \mathbb{N}\}$ is not a well ordered set.

    (b) Show that $S = \{\frac{1}{6^n} | n \in \mathbb{N}\}$ is not a well ordered set.

    (c) Find a subset of $\mathbb{R}$ which is not well-ordered.

    (d) Find a subset of $\mathbb{R}$ which is well-ordered.

15. (a) Prove that the interval $(0, 1)$ does not have a minimal element.

    (b) Prove that $(0, 1)$ is not well-ordered.

    (c) Prove that $(a, b) \cup (c, d)$ is not well-ordered in general for $a, b, c, d \in \mathbb{N}$.

1.2 Equivalence Relations

Definition 1.2.1. *For a given set S and the nonempty subsets S_i, $1 \le i \le n$, of S, the class $\{S_i | 1 \le i \le n\}$ is called a* **partition of** S, *if*

1. $\cup_{1 \le i \le n} S_i = S$ *and*

2. $S_j \cap S_j = \emptyset$ *if* $i \ne j$.

Definition 1.2.1 says that all elements of a partition are disjoint and the union is the whole set.

Example 1.2.1. *1. Let* $A = \{1, 2\}$ *and* $B = \{3, 4, 5\}$*. Then,* $\{A, B\}$ *is a partition of* $S = \{1, 2, 3, 4, 5\}$

2. Let $A = \{a, b, c\}$ *and* $B = \{b, c, d\}$*. Then,* $\{A, B\}$ *is not a partition of* $S = \{a, b, c, d\}$ *since* $A \cap B \neq \emptyset$.

Definition 1.2.2. *An* **equivalence relation** *on a given set* S *is a set* $\mathfrak{E}$ *of ordered pairs of elements of* S *such that:*

1. $(a, a) \in \mathfrak{E}$ *for* $a \in S$ **(Reflexive Property)**.

2. *If* $(a, b) \in \mathfrak{E}$*, then* $(b, a) \in \mathfrak{E}$ **(Symmetry Property)**.

3. *If* (a, b) *and* $(b, c) \in \mathfrak{E}$*, then* $(a, c) \in \mathfrak{E}$ **(Transitive Property)**.

For the equivalence relation $\mathfrak{E}$, an element (a, b) of $\mathfrak{E}$ is denoted by $a \sim b$.

Example 1.2.2. *For a given set* $S = \{a, b, c, d\}$*, let* $\mathfrak{E} = \{(a, a), (b, b), (c, c), (d, d), (a, b), (b, a), (a, c), (c, a), (b, c), (c, b)\}$*. Then* $\mathfrak{E}$ *is an equivalence relation.*

Example 1.2.3. *For a given set* $S = \{a, b, c, d, e\}$*, let* $\mathfrak{E} = \{(a, a), (b, b), (c, c), (d, d), (a, b), (b, a), (a, c), (c, a), (b, c), (c, b), (d, e), (e, d)\}$*. Since* $(e, e) \notin \mathfrak{E}$*,* $\mathfrak{E}$ *is not an equivalence relation.*

For a given equivalence relation $\mathfrak{E}$ on a set S, **the equivalence class** of $a \in S$ is the set of elements that are related to a in S and denoted by $[a]$ or $\overline{a}$, i.e., $[a] = \{x \in S | (a, x) \in \mathfrak{E}\}$.

Example 1.2.4. *In example (1.2.2), the equivalence class is* $[a] = \{a, b, c\}$ *and* $[d] = \{d, e\}$.

Theorem 1.2.1. *Let* $\mathfrak{E}$ *be an equivalent relation on a non-empty set* S *and let* $a, b \in S$. *Then, either* $[a] \cap [b] = \phi$ *or* $[a] = [b]$.

Proof. Let's assume that $[a] \cap [b] \neq \emptyset$. Then there is an element $c \in [a] \cap [b]$. Suppose that $x \in [a]$. Then $x \sim a$. Note that $a \sim c$ and $b \sim c$ since $c \in [a]$ and $c \in [b]$. This implies that $x \sim c$ and with $b \sim c$ we have that $x \sim b$, i.e., $x \in [b]$. This means that $[a] \subset [b]$. Using a similar argument, we can prove that $[b]$ is a subset of $[a]$. This implies that $[a] = [b]$ and this proves the theorem. □

Definition 1.2.3. *For* $a, b \in \mathbb{Z}$ *and a given integer* n, *we say that* a **is congruent to** b **modulo** n *if* n *divides* $a - b$, *i.e.,* $n|(a - b)$.

Since the congruence is an equivalence relation (why?), the equivalence classes for congruence modulo n are called **a residue classes modulo** n and $\mathbb{Z}_n$ denotes the set of residue classes modulo n. It is used frequently enough that $\overline{k}$ is written for $[k]$ with respect to the relation.

Example 1.2.5. *In* $\mathbb{Z}_3 = \{\bar{0}, \bar{1}, \bar{2}\}$,

$$
\begin{aligned}
\bar{0} &= \{\cdots, -3, 0, 3, 6, \cdots\} = 3\mathbb{Z} \\
\bar{1} &= \{\cdots, -2, 1, 4, 7, \cdots\} = 1 + 3\mathbb{Z} \\
\bar{2} &= \{\cdots, -1, 2, 5, 8, \cdots\} = 2 + 3\mathbb{Z}
\end{aligned}
$$

We can see that they are all disjoint and $\bar{0} \cup \bar{1} \cup \bar{2} = \{\cdots, -3, 0, 3, 6, \cdots\} \cup \{\cdots, -2, 1, 4, 7, \cdots\} \cup \{\cdots, -1, 2, 5, 8, \cdots\} = \mathbb{Z}$.

Definition 1.2.4. *For a given set S,* **the cardinality of** S *is the number of elements of S and denoted by $|S|$.*

Example 1.2.6. *If $S = \{1, 2, 3, 4, 7\}$, then $|S| = 5$.*

Note that the cardinality of any of the number sets in p11 is infinity.

Exercises 1.2.

1. (a) Find all the elements of $\mathbb{Z}_5$.

   (b) Find all the elements of $\mathbb{Z}_n$ for a natural number n.

2. For given sets A, B, and C, let us define

$$A \times B = \{(a, b) | a \in A, b \in B\}$$

   So $A \times (B \times C)$ and $(A \times B) \times C$ can be defined similarly.

   (a) Prove that $|A \times (B \times C)| = |(A \times B) \times C|$

   (b) Are $A \times B$ and $B \times A$ the same?

3. (a) For $a, b \in \mathbb{Z}$, let's define a relation $a \sim b$ if $a = \pm b$. Prove that $\sim$ is an equivalence relation and find all the equivalence classes of the relation.

   (b) Let L be the set of all straight lines in the plane and for $l_1, l_2 \in L$, we define a relation $l_1 \sim l_2$ if l_1 is parallel to l_2. Prove that $\sim$ is an equivalence relation and find all the equivalence classes of the relation.

   (c) Let L be the set of all straight lines and for $l_1, l_2 \in L$, if we define a relation $l_1 \sim l_2$ if l_1 is perpendicular to l_2. Prove or disprove that $\sim$ is an equivalence relation.

4. For a given set $S = \{a, b, c, d, e\}$,

   (a) Define an equivalence relation on S.

   (b) Define an equivalence relation on S which is different from the equivalence in (a).

   (c) Find the equivalence classes in (a).

5. Let $\mathfrak{T}$ be the set of all triangles in a plane. We say that two triangles are related if they are similar.

   (a) Prove that the relation is an equivalence relation.

   (b) Find all the equivalence classes of the relation.

6. Prove that the congruence modulo n in Definition 1.2.3 is an equivalence relation.

7. (a) List all the elements of $\mathbb{Z}_6$ and define an equivalence relation on $\mathbb{Z}_6$.

   (b) In general, find all the elements of $\mathbb{Z}_n$ and define an equivalence relation on $\mathbb{Z}_n$.

8. Prove that the following relations are equivalence relations.

   (a) $a \sim b$ if and only if $3|(a-b)$ for $a, b \in \mathbb{Z}$.

   (b) In general, let's fix $c \in \mathbb{N}$ and $a \sim b$ if and only if $c|(a-b)$ for $a, b \in \mathbb{Z}$.

   (c) Find all the equivalence classes.

9. Prove or disprove the following relations are equivalence relations in $\mathbb{R}$.

   (a) $a \sim b$ if and only if $1 + ab \geq 0$.

   (b) $a \sim b$ if and only if $|a - b| > 0$.

(c) $a \sim b$ if and only if $a^2 + b^2 \geq 0$.

(d) $a \sim b$ if and only if $a^4 + b \geq 0$.

10. Let's define a relation $\sim$ on $\mathbb{Z}$ as follows:

$$n_1 \sim n_2 = 0 \text{ if and only if } n_1 - n_2 = 0$$

where $n_1, n_2 \in \mathbb{Z}$.

(a) Prove that $\sim$ is an equivalence relation.

(b) Find the equivalence class $\overline{7} = [7]$.

(c) In genenral, find the equivalence class $\overline{n} = [n]$ for $n \in \mathbb{N}$.

(d) Find the equivalence classes of the relation $\sim$ on $\mathbb{Z}$.

11. Let's use the relation $\sim$ in the previous problem on $\mathbb{Q}$.

(a) Prove that $\sim$ is an equivalence relation on $\mathbb{Q}$.

(b) Find the equivalence class $\overline{8} = [8]$.

(c) Find the equivalence class $\overline{\frac{2}{3}} = [\frac{2}{3}]$.

(d) In genenral, find the equivalence class $\overline{n} = [n]$ for $n \in \mathbb{N}$.

(e) Find all the equivalence classes of the relation $\sim$ on $\mathbb{Q}$.

1.3 Mathematical Induction

The principle of mathematical induction is usually stated as an axiom of the natural numbers. It can be proved in some logical systems. For instance, it can be proved if one assumes the following:

- The set of natural numbers is well-ordered.
- Every natural number is either zero, or $n+1$ for some natural number n.
- For any natural number n, $n+1$ is greater than n.

To derive simple induction from these axioms, we need the following result.

Theorem 1.3.1. *Let $P(n)$ be a proposition predicated on $n \in \mathbb{W}$. If*

1. *$P(0)$ holds and*
2. *$P(k)$ is true implies that $P(k+1)$ is also true,*

then $P(n)$ is ture for all $n \in \mathbb{W}$.

Proof. Take S to be the set of all natural numbers for which $P(n)$ is false. Let us see what happens if we assert that S is nonempty. The Well Ordering Principle tells us that S has a least element, which we will call t. Note that t is not 0 since $P(0)$ is true. Since every natural number is either zero or some $n+1$, there is some natural number n such that $n+1=t$. Now n is less than t, and t is the least element of S. It follows that n is not in S, and so $P(n)$ is true. This means that $P(n+1)$ is true, and so is $P(t)$. This is a contradiction since t was in S. So S is empty and this implies that $P(n)$ is true for all natural number n and we can all $n=0$ by the hypothesis. Therefore we have proven the theorem. □

The way of proving a mathematical statement using Theorem 1.3.1 is called **the mathematical induction** (or simply induction) on $\mathbb{W}$.

We can also apply Theorem 1.3.1 to prove a statement predicated on $n \in \mathbb{N}$. Only difference is the initial statements, $P(0)$ and $P(1)$. The followings are step to use mathematical induction:

Note 1.3.1. *Let $P(n)$ be a mathematical statement. To prove that $P(n)$ is true for all $n \in \mathbb{N}$, we follow these steps:*

Step 1. Prove that $P(1)$ is true.

Step 2. Assume that $P(k)$ is true.

Step 3. Prove that $P(k+1)$ is true.

Then we can say that the statement $P(n)$ is true for any $n \in \mathbb{N}$.

Example 1.3.1. *Prove that*

$$1 + 2 + \cdots + n = \frac{1}{2}n(n+1) \tag{1.2}$$

for any positive integer n using induction.

Solution. Let $P(n)$ be the statement (1.2). Obviously $P(1)$ is truce since the left side of (1.2) is 1 and the right side of (1.2) is also 1. Assume that $P(k)$ is true, i.e., $1 + 2 + \cdots + k = \frac{1}{2}k(k+1)$. If $n = k+1$, then the left side of (1.2) is $1+2+\cdots+k+(k+1) = \frac{1}{2}k(k+1)+k = \frac{(k+1)(k+2)}{2} = \frac{(k+1)(k+1+1)}{2}$ and this is the right side of (1.2) for $n = k+1$. This means that $P(k+1)$ is true. Therefore $P(n)$ is true for any positive integer. □

From Note 1.3.1, we can see that the initial statement doesn't have to be $P(0)$. In fact, it doesn't have to be even $P(1)$. It can be the corresponding statement to the smallest n. Also, note that the Step 2 of Note 1.3.1 can be replaced by the following:

Step 2'. Assume that $P(m)$ is true for $1 \leq m \leq k$ when $k \geq 2$.

Now we'll take a look at more examples.

Example 1.3.2. *Prove de Moivre's Theorem for all non-negative integers:*

$$(\cos\alpha + i\sin\alpha)^n = \cos(n\alpha) + i\sin(n\alpha). \tag{1.3}$$

Proof. Let $P(n)$ be the statement (1.3). For the initial statement, let's consider $P(0)$: $(\cos\alpha + i\sin\alpha)^0 = 1 = \cos(0\cdot\alpha) + i\sin(0\cdot\alpha)$ and it holds. Now assume that $P(k)$ is true, i.e., $(\cos\alpha + i\sin\alpha)^k = \cos(k\alpha) + i\sin(k\alpha)$. Then

$$\begin{aligned}
&(\cos\alpha + i\sin\alpha)^{k+1} = (\cos\alpha + i\sin\alpha)^k(\cos\alpha + i\sin\alpha)\\
&= \cos(k\alpha) + i\sin(k\alpha)(\cos\alpha + i\sin\alpha) \text{ by the induction assumption}\\
&= \cos(k\alpha)\cos\alpha + i\cos(k\alpha)\sin\alpha + i\sin(k\alpha)\cos\alpha) - \sin(k\alpha)\sin\alpha\\
&= \cos(k\alpha)\cos\alpha - \sin(k\alpha)\sin\alpha + i\{\cos(k\alpha)\sin\alpha + \sin(k\alpha)\cos\alpha)\}\\
&= \cos\big((k+1)\alpha\big) + i\sin\big((k+1)\alpha\big).
\end{aligned}$$

So $P(k+1)$ is true. Therefore $P(n)$ is true for any non-negative integer. □

de Moivre's Theorem is a very convenient tool when a complex number is written in a polar form.[1]

Example 1.3.3. *For $n \in \mathbb{N}$, solve the equation $x^n = a$ where $a \in \mathbb{C}$.*

[1]Recall that $a + bi \in \mathbb{C}$, $a, b \in \mathbb{R}$ is a rectangular form and $r(\cos\theta + i\sin\theta)$, $r \in \mathbb{R}$ is a polar form of a complex number.

Solution. Let us put $a = r(\cos\theta) + i\sin\theta)$. Then

$$\begin{aligned} x^n &= \{r(\cos\theta) + i\sin\theta)\}^n \\ &= r^n(\cos\theta) + i\sin\theta)^n \\ &= r^n\{\cos(n\theta) + i\sin(n\theta)\} \end{aligned}$$

by de Moivre's Theorem. This implies that $x = \sqrt[n]{r}\{\cos(\frac{\theta+2k\pi}{n}) + i\sin(\frac{\theta+2k\pi}{n})\}$ are the solutions of the equation where k is a whole number such that $0 \leq k \leq n-1$. □

Exercises 1.3.

1. Solve each of the following equations completely.

   (a) $x^4 - 4 = 0$.

   (b) $(x+3)^4 - 4 = 0$.

   (c) $x^7 + 1 = 0$.

   (d) $(x-5)^7 + 1 = 0$.

   (e) $x^5 - 7 = 0$.

   (f) $x^8 + 5 = 0$.

2. Calculate

   (a) $\{3(\cos\frac{2\pi}{3} + i\sin\frac{2\pi}{3})\}^2$

   (b) $\{3(\cos\frac{2\pi}{3} + i\sin\frac{2\pi}{3})\}^3$

   (c) $\{3(\cos\frac{2\pi}{3} + i\sin\frac{2\pi}{3})\}^6$

   (d) $\{5(\cos\frac{\pi}{3} + i\sin\frac{\pi}{3})\}^7$

3. Prove each of the following equalities using mathematical induction.

   (a) $1^2 + 2^2 + \cdots + n^2 = \frac{n(n+1)(2n+1)}{6}$ for $n \in \mathbb{N}$

   (b) $1^3 + 2^3 + \cdots + n^3 = \frac{n^2(n+1)^2}{4}$ for $n \in \mathbb{N}$

   (c) $1 + 3 + \cdots + (2n-1) = n^2$ for $n \in \mathbb{N}$

   (d) $\frac{1}{1\cdot 2} + \frac{1}{2\cdot 3} + \cdots + \frac{1}{n(n+1)} = \frac{n}{n+1}$ for $n \in \mathbb{N}$

4. Using mathematical induction, prove that $(1+x)^2 \geq 1 + nx$ for any $x \in \mathbb{R}$ and $n \in \mathbb{W}$.

5. (a) Prove that 3 divides one of integers $n+2$, $n+3$, or $n+4$.

   (b) In general, prove that 3 divides one of integers $n+k$, $n+(k+1)$, or $n+(k+2)$ for any $k, n \in \mathbb{W}$.

6. Prove that 15 divides $n^5 - n$ for any $n \in \mathbb{W}$.

7. Prove that $1 + \frac{1}{2} + \cdots + \frac{1}{n}$ is not an integer for all $n \in \mathbb{N}$.

8. Prove that $2n+1$ and $13n+6$ are relatively prime for any $n \in \mathbb{W}$.

9. Prove that 8 divides $7^n + 3^n - 2$ for any $n \in \mathbb{W}$.

10. For $a \in \mathbb{R}$ and $r(\neq 1) \in \mathbb{R}$, prove that

$$a + ar + ar^2 + \cdots + ar^n = \frac{ar^{n+1} - a}{r - 1}$$

    for $n \in \mathbb{N}$ by induction.

11. For a given positive real number x, prove that $1 + nx < (1+x)^n$ where $2 \leq n \in \mathbb{N}$ by induction and induction only.

12. Prove each of the following.

(a) $\overline{(a+bi)^n} = \left(\overline{a+bi}\right)^n$ where $a, b \in \mathbb{R}$ and $n \in \mathbb{W}$

(b) $\overline{(a_1+b_1i)\cdot(a_2+b_2i)} = \overline{(a_1+b_1i)}\cdot\overline{(a_2+b_2i)}$ where $a_1, b_1, a_2, b_2 \in \mathbb{R}$

(c) $\overline{(a_1+b_1i)\cdots(a_n+b_ni)} = \overline{(a_1+b_1i)}\cdots\overline{(a_n+b_ni)}$

where $a_1, b_1, \cdots, a_n, b_2 \in \mathbb{R}$ and $n \in \mathbb{N}$

1.4 Functions

Definition 1.4.1. *Let A and B be sets. A* **map** *(or a* **function** *or a* **mapping***) θ from a set A to a set B is a rule that assigns to each element a of A exactly one element b of B. The element b is called* **the image of a under θ** *denoted by $b = \theta(a)$. The set A is called* **the domain of θ** *and and the set of all images of A under θ is called* **the image of A** *denoted by $\theta(A)$ or $im(\theta)$.*

If θ is a rule that assign to a of A two distinct b_1 and b_2 of B, we say that θ is **not well-defined**[2] (so θ is not a function).

Throughout the book, $\theta : A \to B$ means a map θ from a set A to a set B.

Example 1.4.1. *For $A = \{1, 2, 3\}$ and $B = \{b, c, d\}$, let's define $\theta(1) = b$, $\theta(2) = d$, and $\theta(3) = c$. Then we see that θ is a well-defined map from A to B. We also denote θ as $\begin{pmatrix} 1 & 2 & 3 \\ b & d & c \end{pmatrix}$.*[3]

[2] "Two to one" relation is a function, not "one to two".

[3] It is called an array notation of a permutation and will be studied extensively in Chapter 7

Example 1.4.2. *For* $A = \{1, 2, 3, 4\}$ *and* $B = \{a, b, c, d\}$, *define* $\theta(1) = a$, $\theta(2) = b$, $\theta(3) = c$, *and* $\theta(4) = c$. *Then* θ *is a map from* A *to* B. *If we define* $\theta(1) = a$, $\theta(2) = b$, $\theta(2) = c$, *and* $\theta(3) = c$, *then* θ *is not well-defined so it is not a map from* A *to* B.

Definition 1.4.2. *Let* A, B, *and* C *be sets and let* $\theta : A \to B$ *and* $\psi : B \to C$ *be maps.* **The composition** *(or* **the composite function***)* $\psi \circ \theta = \psi\theta$ *is the map from* A *to* C *defined by* $(\psi\theta)(a) = \psi(\theta(a))$ *for any* $a \in A$. $\circ$ *is called* **the function composition**.

Note that $\psi\theta$ means θ is applied first and ψ later.

Definition 1.4.3. *A map* $\theta : A \to B$ *is called* **one to one**[4] *(or* **injective** *or* **an injection***) if* $\theta(a_1) = \theta(a_2)$ *implies that* $a_1 = a_2$, *for all* $a_1, a_2 \in A$.

The contrapositive statement of Definition 1.4.3 is used often: if $a_1 \neq a_2$, then $\theta(a_1) \neq \theta(a_2)$ for all $a_1, a_2 \in A$.

Definition 1.4.4. *A map* $\theta : A \to B$ *is called* **onto** *(or* **surjective** *or* **a surjection***) if for each element* b *of* B, *there is an element* a *such that* $\theta(a) = b$

Definition 1.4.4 means that there exists a pre-image of every element in the set B. It also implies that $\theta(A) = B$ or $im(\theta) = B$.

[4] If a function is not one to one, it might be a two to one, three to one, or even many to one function. Since these are all two to one partially, one best way to remember an one to one function is "not a two to one" function.

Example 1.4.3. *Let $A = \{a, b, c, d\}$, $B = \{1, 2, 3\}$, and $C = \{k, l, m, n, p\}$. If θ is a map from A to B such that $\theta(a) = 1$, $\theta(b) = 2$, $\theta(c) = 3$, $\theta(d) = 3$ and ψ from B to C such that $\psi(1) = k$, $\psi(2) = l$, $\psi(3) = m$, then find $\psi\theta$.*[5]

Solution. We have that $\psi\theta(a) = \psi(\theta(a)) = \psi(1) = k$, $\psi\theta(b) = \psi(\theta(b)) = \psi(2) = l$, $\psi\theta(c) = \psi(\theta(c)) = \psi(3) = m$, and $\psi\theta(d) = \psi(\theta(d)) = \psi(3) = m$. The following graph explains how this composite function works.

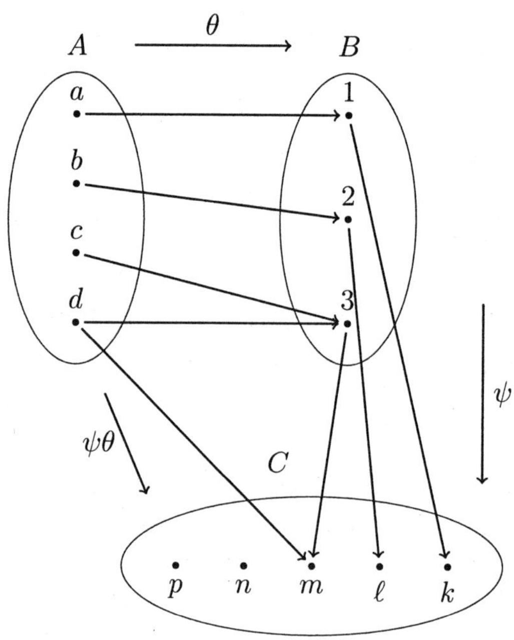

Figure 1.4.1.

Example 1.4.4. *If we define a map θ from $\{a, b, c, d\}$ to $\{1, 2, 3, 4, 5\}$ by $\theta(1) = a$, $\theta(2) = b$, $\theta(3) = c$, and $\theta(4) = d$, then we have the following graph:*

[5]"Find the function" means that the rule of the function needs to be explained. Typically it is enough to find the image of an arbitrary input under the function.

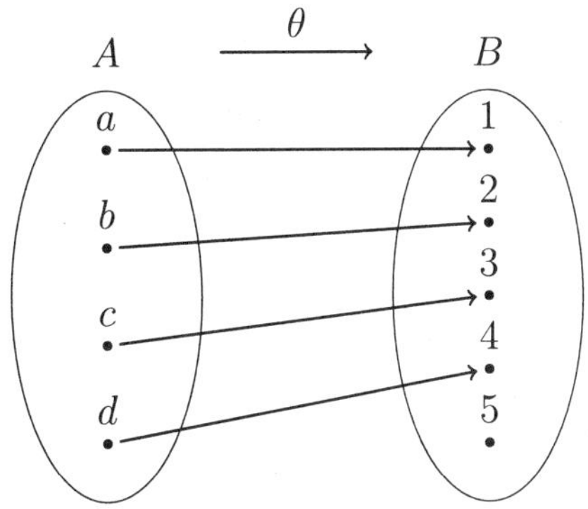

Figure 1.4.2.

We can see that θ is an injection.

Example 1.4.5. *If we define a map θ from $\{a, b, c, d\}$ to $\{1, 2, 3, 4, 5\}$ by $\theta(1) = a$, $\theta(2) = b$, $\theta(3) = c$, and $\theta(4) = d$, then we have the following graph:*

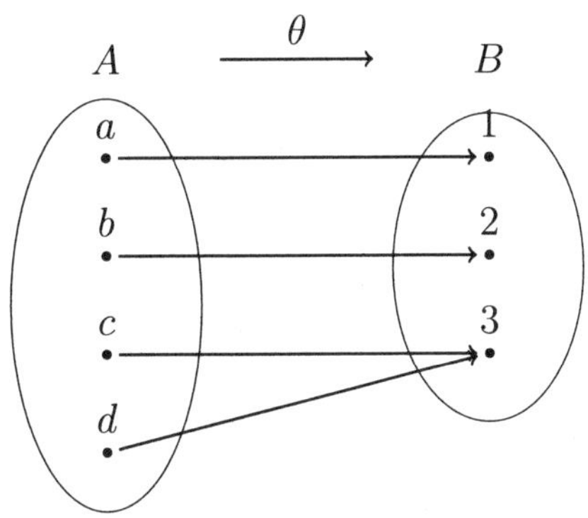

Figure 1.4.3.

Note that θ is not an injection but a surjection.

If a function is one to one and onto, it is called **bijective** or **a bijection**.

Definition 1.4.5. *For a given set A, the mapping I_A is called* **the identity map from A to itself** *if $I_A(a) = a$ for any $a \in A$. Generally, we will denote I_A as I (or simply 1). A map θ from A to itself has* **the inverse function** *(or simply* **the inverse***) if there exists $\psi : A \to A$ such that $I_A = \theta\psi = \psi\theta$. The* *inverse of θ is denoted as θ^{-1} and we say that θ is* **invertible** *if there exists θ^{-1}.*

Proposition 1.4.1. *Let $\theta : A \to B$, $\psi : B \to C$, and $\gamma : C \to D$ be functions. Then we have*

1. $\gamma(\psi\theta) = (\gamma\psi)\theta$ **(associativity of composition)**
2. *If θ and ψ are one to one, then $\psi\theta$ is one to one.*
3. *If θ and ψ are onto, then $\psi\theta$ is onto.*
4. *If θ is bijective, then there is a map θ^{-1} such that $\theta^{-1}\theta = I_A$ and $\theta\theta^{-1} = I_B$.*

Proof.

1. For any $a \in A$, $\gamma(\psi\theta(a)) = \gamma(\psi(\theta(a))$. On the other hand, $(\gamma\psi)(\theta(a)) = \gamma(\psi(\theta(a)))$. Thus $\gamma(\psi\theta) = (\gamma\theta)\phi$.

2. Let's assume that $(\psi\theta)(a_1) = (\psi\theta)(a_2)$ for $a_1, a_2 \in A$. This implies that $\psi(\theta(a_1)) = \psi(\theta(a_2))$. Since ψ is one to one, $\theta(a_1) = \theta(a_2)$. Since θ is one to one, $a_1 = a_2$. This implies that $\psi\theta$ is one to one.

3. For c in C, there is $b \in B$ such that $\psi(b) = c$ since ψ is surjective. Also, there is $a \in A$ such that $\theta(a) = b$ since θ is surjective. This implies that $(\psi\theta)(a) = \psi(\theta(a)) = \psi(b) = c$ and it proves that $\psi\theta$ is surjective.

4. For any $a \in A$, let us put $\theta(a) = b$. If we define $\theta^{-1} : B \to A$ as $\theta^{-1}(b) = a$. Then $(\theta^{-1}\theta)(a) = \theta^{-1}(\theta(a)) = \theta^{-1}(b) = a = I_A(a)$. Similarly we can prove that $\theta\theta^{-1} = I_B$. Thus we have proven 4.

Example 1.4.6. *Let* $A = \{a, b, c, d\}$ *and* $B = \{1, 2, 3, 4, 5\}$*, and* θ *be a map from* A *to* B *such that* $\theta(a) = 1$*,* $\theta(b) = 2$*,* $\theta(c) = 3$*, and* $\theta(d) = 4$*. If we define* $\psi : \{1, 2, 3, 4, 5\} \to \{a, b, c, d\}$ *by* $\psi(1) = a$*,* $\psi(2) = b$*,* $\psi(3) = c$*,* $\psi(4) = d$*, and* $\psi(5) = d$*, then we can see that* $\psi\theta = I_A$ *by the following graph:*

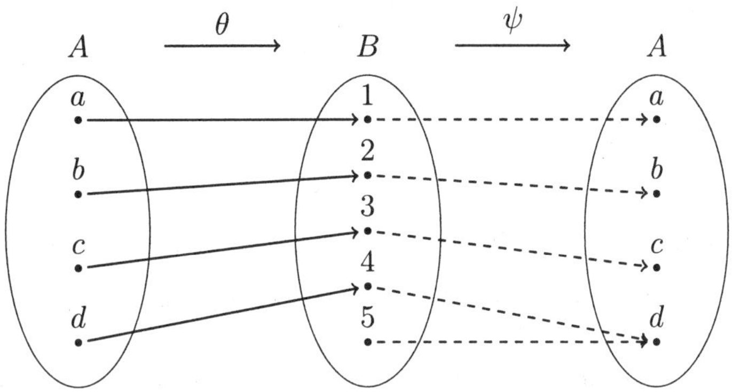

Figure 1.4.4.

Proposition 1.4.2. *If* $\theta : A \to B$ *is injective, then there is a mapping* $\psi : B \to A$ *such that* $\psi\theta = I_A$ *and* ψ *is called* **the left inverse of** ϕ**.**

Proof. Let us define $\psi : B \to A$ so that if $y \in im(\theta) = \theta(A)$, then $\psi(y)$ is the unique x such that $\theta(x) = y$ and if $y_1 \notin im(\theta)$, then $\psi(y_1) = a$ for some fixed $a \in A$. This implies that $\psi\theta(x) = \theta(y) = x = I_A(x)$ for any $x \in A$. Thus $\psi\theta = I_A$. □

Proposition 1.4.2 implies that ψ is not unique if $\theta(A) \neq B$.

Example 1.4.7. *Let* $A = \{a, b, c, d\}$ *and* $B = \{1, 2, 3\}$*, and* θ *be a map from* A *to* B *such that* $\theta(a) = 1$*,* $\theta(b) = 2$*,* $\theta(c) = 3$ *and* $\theta(d) = 3$*. If we define* $\psi : \{1, 2, 3\} \to \{a, b, c, d\}$ *by* $\psi(1) = a$*,* $\psi(2) = b$*, and* $\psi(3) = c$*, then we can see that* $\theta\psi = I_B$ *from the following graph:*

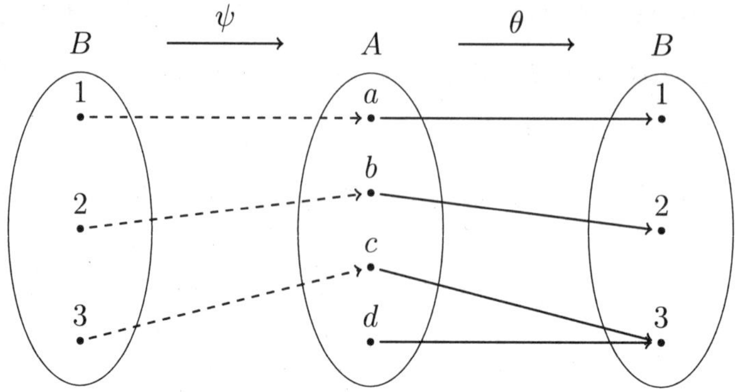

Figure 1.4.5.

Proposition 1.4.3. *If $\theta : A \to B$ is surjective, then there is a map $\psi : B \to A$ such that $\theta\psi = I_B$ and ψ is called* **the right inverse of θ.**

Proof. Let us define a map $\psi : B \to A$ such that for any $y \in B$, pick an element $x \in A$ such that $\theta(x) = y$ and define $\psi(y) = x$. Then $\theta\psi(y) = \psi(x) = y = I_B(y)$ and this means $\theta\psi = I_B$. □

Corollary 1.4.1. *ψ is the inverse of θ if and only if ψ is the left inverse and the right inverse of θ in Proposition 1.3.3.*

Corollary 1.4.2. *If θ is invertible, θ^{-1} is unique in Definition 1.3.5.*

Proof. Suppose ψ_1 and ψ_2 are two inverses of θ. Then for any $a \in A$, we have that

$$\begin{aligned} \psi_1(a) &= \psi_1\Big(\theta\psi_2(a)\Big) && \text{since } \theta \circ \psi_2 = I \\ &= (\psi_1\theta)\psi_2(a) && \text{by Proposition 1.4.1} \\ &= \psi_2(a) && \text{since } \psi_1 \circ \theta = I \end{aligned}$$

This shows that $\psi_1 = \psi_2$ and it proves the corollary. □

Exercises 1.4.

1. (a) Let θ be a map from $\{1,2,3\}$ to $\{a,b,c,d\}$ such that $\theta(1)=a, \theta(2)=b$, and $\theta(3)=d$. Find the left inverse of θ.

   (b) Let ϕ be a map from $\{a,b,c,d\}$ to $\{1,2,3\}$ such that $\phi(a)=1, \phi(b)=2$, and $\phi(c)=3$, and $\phi(d)=1$. Find the right inverse of ϕ.

2. Let $A=\{1,2,3,4\}$ and $B=\{a,b,c,d,e\}$.

   (a) Define an injection from A to B and find the left inverse of your map.

   (b) Is your map surjective? Why or why not?

3. Let $A=\{1,2,3,4,5\}$ and $B=\{a,b,c,d\}$.

   (a) Define a surjection from A to B and find the right inverse of your map.

   (b) Is your map injective? Why or why not?

4. If any, find a map $\theta:\{a,b,c,d\}\to\{1,2,3,4,5\}$ such that:

   (a) θ is injective, not surjective. Also find the left inverse of θ.

   (b) θ is surjective, not injective. Also find the right inverse of θ.

5. Let θ, ψ, and σ be functions from $\mathbb{Z}$ to $\mathbb{Z}$ defined by

$$\theta(x)=2x, \quad \psi(x)=x+1, \quad \text{and} \quad \sigma(x)=x^2$$

   for $x\in\mathbb{Z}$.

   (a) Find the composition $\theta\psi$.

   (b) Find the composition $\psi\theta$.

   (c) Are $\theta\psi$ and $\psi\theta$ the same?

(d) Is θ surjective? Injective?

(e) Is ψ surjective? Injective?

(f) Is σ surjective? Injective?

6. Prove that θ is invertible if and only if θ is bijective.

7. For given non-empty sets A, B, and C, we define

$$A \times B = \{(a,b) | a \in A, b \in B\}$$

(a) Find a bijection from $A \times B$ to $B \times A$.

(b) Find a bijection from $A \times (B \times C)$ to $(A \times B) \times C$.

(c) Find a surjection from $A \times B$ to B.

8. Let f and g be maps from $\mathbb{R}$ to $\mathbb{R}$ such that

$$f(x) = 3x^2 - 5 \quad \text{and} \quad g(x) = \frac{x}{x^2+1}$$

for $x \in \mathbb{R}$.

(a) Find fg and gf. Are they the same?

(b) Is f injective?

(c) Is g surjective?

9. Let $A = \{1, 2, 3, 4, 5\}$ and $B = \{a, b, c, d, e\}$.

(a) Define an injection from A to B.

(b) Is your map θ surjective? If yes, find its inverse.

Chapter 2

Basic Group Theory

When we say that a set has a structure, in particular an algebraic structure, it means that a operation among the elements of the set is well-defined. Examples of operations are an addition, a subtraction, a multiplication, etc. If it is well-defined, we called it **a binary operation**.

Formally, a binary operation on a set G is a function $\phi : G \times G \to G$ where $G \times G = \{(g_1, g_2) | g_1, g_2 \in G\}$. For example, a usual addition among integers is a binary operation and several properties of this binary operation include $(a + b) + c = a + (b + c)$ and $a + 0 = 0 + a = a$ for $a, b, c, 0 \in \mathbb{Z}$. Also, there is an element $-a$ such that $a + (-a) = (-a) + a = 0$ for any $a \in \mathbb{Z}$. These type of additive properties of $\mathbb{Z}$ will be generalized to an abstract definition of a group.

2.1 Definition and Examples

Definition 2.1.1. *A* **group** *is a non-empty set G with a binary operation $\cdot$ satisfying the following:*

1. $g_1 \cdot g_2 \in G$ *for any* $g_1, g_2 \in G$ **(closedness** *or* **closure property***)*[1]

2. $g_1 \cdot (g_2 \cdot g_3) = (g_1 \cdot g_2) \cdot g_3$ *for any* $g_1, g_2, g_3 \in G$ **(associativity)**

3. *There is an element* e *of* G *such that* $e \cdot g = g \cdot e = g$ *for any* $g \in G$. *This element* e *is called* **the identity of** G.

4. *There is an element* $g^{-1} \in G$ *such that* $g \cdot g^{-1} = g^{-1} \cdot g = e$ *for any* $g \in G$. *This element* g^{-1} *is called* **the inverse of** g.[2]

A set G with a binary operation $\cdot$ is compactly written as $(G, \cdot)$ and these four axioms in Definition 2.1.1 are called **the group axioms**. So we can say that $(G, \cdot)$ is a group if it satisfies the group axioms.

Example 2.1.1. *Prove that* $\mathbb{Z}_n$ *is a group with the addition* $+$ *modulo* n.

Solution. For $\overline{a}, \overline{b} \in \mathbb{Z}_n$, let us define the addition as $\overline{a} + \overline{b} = \overline{a+b} \in \mathbb{Z}_n$. We need to show that the addition is well defined. For $\overline{a} = \overline{b}$ and $\overline{c} = \overline{d}$ of $\mathbb{Z}_n$, we need to show that $\overline{a} + \overline{c} = \overline{c} + \overline{d}$. Since $\overline{a} = \overline{b}$, we have that $a = b + k_1 n$ for $k_1 \in \mathbb{Z}$ and similarly we have that $c = d + k_2 n$ for $k_2 \in \mathbb{Z}$. This implies that $\overline{a} + \overline{c} = \overline{a+c} = \overline{b + k_1 n + d + k_2 n} = \overline{b+d} = \overline{b} + \overline{d}$. Thus the addition is well defined. For $\overline{a}, \overline{b}, \overline{c} \in \mathbb{Z}_n$, we have that

$$\overline{a} + (\overline{b} + \overline{c}) = \overline{a} + \overline{b+c} = \overline{a+b+c} = (\overline{a} + \overline{b}) + \overline{c}.$$

This shows that the addition is associative. Obviously $\overline{0}$ is the additive identity of $\mathbb{Z}_n$. For any $\overline{a} \in \mathbb{Z}_n$, $-\overline{a}$ is the inverse of $\overline{a}$. Therefore $\mathbb{Z}_n$ is a group under addition. □

[1] It seems redundant to mention a binary operation and the closure property of a group at the same time but we do that here since some texts distinguish them

[2] The inverse function of a function is also called the inverse since a function can be an element of some group.

We also say that $\mathbb{Z}_n$ is an additive group or $(\mathbb{Z}_n, + \bmod n)$ is a group.

Example 2.1.2. *The set $\mathbb{Q}$ of all rational numbers is a group under addition $+$, and so are $\mathbb{R}$ and $\mathbb{C}$.*

Example 2.1.3. *$\mathbb{Q}$, $\mathbb{R}$, and $\mathbb{C}$ are not groups under multiplication.*

Example 2.1.4. *The set $\mathbb{Q}^*$ of all non-zero rational numbers is a group under multiplication $\cdot$. Also $(\mathbb{R}^*, \cdot)$ and $(\mathbb{C}^*, \cdot)$ are multiplicative groups.*[3]

Example 2.1.5. *If p is prime, show that $\mathbb{Z}_p^*$ is a multiplicative group.*

Solution. For any $\overline{a}, \overline{b} \in \mathbb{Z}_p^*$, let us define $\overline{a}\overline{b} = \overline{ab}$. Let us show that the multiplication is well defined. For $\overline{a}_1 = \overline{a}_2$ and $\overline{b}_1 = \overline{b}_2$, we need to show that $\overline{a}_1\overline{b}_1 = \overline{a}_2\overline{b}_2$. Since $\overline{a}_1 = \overline{a}_2$, we have that $a_1 = k_1p + a_2$ and $b_1 = k_2p + b_2$ for $k_1, k_2 \in \mathbb{Z}$. This makes $a_1b_1 = (k_1p+a_2)(k_2p+a_2) = k_1k_2p^2+k_1pb_2+a_2k_2p+a_2a_2$ and if we apply the modulo p arithmetics, then $\overline{a_1b_1} = \overline{(k_1p + a_2)(k_2p + a_2)} = \overline{k_1k_2p^2 + k_1pb_2 + a_2k_2p + a_2a_2} = \overline{a_2a_2}$. Thus the multiplication is well defined. Since $(a, p) = 1$ and $(b, p) = 1$, we have that $(ab, p) = 1$. Thus $\overline{a_1b_1} \in \mathbb{Z}_p^*$. Thus it is closed. For any $\overline{c} \in \mathbb{Z}_p^*$, we have that $(\overline{1})(\overline{c}) = \overline{c} = (\overline{c})(\overline{1})$, i.e., $\overline{1}$ is the identity. Note that for any $\overline{d} \in \mathbb{Z}_p^*$, $(d, p) = 1$ so there are $s, t \in \mathbb{Z}$ such that $ds + pt = 1$. This makes $\overline{ds + pt} = \overline{1}$ and this implies that $\overline{d}\overline{s} = \overline{1}$. So, $\overline{d}$ has the multiplicative inverse and we have showed that $(\mathbb{Z}_p^*, \cdot \bmod n)$ is a group. □

Example 2.1.6. *The set of all $n \times m$ matrices $M_{n\times m}(\mathbb{R})$ over $\mathbb{R}$ is an additive group and so is the set of all $n \times n$ square matrices $M_n(\mathbb{R})$ over $\mathbb{R}$.*

[3]In general, the set of all nonzero elements of G is denoted by $G^* = G\backslash\{0\}$

Check the Appendix for more information about matrices.

Example 2.1.7. *Let* $GL_n(\mathbb{R}) = \{A \in M_n(\mathbb{R}) | det(A) \neq 0\}$*. Then* $GL_n(\mathbb{R})$ *is the set of all invertible* $n \times n$ *matrices over* $\mathbb{R}$ *and* $(GL_n(\mathbb{R}), \cdot)$ *is a multiplicative group.*[4]

Example 2.1.8. *Let* A *be a non-empty set and* $Map(A, A) =: \{\theta | \theta : A \to A, \theta$ *is one to one and onto*$\}$*. Then the composition* $\circ$ *of maps is a binary operation on* $Map(A, A)$*. So we say that* $\big(Map(A, A), \circ\big)$ *is a group.*

Example 2.1.9. Klein 4 **group** K_4

Let $K_4 = \{e, a, b, c\}$*. If we define a binary multiplication* $\cdot$ *as the following table,* $(K_4, \cdot)$ *becomes a group called* **the Klein** 4 **group***:*

Figure 2.1.1.

$\cdot$	e	a	b	c
e	e	a	b	c
a	a	e	c	b
b	b	c	e	a
c	c	b	a	e

This table is called **the Cayley table** of the group K_4.

Example 2.1.10. *Let* $U(n) := \{\overline{x} \in \mathbb{Z}_n | x$ *and* n *are relatively prime*$\}$*. Then* $U(n)$ *is a multiplicative group.*

[4] $GL_n(\mathbb{R})$ is called **the general linear group of degree** n **over** $\mathbb{R}$.

Solution. Like the addition of $\mathbb{Z}_n$, the multiplication is well defined. For any $\overline{x}, \overline{y} \in \mathbb{Z}_n$, we have that $(x,n) = 1 = (y,n)$. This implies that $(xy,n) = 1$. So we have that $\overline{xy} \in U(n)$. Obviously $\overline{1} \in U(n)$. For $\overline{x}$, we have that $(x,n) = 1$. Thus there are $s,t \in \mathbb{Z}$ so that $sx + nt = 1$. This implies that $\overline{sx+nt} = \overline{sx} = \overline{1}$, i.e., $\overline{s} \cdot \overline{x} = \overline{1}$ so we derive that $\overline{x}^{-1} = \overline{s}$. Therefore $\big(U(n), \cdot \bmod n\big)$ is a group. □

Note that if p is prime, then $\mathbb{Z}_p^*$ is $U(p)$.

Example 2.1.11. *1. Since there is no additive identity, the set of all non-negative integers* $\mathbb{N}$ *is not a group under addition.*

2. Since the multiplicative inverse of 3 *is* $\frac{1}{3}$, *the set of all non-negative integers* $\mathbb{N}$ *is not a group under multiplication.*

Example 2.1.12. *Let* $\mathbf{Q} = \{\pm 1, \pm \mathbf{i}, \pm \mathbf{j}, \pm \mathbf{k}\}$ *with the following multiplications* $\mathbf{ii} = \mathbf{jj} = \mathbf{kk} = 1$, $\mathbf{ij} = \mathbf{k}, \mathbf{jk} = \mathbf{i}, \mathbf{ki} = \mathbf{j}$, $\mathbf{ji} = -\mathbf{k}, \mathbf{kj} = -\mathbf{i}$, *and* $\mathbf{ik} = -\mathbf{j}$. *Then, it becomes a group and is called* **the quaternion group**

2.2 Basic Properties

When there is no ambiguity, the binary operation $\cdot$ or $+$ is omitted so sometimes we say that a set G is a group.

Theorem 2.2.1. (Uniqueness of the Identity and the Inverse)
In a group, the identity and the inverse of each element are unique.

Proof. Suppose e and e' are two identity elements of a group G. Then since e and e' are identity elements of G, we have that $e = ee' = e'$. This implies that the identity element of G is unique. Now suppose g_1 and g_2 are inverses of g. Then $g_1 = g_1(gg_2) = (g_1g)g_2 = g_2$ and this implies that $g_1 = g_2$. Therefore the inverse of g is unique. $\square$

Theorem 2.2.2. *For any integer $n \in \mathbb{Z}$ and g of a group G, $(g^n)^{-1} = (g^{-1})^n$.*

Proof. Note that the inverse of g^n is $(g^n)^{-1}$. We also have that $(g^{-1})^n g^n = g^{-1} \cdots g^{-1} g \cdots g = e$. Since the inverse of g^n is unique, we have that $(g^n)^{-1} = (g^{-1})^n$. $\square$

In Theorem 2.2.2, the element $(g^{-1})^n$ is denoted as g^{-n}.

Lemma 2.2.1. (Cancelation Law)
Let a, b, and c be elements of a group G. If $ab = ac$ or $ba = ca$, then $b = c$.

Proof. Suppose $ab = ac$. Since $a \in G$, $a^{-1} \in G$, we can derive that $a^{-1}(ab) = a^{-1}(ac)$. This implies that $(a^{-1}a)b = (a^{-1}a)c$. So $b = c$. The other case is similar. $\square$

Definition 2.2.1. *A group G is called* **Abelian** *(or commutative) if $ab = ba$ for any $a, b \in G$. Otherwise, it is called* **non-Abelian** *(or non-commutative).*

Example 2.2.1. *$\mathbb{Z}$, $\mathbb{Q}$, $\mathbb{R}$, and $\mathbb{C}$ are all Abelian groups under the addition.*

Example 2.2.2. *We saw that $\mathbb{Z}_n$ is an additive group in Example 2.1.1 and it is an Abelian group because $\overline{a} + \overline{b} = \overline{b} + \overline{a}$ for $\overline{a}, \overline{b} \in \mathbb{Z}_n$.*

Example 2.2.3. *If p is prime, then $\mathbb{Z}_p^*$ is a multiplicative Abelian group.*

Example 2.2.4. *$M_{n,m}(\mathbb{R})$ and $M_n(\mathbb{R})$ are additive Abelian groups.*

Example 2.2.5. *Since $\begin{pmatrix} 1 & 4 \\ 0 & 2 \end{pmatrix}\begin{pmatrix} 1 & 1 \\ 4 & 3 \end{pmatrix} \neq \begin{pmatrix} 1 & 1 \\ 4 & 3 \end{pmatrix}\begin{pmatrix} 1 & 4 \\ 0 & 2 \end{pmatrix}$, $GL_2(\mathbb{R})$ is a non-Abelian group under the matrix multiplication. In general, $GL_n(\mathbb{R}) = \{A \in M_n(\mathbb{R}) | det(A) \neq 0\}$ is a non-Abelian multiplicative group.*[5]

Example 2.2.6. *Let A be a non-empty set. Then $Map(A, A)$ in Example 2.1.8 is a non-Abelian group (why?) under the composition of maps.*

Definition 2.2.2. *A non-empty subset H of a group G is called* **a subgroup of** *G if H is a group under the binary operation of G.*

Obviously a subgroup of an Abelian group is Abelian. Also, the following is a good example of an Abelian subgroup.

Example 2.2.7. *For an element g of a group G, we define $< g >:= \{g^n | n \in \mathbb{Z}\}$. Note that $g^0 = 1 \in \mathbb{Z}$ is the identity of G and $(g^n)^{-1} = g^{-n} \in \mathbb{Z}$ by Theorem 2.2.2. Therefore, $< g >$ is a subgroup of G called* **the cyclic subgroup of** *G* **generated by** *g.*

Example 2.2.8. *1. For $\overline{2} \in \mathbb{Z}_8$, $< \overline{2} >= \{\overline{2}, \overline{2} + \overline{2} = \overline{4}, \overline{2} + \overline{2} + \overline{2} = \overline{6}, \overline{2} + \overline{2} + \overline{2} + \overline{2} = \overline{0}\} = \{\overline{0}, \overline{2}, \overline{4}, \overline{6}\}$ is a subgroup of $\mathbb{Z}_8$.*

[5] See the Appendix for more information about matrices.

2. For $3 \in \mathbb{Z}$, $<3>=\{3n|n \in \mathbb{Z}\}$ *is a subgroup of* $\mathbb{Z}$.

Note that the subgroups in 1. and 2. of Example 2.2.8 can be written as $\overline{2}\mathbb{Z}_8$ and $3\mathbb{Z}$ respectively.

Definition 2.2.3. *The cardinality of a group* G *is called* **the order** *of* G *and denoted by* $|G|$.

Example 2.2.9. $\mathbb{Z}_{12}$ *is an additive group with* $|\mathbb{Z}_{12}| = 12$. *In general, the order of an additive group* $\mathbb{Z}_n$ *is* n.

Example 2.2.10. $|\mathbb{Q}| = |\mathbb{R}| = |\mathbb{C}| = \infty$.[6]

Note that $|<g>|$ can be finite or infinite and it depends on the nature of g. For example, if $g = 2 \in \mathbb{Z}_4$, then $|<g>| = 2$ but if $g = 2 \in \mathbb{Z}_6$, then $|<g>| = 3$. Also, if $g = 2 \in \mathbb{Z}$, then $|<g>| = \infty$.

Definition 2.2.4. *A group is called* **an infinite group** *if the group order is infinity. If the order is finite, it is called* **a finite group**.

Definition 2.2.5. *Let* $(G_1, *_1)$ *and* $(G_2, *_2)$ *be groups. We define* **the external direct product** *of two groups* G_1 *and* G_2 *as the set* $G_1 \times G_2 = \{(g_1, g_2|g_1 \in G_1, g_2 \in G_2\}$ *with the binary* $*$ *defined as follows:*

$$(g_1, g_2) * (g_1', g_2') = (g_1 *_1 g_1', g_2 *_2 g_2')$$

for $(g_1, g_2), (g_1', g_2') \in G_1 \times G_2$.

[6] $|\mathbb{Q}|$ and $|\mathbb{R}|$ are different infinities though.

It is easy to prove that $G_1 \times G_2$ is a group under $*$. Sometimes $G_1 \oplus G_2$ is used for the external direct product.[7]

Example 2.2.11. *Note that* $\mathbb{Z}_3 \times \mathbb{Z}_5 = \{(a,b) | a \in \mathbb{Z}_3 \text{ and } b \in \mathbb{Z}_5\}$. *For* $(a_1,b_1),(a_2,b_2) \in \mathbb{Z}_3 \times \mathbb{Z}_5$, *if we define*

$$(a_1,b_1)+(a_2,b_2) = \big((a_1+a_2)\ mod\ 3, (b_1+b_2)\ mod\ 5\big),$$

then $\mathbb{Z}_3 \times \mathbb{Z}_5$ *becomes a group.*

Chapter 2 Exercises.

1. (a) Show that $\mathbb{C}$ is not a multiplicative group.

   (b) Prove that $\mathbb{C}$ is an additive Abelian group.

   (c) Prove that $\mathbb{C}^*$ is a multiplicative Abelian group.

2. Let $a,b,c \in \mathbb{R}$ and define a map $T_{a,b,c} : \mathbb{R}^3 \to \mathbb{R}^3$ as follows:

$$T_{a,b,c}\big((x_1,x_2,x_3)\big) = (a+x_1, b+x_2, c+x_3)$$

   for $(x_1,x_2,x_3) \in \mathbb{R}^3$. Prove that the set $T = \{T_{a,b,c} | a,b,c \in \mathbb{R}\}$ is an additive group.[8] Is it Abelian?

3. (a) Prove that $\mathbb{Q}[\sqrt{7}] = \{a+b\sqrt{7} | a,b \in \mathbb{Q}\}$ is an additive Abelian group.

   (b) Prove that $\mathbb{Q}[\sqrt{7}]^* = \{a+b\sqrt{7} | a,b \in \mathbb{Q}, a+b\sqrt{7} \neq 0\}$ is a multiplicative Abelian group.

[7] Besides the external direct product, the internal direct product G_1G_2 can be defined. Both G_1 and G_2 are subsets of G_1G_2 but they are not physically subsets of $G_1 \oplus G_2$.

[8] $T_{a,b,c}$ is called a translation in $\mathbb{R}^3$.

4. Prove that $(M_2(\mathbb{Z}_5), +)$ is an additive group. Is it Abelian?

5. Find the inverse $\begin{pmatrix} \overline{1} & \overline{1} \\ \overline{3} & \overline{2} \end{pmatrix}$ in the multiplicative group $(GL_2(\mathbb{Z}_5), \cdot)$.

6. Let $SL_n(\mathbb{R}) = \{A \in M_n(\mathbb{R}) | det(A) = 1\}$.[9] Prove that $SL_n(\mathbb{R})$ is a multiplicative group. Is it Abelian?

7. Identify all the elements of the following groups.

   (a) $GL_2(\mathbb{Z}_2)$.

   (b) $SL_2(\mathbb{Z}_2)$.

   (c) $GL_2(\mathbb{Z}_3)$.

   (d) $SL_2(\mathbb{Z}_3)$.

8. Let A be a proper subset of a finite set S. Prove that

$$S_A = \{\theta : S \to S | \theta \text{ is a bijection such that } \theta(a) = a \text{ for any } a \in A\}$$

   is a group under the composition of maps. Is it Abelian?

9. Prove that $\mathbb{Z}_5 \times \mathbb{Z}_7$ is a group and find its order. Is it Abelian?

10. Find all the elements of each of the following U-groups and their group orders.

    (a) $U(6)$.

    (b) $U(7)$.

    (c) $U(36)$.

    (d) $U(p)$ where p is prime.

[9] $SL_n(\mathbb{R})$ is called **the special linear group of degree n over $\mathbb{R}$** and it will be studied in the next chapter.

(e) $U(pq)$ where p and q are prime.

11. Prove that the following four matrices form an Abelian group:

$$S = \left\{ \begin{pmatrix} 1 & 0 \\ 0 & 1 \end{pmatrix}, \begin{pmatrix} -1 & 0 \\ 0 & 1 \end{pmatrix}, \begin{pmatrix} 1 & 0 \\ 0 & -1 \end{pmatrix}, \begin{pmatrix} -1 & 0 \\ 0 & -1 \end{pmatrix} \right\}$$

12. Let H and K are subgroups of a group G and let $H \vee K$ be the smallest subgroup of G that contains $H \cup K$. Find $H \vee K$ for the following:

(a) $H = 6\mathbb{Z}$, $K = 8\mathbb{Z}$, and $G = \mathbb{Z}$.

(b) $H = 12\mathbb{Z}$, $K = 8\mathbb{Z}$, and $G = \mathbb{Z}$.

13. The dihedral group D_n is a group generated by x and y satisfying the following equations:

$$x^2 = y^n = e, \quad n > 2,$$
$$xy = y^{-1}x.$$

(a) Represent $(x^i y^j)(x^k y^l)$ as $x^a y^b$.

(b) Prove that D_n is a non Abelian group of order $2n$.

14. Let $G = \left\{ \begin{pmatrix} a & b \\ -b & a \end{pmatrix} | a, b \in \mathbb{Q}, a^2 + b^2 \neq 0 \right\}$. Prove that G is a group under the usual matrix multiplication. Is G Abelian?

15. Let $G = \left\{ \begin{pmatrix} a & b \\ -b & a \end{pmatrix} | a, b \in \mathbb{R}, a^2 + b^2 \neq 0 \right\}$. Prove that G is a group under the usual matrix multiplication. Is G Abelian?

16. Let $G = \left\{ \begin{pmatrix} a & b \\ -b & a \end{pmatrix} | a, b \in \mathbb{Z}_p, a^2 + b^2 \neq 0 \right\}$. Prove that G is a group

under the usual matrix multiplication. Is G Abelian?

17. Let $G = \left\{ \begin{pmatrix} a & 0 \\ 0 & a^{-1} \end{pmatrix} | a \in \mathbb{Q}, a \neq 0 \right\}$. Prove that G is an Abelian group under the usual matrix multiplication.

18. Suppose G is a group such that $g = g^{-1}$ for any $g \in G$. Prove that G is Abelian.

19. Suppose G is a group such that $(ab)^2 = a^2b^2$ any $a, b \in G$. Prove that G is Abelian.

20. Let $\mathbb{R}^+$ be the set of all positive real numbers.

    (a) Is $\mathbb{R}^+$ be an additive group?

    (b) Is $\mathbb{R}^+$ be a multiplicative group?

Chapter 3

Subgroups

We have defined a subgroup of a group in the previous chapter. It is beneficial to study properties of subgroups for investigating a group itself. Let us start with several examples of a subgroup. If H is a subgroup of G, we denote $H < G$ or $H \leq G$.

3.1 Examples

Example 3.1.1. *For any group, the identity group* $\{e\}$ *and the group itself are always subgroups. These are called* **trivial subgroups**. *If there is other subgroup other than trivial subgroups, it is called* **a non-trivial subgroup.**

Example 3.1.2. *There are 4 subgroups of* $\mathbb{Z}_6$*, 2 trivial and 2 non-trivial ones:* $\{\overline{0}\}$, $\{\overline{0}, \overline{3}\}$, $\{\overline{0}, \overline{2}, \overline{4}\}$, *and* $\mathbb{Z}_6$ *itself. All of them are cyclic.*

Example 3.1.3. $\mathbb{Z}$ *is a subgroup or* $\mathbb{Q}$, $\mathbb{Q}$ *is a subgroup of* $\mathbb{R}$, *and* $\mathbb{R}$ *is a subgroup of* $\mathbb{C}$, *i.e.,* $\mathbb{Z} < \mathbb{Q} < \mathbb{R} < \mathbb{C}$. *They are all under addition.*

Example 3.1.4. $\mathbb{Q}^* < \mathbb{R}^* < \mathbb{C}^*$. *They are all under multiplication.* $\mathbb{Z}^*$ *is not a group.*

Example 3.1.5. $SL_n(\mathbb{R}) = \{A \in GL_n(\mathbb{R}) | det(A) = 1\}$ *is a subgroup of the general linear group* $GL_n(\mathbb{R})$ *and it is called* **the special linear group of degree** n **over** $\mathbb{R}$. $SL_n(\mathbb{Q})$ *is a subgroup of* $SL_n(\mathbb{R})$ *or* $GL_n(\mathbb{R})$.

Definition 3.1.1. *Let* g *be an element of a group* G. **The order of** g *is the minimum positive integer* m *such that* $g^m = e$ *where* e *is the identity of* G *and it is denoted by* $|g|$.

If there is no such a positive integer m such that $g^m = e$, we say that g has **infinite order** (or $|g| = \infty$). Otherwise, g has **finite order.**

Example 3.1.6. *For the additive group* $\mathbb{Z}_4$ *and* $\overline{2} \in \mathbb{Z}_4$, $|\overline{2}|$ *is* 2.

Example 3.1.7. *For the additive group* $\mathbb{Z}_6$, *we have that* $|\overline{0}| = 1$, $|\overline{1}| = 1$, $|\overline{2}| = 3$, $|\overline{3}| = 2$, $|\overline{4}| = 3$, *and* $|\overline{5}| = 6$.

Example 3.1.8. *Every element in* $\mathbb{Z}$ *has infinite order except* 0 *and* $|0| = 1$ *in* $\mathbb{Z}$. *It is the same for* $\mathbb{Q}$, $\mathbb{R}$, *or* $\mathbb{C}$.

The following lemma is stated here again with a property of the orders. The 1st part will be proved in Theorem 3.2.1.

Lemma 3.1.1. *Let* g *be an element of a group* G. *Then*

1. $< g >$ *is a subgroup of* G.

2. $| < g > | = |g|$.

Elements of $\mathbb{Z}_n$ is related to their orders and it can be seen through the following lemmas. We will prove the following lemmas in Theorem 6.0.16.

Lemma 3.1.2. *Let* g *be an element of a group* G *with* $|G| = n$. *Then*

1. $g^n = e$.

2. *If* $|g| = m$, $m = \frac{n}{(n,m)}$.

Lemma 3.1.3. *Let* $\overline{k} \in \mathbb{Z}_n$. *Then the order of* $\overline{k}$ *is* $\frac{n}{(n,k)}$, *i.e.,* $|\overline{k}| = \frac{n}{(n,k)}$.

Example 3.1.9. *For* Z_{12}, $|\overline{8}| = \frac{12}{(12,8)} = \frac{12}{4} = 3$ *and* $|\overline{7}| = \frac{12}{(12,7)} = \frac{12}{1} = 12$. *Note that* $Z_{12} =< \overline{7} >$.

3.2 Subgroup Tests

The following lemma is a convenient tool to check if a subset is a subgroup and it is called the subgroup test lemma. If H is a subgroup of G, sometimes we used the notation $H \leq G$.

Lemma 3.2.1. *Let* H *be a non-empty subset of a group* G. *Then the following are equivalent.*

1. H *is a subgroup of* G, *i.e.,* $H \leq G$.

2. For any $g_1, g_2 \in H$, $g_1 g_2^{-1} \in H$.

3. For any $g_1, g_2 \in H$, $g_1 g_2 \in H$ *and* $g_1^{-1} \in H$.

Proof.

1. $\Rightarrow$ 2.: It is obvious since H itself is a group.

2. $\Rightarrow$ 3.: Let us assume that 2. holds. Since H is not empty, there exists $g \in H$. If we apply $g, g \in H$ (not $g_1, g_2 \in H$), $gg^{-1} = e \in H$ by the assumption. Then since $e, g \in H$, we have that $eg^{-1} = g^{-1} \in H$. Suppose $g_1, g_2 \in H$. Then $g_1, g_2^{-1} \in H$. This implies that $g_1(g_2^{-1})^{-1} = g_1 g_2 \in H$ and we can conclude the part 3. since the existence of the inverse is already proved.

3. $\Rightarrow$ 1.: Let us assume that 3. holds. We only need to show the existence of the identity in H. Since H is not the empty set, there is an element $g \in H$. This implies that $g^{-1} \in H$ and this makes $gg^{-1} = e \in H$. Thus H is a subgroup of G. □

2. is called **the one-step subgroup test** and 3. is called **the two-step subgroup test.**

Theorem 3.2.1. *For a given element* g *of a group* G, $< g >= \{g^n | n \in \mathbb{Z}\}$ *is a subgroup of* G.

Proof. Note that $g^0 = e$ is the identity of G. For any $g^n, g^m \in < g >$, we have that $g^n g^{-m} = g^{n-m} \in < g >$. This implies that $< g >$ is a subgroup of G by the one-step subgroup test. □

Example 3.2.1. *Let n be a positive integer and G be an Abelian group. Prove that nth power $G^n := \{g^n | g \in G\}$ is a subgroup of G.*

Solution. For any $g_1^n, g_2^n \in G^n$, we have that $g_1^n(g_2^n)^{-1} = g_1^n g_2^{-n} \in G^n$. This shows that G^n is a subgroup of G by the one-step subgroup test. □

The following is called **the finite subgroup test.**

Lemma 3.2.2. *Let H be a non-empty finite subset of a group G. Then H is a subgroup of G if $g_1, g_2 \in H$ for any $g_1, g_2 \in H$.*

Proof. By Lemma 3.2.1, it is enough to prove that $g^{-1} \in H$ for $g \in H$. If $g = e$, then there is nothing to prove. Let's assume that $g \neq e$. We have that $g^2 \in H$ and inductively we derive that $g^n \in H$ for any positive integer n. Since H is a finite set, there are integers n and m such that $g^n = g^m$ with $n > m$. If $n = m + 1$, $g = e$, a contradiction. So $n > m + 1$, i.e. $n - m - 1 > 0$. Since $g^{n-m} = e$, we derive that $g^{n-m-1}g = e$. This implies that $g^{-1} = g^{n-m-1} \in H$ since $n - m - 1$ is a positive integer. Thus we have proven the lemma. □

Proposition 3.2.1. *Let H_1 and H_2 be subgroups of a group G, then $H_1 \cap H_2$ is a subgroup of G.*

Proof. Since the identity of e of G is an element of $H_1 \cap H_2$, $H_1 \cap H_2$ is not empty. For any $g_1, g_2 \in H_1 \cap H_2$, we have that $g_1, g_2 \in H_1$ and $g_1, g_2 \in H_2$. By Lemma 3.2.1, we have that $g_1^{-1}g_2 \in H_1$ and $g_1^{-1}g_2 \in H_2$ and this implies that $g_1^{-1}g_2 \in H_1 \cap H_2$. So, by Lemma 3.2.1 again, $H_1 \cap H_2$ is a subgroup of G. □

Definition 3.2.1. The center $\mathfrak{Z}$ of a group G *is the set*

$$\{g' \in G | gg' = g'g, \text{ for any } g \in G\}.$$

Example 3.2.2. *The center* $\mathfrak{Z}$ *of the quaternion group* $\mathbf{Q}$ *is* $\{1, -1\}$.[1]

Theorem 3.2.2. *The center* $\mathfrak{Z}$ *of the group* G *is a subgroup of* G.

Proof. Since $e \in \mathfrak{Z}$, $\mathfrak{Z}$ is not the empty set. For any $g \in G$ and $g_1, g_2 \in \mathfrak{Z}$, we have that $g_1g_2(g) = g_1(g_2g) = g_1(gg_2) = (g_1g)g_2 = gg_1g_2$. This implies that $g_1g_2 \in \mathfrak{Z}$. By $g_1g = gg_1$, we have that $gg_1^{-1} = g_1^{-1}g$. This implies that $g_1^{-1} \in \mathfrak{Z}$. Thus the center $\mathfrak{Z}$ is a subgroup of G by the two-step subgroup test. □

It is clear that the center of a group is Abelian by the definition.

Definition 3.2.2. *For a given element* g *of a group* G, $\mathfrak{Z}(g) = \{g_1 \in G | gg_1 = g_1g\}$ *is called* **the centralizer of** g.

Example 3.2.3. *The centralizer* $\mathfrak{Z}(\mathbf{i})$ *of the element* $\mathbf{i}$ *of the quaternion group* $\mathbf{Q}$ *is* $\{1, -1\}$.[2]

Theorem 3.2.3. *The centralizer* $\mathfrak{Z}(g)$ *of a group* G *is a subgroup of* G *and the center* $\mathfrak{Z}$ *is a subgroup of* $\mathfrak{Z}(g)$, *i.e.,* $\mathfrak{Z} \leq \mathfrak{Z}(g) \leq G$.

Proof. Let $g \in G$. Since the identity of G is an element of $\mathfrak{Z}(g)$, $\mathfrak{Z}(g)$ is not empty. For any $g_1, g_2 \in \mathfrak{Z}(g)$, we have that $g_1g_2(g) = g_1(g_2g) = g_1(gg_2) = (g_1g)g_2 = gg_1g_2$. This implies that $g_1g_2 \in \mathfrak{Z}(g)$. Since $g_1g = gg_1$ implies that $gg_1^{-1} = g_1^{-1}g$. This implies that $g_1^{-1} \in \mathfrak{Z}(g)$. Thus the centralizer $\mathfrak{Z}(g)$ is a subgroup of G by the two-step subgroup test and $\mathfrak{Z} \leq \mathfrak{Z}(g)$ is obvious. □

---

[1]See Example 2.1.12
[2]See Example 2.1.12

Example 3.2.4. *Show that the center of the dihedral group D_3 is the trivial group $\{e\}$.*

Proof. The dihedral group D_3 is the set of all bijections from $\{1,2,3\}$ to itself. For example, if $\sigma_1(1) = 1$, $\sigma_1(2) = 2$, and $\sigma_1(3) = 3$, then $\sigma_1 \in D_3$. The element σ_1 is denoted as $\begin{pmatrix} 1 & 2 & 3 \\ 1 & 2 & 3 \end{pmatrix}$. So the other elements of D_3 are

$\sigma_2 = \begin{pmatrix} 1 & 2 & 3 \\ 3 & 1 & 2 \end{pmatrix}$, $\sigma_3 = \begin{pmatrix} 1 & 2 & 3 \\ 2 & 3 & 1 \end{pmatrix}$, $\sigma_4 = \begin{pmatrix} 1 & 2 & 3 \\ 3 & 2 & 1 \end{pmatrix}$, $\sigma_5 = \begin{pmatrix} 1 & 2 & 3 \\ 2 & 1 & 3 \end{pmatrix}$, and $\sigma_6 = \begin{pmatrix} 1 & 2 & 3 \\ 1 & 3 & 2 \end{pmatrix}$.

Note that the multiplication on D_3 is the composition of maps. For example,

$$\sigma_5\sigma_3(1) = \sigma_5(2) = 1,$$

$$\sigma_5\sigma_3(2) = \sigma_5(3) = 3,$$

$$\sigma_5\sigma_3(3) = \sigma_5(1) = 2,$$

i.e., $\sigma_5\sigma_3 = \begin{pmatrix} 1 & 2 & 3 \\ 1 & 3 & 2 \end{pmatrix} = \sigma_6$. Since for any $\sigma_i \in D_3$, there is an element $\sigma_j \in D_3$ such that $\sigma_i\sigma_j \neq \sigma_j\sigma_i$ where $i \neq 1$. Thus the center of the dihedral group D_3 is the trivial group $\{e\}$. □

One can construct the Cayley table for D_3 to confirm the argument in Example 3.2.4, which is not so obvious. The following is obvious though.

Example 3.2.5. *The center of an Abelian group is the group itself. Also the centralizer of any element of an Abelian group is the group itself.*

Proposition 3.2.2. *Show that the center* $\mathfrak{Z}$ *of* $GL_2(\mathbb{R})$ *is* $\left\{ \begin{pmatrix} a & 0 \\ 0 & a \end{pmatrix} \mid a \in \mathbb{R}^* \right\}$.

Proof. Suppose $g = \begin{pmatrix} \alpha & \beta \\ \gamma & \delta \end{pmatrix} \in \mathfrak{Z}$. Then we have that

$$\begin{pmatrix} \alpha & \beta \\ \gamma & \delta \end{pmatrix} \begin{pmatrix} 1 & 2 \\ 1 & 0 \end{pmatrix} = \begin{pmatrix} 1 & 2 \\ 1 & 0 \end{pmatrix} \begin{pmatrix} \alpha & \beta \\ \gamma & \delta \end{pmatrix}$$

and this implies that

$$\begin{pmatrix} \alpha+\beta & 2\alpha \\ \gamma+\delta & 2\gamma \end{pmatrix} = \begin{pmatrix} \alpha+2\gamma & \beta+2\delta \\ \alpha & \beta \end{pmatrix}.$$

So we derive that $\beta = 2\gamma$, i.e., $g = \begin{pmatrix} \alpha & 2\gamma \\ \gamma & \delta \end{pmatrix}$. Next, using $g \begin{pmatrix} 1 & 2 \\ 3 & 4 \end{pmatrix} = \begin{pmatrix} 1 & 2 \\ 3 & 4 \end{pmatrix} g$, we derive that $\alpha = \delta$ and $\gamma = 0$. This implies that $g = \begin{pmatrix} \alpha & 0 \\ 0 & \alpha \end{pmatrix}$ and this type of matrices commute with any 2×2 matrix for any $\alpha \in \mathbb{R}$. Note that if $\alpha = 0$, g is not an element of $GL_2(\mathbb{R})$. □

We can also write the center of $GL_2(\mathbb{R}) = \{aI_2 | a \in \mathbb{R}^*\} = \mathbb{R}I_2$ where I_2 is the 2×2 identity matrix.

Corollary 3.2.1. *The center* $\mathfrak{Z}$ *of* $SL_2(\mathbb{R})$ *is* $\{I_2, -I_2\}$.

Proof. It is obvious by Proposition 3.2.2 since the determinants are ± 1. □

Example 3.2.6. $\mathbb{Z}_{15}$ *has three proper subgroups* $\{\overline{0}\}$, $< 3 >= \{\overline{0}, \overline{3}, \overline{6}, \overline{9}, \overline{12}\}$, *and* $< \overline{5} >= \{\overline{0}, \overline{5}\}$. *Using them, we have the following figure and it is called* a

subgroup lattice.

Figure 3.2.1.

$$\mathbb{Z}_{15}$$

$$\nearrow \qquad\qquad \nwarrow$$

$$<\bar{3}> \qquad\qquad <\bar{5}>$$

$$\nwarrow \qquad\qquad \nearrow$$

$$<\bar{0}>$$

Chapter 3 Exercises.

1. (a) Find the orders of all elements in $\mathbb{Z}_{16}$.

   (b) Find all the subgroups of $\mathbb{Z}_{16}$ and draw its subgroup lattice.

   (c) Find all the subgroups of $\mathbb{Z}_{15}$ and draw its subgroup lattice.

2. (a) Prove that $< 4 >$ is a subgroup of $< 2 >$.

   (b) Prove that if n divides m, then $m\mathbb{Z}$ is a subgroup of $n\mathbb{Z}$.

3. (a) Prove that $H_1 \cap H_2$ is a subgroup of a group G if H_1 and H_2 are subgroups of G.

   (b) Prove that $\cap_{1\le i\le n} H_i$ is a subgroup of a group G if H_1, H_2, $\cdots$, and H_n are subgroups of G.

   (c) Find subgroups H_1 and H_2 of $\mathbb{Z}_{24}$ such that $H_1 \cup H_2$ is not a subgroup of $\mathbb{Z}_{24}$. (This shows that $H_1 \cup H_2$ might not be a subgroup of a group G even though H_1 and H_2 are subgroups of G.)

4. For the additive group $G = (\mathbb{Z}, +)$, find $G^2 = \{x + x = 2x | x \in \mathbb{Z}\}$ and G^3.

5. (a) Is $\begin{pmatrix} \overline{3} & \overline{1} \\ \overline{1} & \overline{4} \end{pmatrix}$ an element of $GL_2(\mathbb{Z}_5)$? What about $\begin{pmatrix} \overline{2} & \overline{3} \\ \overline{1} & \overline{3} \end{pmatrix}$?

   (b) Find the multiplicative inverse of $\begin{pmatrix} \overline{1} & \overline{2} \\ \overline{3} & \overline{5} \end{pmatrix} \in GL_2(\mathbb{Z}_5)$.

6. Let G be a group with identity e.

   (a) If $g^2 = e$, prove that $g = g^{-1}$.(Such a g is called an **idempotent** element.)

   (b) If $g^2 = e$ for all $g \in G$, prove that G is Abelian. (Hint: Show that $gh = hg$ for all $g, h \in G$ using $(gh)^2 = e$.)

7. Let H and K be subgroups of an Abelian group G.

   (a) Prove that $HK = \{hk | h \in H, k \in K\}$ is a subgroup of G.

   (b) Prove that H and K are subgroups of HK.

   (c) Show that HK is not a subgroup in general, i.e. if G is not Abelian.

8. If H is a subgroup of a group G with the center $\mathfrak{Z}$, prove that $H\mathfrak{Z}$ is a subgroup of G.

9. Let $G_1 = \left\{ \begin{pmatrix} a & b \\ -b & a \end{pmatrix} | a, b \in \mathbb{Q}, a^2 + b^2 \neq 0 \right\}$ and $G_2 = \left\{ \begin{pmatrix} a & b \\ -b & a \end{pmatrix} | a, b \in \mathbb{R}, a^2 + b^2 \neq 0 \right\}$. Prove that G_1 is a subgroup of G_2 under the usual matrix multiplication.

10. Let $G = \left\{ \begin{pmatrix} a & b \\ 0 & c \end{pmatrix} | a, b, c \in \mathbb{Q}, ac \neq 0 \right\}$. Prove that G is a subgroup of $GL_2(\mathbb{Q})$ under the usual matrix multiplication.

11. Let $G = \left\{ \begin{pmatrix} a & 0 \\ 0 & a^{-1} \end{pmatrix} |a \in \mathbb{Z}_p, a \neq 0 \right\}$ where p is a prime. Prove that G is an Abelian subgroup of $GL_2(\mathbb{Z}_p)$ under the usual matrix multiplication.

12. Prove that $< g >$ is an Abelian subgroup of a group G if $g \in G$.

13. Prove that the center $\mathfrak{Z}$ is a subgroup of the centralizer $\mathfrak{Z}(g)$ for any g in a group G. Is $\mathfrak{Z}(g)$ Abelian?

14. (a) Construct a group of order 3 in general, make its Cayley table, and prove that it is Abelian.

    (b) Construct a group of order 4 in general, make its Cayley table, and prove that it is Abelian.

    (c) Construct a group of order 5 in general, make its Cayley table, and prove that it is Abelian.

15. Make the Cayley table of D_3.

Chapter 4

Normal Subgroups and Lagrange Theorem

Suppose H is a subgroup of a group G. For $g \in G$, $Hg = \{hg|h \in H\}$ is called **the right coset of H in G**. Similarly, $gH = \{gh|h \in H\}$ is called **the left coset of H in G**. Here, g is called **a coset representative of** Hg (or gH). Note that if G is Abelian, a left coset of G is a right coset. Using these cosets, we define $G/H := \{Hg|g \in G\}$, the set of all right cosets and this set becomes a group if H is so called "a normal subgroup". We will define a normal subgroup of a group in this chapter and we will see that the normality of a group is a fundamental concept to study the group structure.

4.1 Cosets and their operations

Example 4.1.1. *Find all left cosets of* $3\mathbb{Z}$ *in* $\mathbb{Z}$.

Solution. There are three distinct left cosets: $0+3\mathbb{Z} = 3\mathbb{Z} = \{\cdots, -3, 0, 3, \cdots\}$, $1 + 3\mathbb{Z}$, and $2 + 3\mathbb{Z}$. $\square$

Note that these cosets are also right cosets. Also note that they are all disjoint and $\mathbb{Z} = (0 + 3\mathbb{Z}) \cup (1 + 3\mathbb{Z}) \cup (2 + 3\mathbb{Z})$.

Example 4.1.2. *Find all the cosets of* $H = \{1, -1, \mathbf{i}, -\mathbf{i}\}$ *in the quaternion group. Recall that* $\mathbf{Q} = \{1, -1, \mathbf{i}, \mathbf{j}, \mathbf{k}, -\mathbf{i}, -\mathbf{j}, -\mathbf{k}\}$ *is the quaternion group.*

Solution. Note that H is a subgroup of $\mathbf{Q}$. Also, note that $1H = H$, $-H = H$, $iH = H$, and $-iH = H$. Then we have that $\mathbf{j}H = \{\mathbf{j}(1), \mathbf{j}(-1), \mathbf{j}\mathbf{i}, \mathbf{j}(-\mathbf{i})\} = \{\mathbf{j}, -\mathbf{j}, -\mathbf{k}, \mathbf{k}\}$, $-\mathbf{j}H = \{-\mathbf{j}, \mathbf{j}, \mathbf{k}, -\mathbf{k}\}$, $\mathbf{k}H = \{\mathbf{k}, -\mathbf{k}, -\mathbf{j}, \mathbf{j}\}$, and $-\mathbf{k}H = \{-\mathbf{k}, \mathbf{k}, \mathbf{j}, -\mathbf{j}\}$. Therefore, there are only two disjoint cosets of H in Q: $H = -H = \mathbf{i}H = -\mathbf{i}H$ and $\mathbf{j}H = -\mathbf{j}H = \mathbf{k}H = -\mathbf{k}H$. $\square$

Lemma 4.1.1. *Suppose* H *is a subgroup of a group* G. *Then* $Hg_1 = Hg_2$ *if and only if* $g_1 g_2^{-1} \in H$ *where* $g_1, g_2 \in G$.

Proof. Assume that $Hg_1 = Hg_2$. Since $e \in H$, $eg_1 \in Hg_1$ so $g_1 \in Hg_2$. Then $g_1 = hg_2$ for some $h \in H$ and this implies that $g_1 g_2^{-1} = h \in H$. Conversely let's assume that $g_1 g_2^{-1} \in H$ for some $g_1, g_2 \in G$. This means that $g_1 g_2^{-1} = h$ for some $h \in H$. Suppose $g \in Hg_1$. Then $g = h_1 g_1$ for some $h_1 \in H$. This implies that $g = h_1 h g_2$ and this makes $g \in Hg_2$ so $Hg_1 \subset Hg_2$. Similarly we prove that $Hg_2 \subset Hg_1$. Therefore, $Hg_1 = Hg_2$ and this proves the lemma. $\square$

Example 4.1.3. *Show that* $x + 7\mathbb{Z} = 7\mathbb{Z}$ *for any* $x \in 7\mathbb{Z}$.

Solution. Note that $7\mathbb{Z} = \{\cdots, -7, 0, 7, \cdots\}$ is a subgroup of $\mathbb{Z}$. Suppose $x \in 7\mathbb{Z}$, i.e. $x = 7k$ for some $k \in \mathbb{Z}$. This means that $x + 7\mathbb{Z} = 7k + \{\cdots, -7, 0, 7, \cdots\} = \{\cdots, -7, 0, 7, \cdots\} = 7\mathbb{Z}$. $\square$

If we apply $x = 7$ to Example 4.1.3 and subtract 5 from both sides , we also derive that $2 + 7\mathbb{Z} = -5 + 7\mathbb{Z}$ for example.

Example 4.1.4. *Show that* $38 + 6\mathbb{Z} = 2 + 6\mathbb{Z} \in \mathbb{Z}/(6\mathbb{Z})$.

Solution. Note that $6\mathbb{Z}$ is a subgroup of $\mathbb{Z}$ and since $\mathbb{Z}$ is Abelian, $6\mathbb{Z} \lhd \mathbb{Z}$. So $38 + 6\mathbb{Z}6, \mathbb{Z} \in \mathbb{Z}/(6\mathbb{Z})$ and since $36 \in 6\mathbb{Z}$, $36 + 6\mathbb{Z} = 6\mathbb{Z}$. If we add 2 on both sides, we derive the conclusion. □

Theorem 4.1.1. *Let H be a subgroup of a group G. Then $|Hg_1| = |Hg_2|$ for $g_1, g_2 \in G$. Also any two cosets are identical or disjoint, i.e. either $Hg_1 \cap Hg_2 = \phi$ or $Hg_1 = Hg_2$.*

Proof. For any $g \in G$, it is enough to prove that $|H| = |Hg|$. We define a map $\theta_g : H \to Hg$ as $\theta_g(h) = hg$ for $h \in H$. First we show that the map is one to one. Suppose that $\theta_g(h_1) = \theta_g(h_2)$. Then we have $h_1 g = h_2 g$ and this implies that $h_1 = h_2$. So θ_g is injective. Note that $\theta_g(h) = hg$ for any $hg \in Hg$. This implies that θ_g is surjective. Since this exists a bijection from H to Hg, $|H| = |Hg|$. Now suppose that $Hg_1 \cap Hg_2 \neq \phi$. Then there must be an element $g \in Hg_1 \cap Hg_2$. This implies that $g = h_1 g_1 = h_2 g_2$ for some $h_1, h_2 \in H$. Then we can derive that $h_1^{-1} g = g_1$ and $g^{-1} h_2 = g_2^{-1}$. This makes that $g_1 g_2^{-1} = h_1^{-1} g g^{-1} h_2 = h_1^{-1} h_2 \in H$. Thus, $Hg_1 = Hg_2$ by Lemma 4.1.1 and it proves the lemma. □

Now we state an important relation on orders of a group and its subgroup.

Theorem 4.1.2. (Lagrange Theorem)

If H is a subgroup of a finite group G, $|H|$ divides $|G|$. Moreover, the number of distinct right (or left) cosets of H is $|G|/|H| = [G : H]$ (or $|G| = |H|[G : H]$).

Proof. For any $g \in G$, $g \in Hg$. This implies that $G = \cup_{g_1 \in G} Hg_1$ and the rest of the proof is straightforward by Theorem 4.1.1. $\square$

Next one is an example of Theorem 4.1.2 using a cyclic subgroup.

Corollary 4.1.1. *For an element g of a finite group G, $| < g > |$ divides $|G|$ and $a^{|G|} = e$.*

Proof. For the subgroup $< g >$ of G, by Theorem 4.0.7, we have that $| < g > | [G :< g >] = |G|$. Obviously $g^{|G|} = e$. This implies that $g^{|G|} = g^{|<g>|[G:<g>]} = e^{[G:<g>]} = e$ and we see that $| < g > |$ divides $|G|$. $\square$

The following is a relation between a number and its prime power.

Corollary 4.1.2. (Fermat's Little Theorem)
For $a \in \mathbb{Z}$ and for a prime p, $a^p \equiv a \ mod \ p$.

Proof. For given a and p of the corollary, we have that $a = pm + r$ where $0 \leq r < p$. This implies that $a \equiv r \ mod \ p$. If $r = 0$, then $a^p \equiv a \ mod \ p$ holds trivially. Now, we assume that r is an element of $U(p)$, the set of positive integers less than or equal to p and relative prime to p. Since $|U(p)| = p - 1$, by Corollary 4.1.1, we have that $r^{p-1} \equiv 1 \ mod \ p$, i.e., $r^p \equiv r \ mod \ p$ and this proves the corollary. $\square$

4.2 Normal Subgroup and Factor Group

In this section, we'd like to look at the set of all cosets using a special subgroup. This type of subgroup makes the set of all cosets a group.

Definition 4.2.1. *A subgroup N of a group G is called* **normal** *if $gng^{-1} \in N$ for any $g \in G$ and $n \in N$, and it is denoted by $N \lhd G$.*

Note that $gng^{-1} \in N$ means that $gng^{-1} = n_1$ for some $n_1 \in N$.

Theorem 4.2.1. *For a subgroup N of G, the following are equivalent.*

1. *N is a normal subgroup of G.*
2. *For any $g \in G$, $gN = Ng$.*
3. *For any $g \in G$, $gNg^{-1} = N$.*

Proof. 1. $\to$ 2. : Suppose that N is a normal subgroup of G. For all $x \in gN$, we have that $x = gn_1$ for some $n_1 \in N$. Since N is a normal subgroup of G, we have that $gn_1 = n_2g$ for some $n_2 \in N$. This implies that $x = n_2g \in Ng$, i.e., $x \in Ng$. So $gN \subset Ng$ for any $g \in G$. For example, $g^{-1}N \subset Ng^{-1}$ holds as well. Then if we multiply g on the left and on the right of the inclusion, we also derive that $Ng \supset gN$ and this makes 2. $gN = Ng$.
2. $\to$ 3. : Suppose that $gN = Ng$ for $g \in G$. This implies that $gn = n_1g$ for $n, n_1 \in N$. For any $gng^{-1} \in gNg^{-1}$, we have that $gng^{-1} = n_1g^{-1}g = n_1 \in N$. This implies that $gNg^{-1} \subset N$. Conversely for $n \in N$, we have that $n = ngg^{-1} = gn_2g^{-1}$ for $n_2 \in N$. This implies that $n = gn_2g^{-1} \in gNg^{-1}$. Thus N is a subset of gNg^{-1}. Thus $gNg^{-1} = N$.
3. $\to$ 2. : Suppose that for any $g \in G$, $gNg^{-1} = N$. This means that $gng^{-1} = n_1 \in N$ for any $n \in N$ and this is the definition of a normal subgroup.
□

By simply multiplying g^{-1} on the right of the equation in 2, we can derive 3 as well.

Example 4.2.1. *If G is an Abelian group, then every subgroup of G is normal.*

Proof. Let H be a subgroup of G. For any $g \in G$ and for any $h \in H$, we have that $g^{-1}hg = g^{-1}gh = h \in H$. This implies that H is a normal subgroup of G. □

Example 4.2.2. *In Exercise 3.2.4, the subgroup $N = \{\sigma_1, \sigma_2, \sigma_3\}$ of D_3 is a normal subgroup of D_3.*

Hint. Check $\sigma_i N \sigma_i^{-1} \in N$ for all $\sigma_i \in D_3$ where $1 \leq i \leq 6$. □

Now we'll make new groups using normal subgroups. Suppose N is a normal subgroup of a group G and we denote G/N, the set of all left cosets of N in G, i.e.,

$$G/N = \{Ng | g \in G\}$$

Then we define an operation on G/N as

$$Ng_1 \cdot Ng_2 = Ng_1g_2$$

for Ng_1 and $Ng_2 \in G/N$. First we show that this operation is well-defined. Suppose $Ng_1 = Ng_1'$ and $Ng_2 = Ng_2'$. Then $g_1 = n_1g_1'$ and $g_2 = n_2g_2'$ for some $n_1, n_2 \in N$ and this implies that $Ng_1Ng_2 = Ng_1g_2 = Nn_1g_1'n_2g_2' = Ng_1'n_2g_2' = Nn_3g_1'g_2' = Ng_1'g_2'$ for $n_3 \in N$. This shows that the operation on G/N is well-defined. Obviously the associativity holds on G/N and $N = Ne$ is the identity of G/N. Also, for any $Ng \in G/N$, Ng^{-1} is the inverse of Ng. Therefore, G/N is a group under the operation and it is called the **factor group** (or **quotient group**) of G by N.

Example 4.2.3. *Find* $\mathbb{Z}/5\mathbb{Z}$.

Solution. $\mathbb{Z}/5\mathbb{Z} = \{0 + 5\mathbb{Z} = 5\mathbb{Z}, 1 + 5\mathbb{Z}, 2 + 5\mathbb{Z}, 3 + 5\mathbb{Z}, 4 + 5\mathbb{Z}\}$. □

Note that $5\mathbb{Z}$ acts as the identity. Also, note that $2 + 3 \in 5\mathbb{Z}$ and it makes $2 + 3 + 5\mathbb{Z} = 5\mathbb{Z}$. This also implies that $3 + 5\mathbb{Z} = -2 + 5\mathbb{Z}$ and the inverse of $3 + 5\mathbb{Z}$ is $2 + 5\mathbb{Z}$. In general, the inverse of $m + 5\mathbb{Z}$ is $(5 - m) + 5\mathbb{Z}$ so we can look at the operation on $\mathbb{Z}/5\mathbb{Z}$ as the addition modulo $5\mathbb{Z}$.

Example 4.2.4. *Make a factor group of* $\mathbb{Z}_{12}$.

Solution. First, we need a normal subgroup of $\mathbb{Z}_{12}$ but all subgroups are normal since $\mathbb{Z}_{12}$ is Abelian. For example, since $< \overline{4} > \lhd \mathbb{Z}_{12}$, $\mathbb{Z}_{12}/< \overline{4} > = \{\overline{0}+ < \overline{4} >, \overline{1}+ < \overline{4} >, \overline{2}+ < \overline{4} >, \overline{3}+ < \overline{4} >\}$. □

An easy way to check if a set is a group is to make the Cayley table so you will see that $\mathbb{Z}_{12}/< \overline{4} >$ is a well-defined group from its Cayley table. Also note that $< \overline{4} >= \overline{0}+ < \overline{4} >= \overline{4}+ < \overline{4} >= \overline{8}+ < \overline{4} >$, $\overline{1}+ < \overline{4} >= \overline{5}+ < \overline{4} >= \overline{9}+ < \overline{4} >$, $\overline{2}+ < \overline{4} >= \overline{6}+ < \overline{4} >= \overline{10}+ < \overline{4} >$, and $\overline{3}+ < \overline{4} >= \overline{7}+ < \overline{4} >= \overline{11}+ < \overline{4} >$ in $\mathbb{Z}_{12}/< \overline{4} >$. This means that a coset representative is not unique but the operation on a factor group is still well-defined.

Theorem 4.2.2. *Let H and L be normal subgroups of a group G.*

1. *If L is a subgroup of H, then L is a normal subgroup of H.*

2. *$H \cap L$ is a normal subgroup of L and H.*

Proof. 1. Since L is a normal subgroup of G, for any $h \in H \subset G$ and $l \in L$, $h^{-1}lh \in L$. This implies that L is a normal subgroup of H.
2. For any $x \in H \cap L$, we have that $x \in H$ and $x \in L$. Since H and L be

normal subgroups of a group G, for all $l \in L$, $l^{-1}xl \in H$ and $l^{-1}xl \in L$. This implies that $l^{-1}xl \in L \cap H$. Thus $H \cap L$ is a normal subgroup of L. Similarly, $H \cap L \lhd H$. □

If $N \lhd G$, N commutes with all elements in G and if $N \not\lhd G$, N commutes with "some" elements in G. This leads us the following useful definition.

Definition 4.2.2. *Let H be a subgroup of a group G. Then $\{g \in G | gHg^{-1} = H$, i.e., $gH = Hg\}$ is called the* **normalizer** *and denoted by $N_G(H)$.*

It is obvious that $N_G(H) = G$ if $N \lhd G$. The following is for a relation between a subgroup and its normalizer.

Theorem 4.2.3. *If $H \leq G$, $H \lhd N_G(H) \leq G$.*

Proof. First we prove that $N_G(H)$ is a subgroup of G. Since the identity of a group is an element of $N_G(H)$, $N_G(H)$ is not empty. For any $x \in N_G(H)$, we have that $xH = Hx$. This implies that $Hx^{-1} = x^{-1}H$, i.e., $x^{-1} \in N_G(H)$. For any $x, y \in N_G(H)$, we have that $xy^{-1}h(xy^{-1})^{-1} = xy^{-1}hyx^{-1} = xh_1x^{-1} \in H$ for $h_1 = y^{-1}hy \in H$. So $N_G(H) < G$. For any $x \in N_G(H)$, we have that $x^{-1}hx \in H$. This implies that H is a normal subgroup of $N_G(H)$ and it completes the proof. □

The following is another example of normal subgroups and it is used in many parts of mathematics and sciences.

Definition 4.2.3. *Suppose a, b are elements of a group G. We call $a^{-1}b^{-1}ab$* **the commutator of** *a* **and** *b and it is denoted by $[a, b]$. We also call the set* *of all commutators of elements in G* **the commutator subgroup of** *G and it is denoted by G'.*

Clearly G' is a subgroup of G. For example, for any $g_1^{-1}g_2^{-1}g_1g_2 \in G'$, the inverse $(g_1^{-1}g_2^{-1}g_1g_2)^{-1} = g_2^{-1}g_1^{-1}g_2g_1$ of $g_1^{-1}g_2^{-1}g_1g_2 \in G'$ is an element of G'. We also can say that G' is generated by all the commutators of G, i.e.,

$$G' = \{x_1 \cdots x_n | x_i \text{ is a commutator of } G, 1 \leq i \leq n\}.$$

Proposition 4.2.1. *$G' \lhd G$.*

Proof. It is enough to prove that for any $a \in G$ and $g_1^{-1}g_2^{-1}g_1g_2 \in G'$, $a^{-1}g_1^{-1}g_2^{-1}g_1g_2a$ is an element of G' and it is true since

$$\begin{aligned} a^{-1}g_1^{-1}g_2^{-1}g_1g_2a &= a^{-1}g_1^{-1}aa^{-1}g_2^{-1}aa^{-1}g_1aa^{-1}g_2a \\ &= (a^{-1}g_1a)^{-1}(a^{-1}g_2a)^{-1}a^{-1}g_1aa^{-1}g_2a \in G' \quad \square \end{aligned}$$

Example 4.2.5. *If G is Abelian, $G' = \{e\}$.*

Chapter 4 Exercises.

1. Suppose $H < G$ and $x \in G$. Then prove the following

   (a) $xHx^{-1} < G$

   (b) $|xHx^{-1}| = |H|$

   (c) If H is Abelian, xHx^{-1} is Abelian as well.

2. Suppose $H < G$ and let's define a relation $\sim$ on G as follows: $a \sim b$ if and only if $a^{-1}b \in H$ for any $a, b \in G$.

   (a) Prove that $\sim$ is an equivalence relation.

   (b) Find the set of all the equivalence classes.

3. If $H < G$ and $N \triangleleft G$, prove that $HN < G$. Is $HN \triangleleft G$?

4. Suppose $N_1, N_2 \triangleleft G$. Then prove the following

   (a) $N_1 \cap N_2 \triangleleft G$

   (b) $N_1N_2 \triangleleft G$

   (c) If $N_1 \cap N_2 = \{e\}$, $n_1n_2 = n_2n_1$ for any $n_1 \in N_1$ and $n_2 \in N_2$

5. Suppose $N_1, \cdots, N_n$ are normal subgroups of a group G. Then prove the following.

   (a) $\cap_{1 \le i \le n} N_i \triangleleft G$

   (b) $N_1 \cdots N_n \triangleleft G$

6. Prove that if every proper subgroup of a group is Abelian, then the group is Abelian.

7. If $H < G$ with $[G : H] = 2$, prove that $H \triangleleft G$.

8. Prove that if a group G has only two subgroups (trivial ones), then $|G|$ is prime.

9. Suppose $H < G$ and let $N = \cap_{g \in G} gHg^{-1}$. Then prove that $N \triangleleft G$.

10. If g and h are elements of a group such that $g^5 = e$ and $ghg^{-1} = h^2$, then compute $|h|$.

11. Prove that $SL_n(\mathbb{R})$ is a normal subgroup of $GL_n(\mathbb{R})$.

12. Let $G = \left\{ \begin{pmatrix} a & b \\ 0 & c \end{pmatrix} | a, b, c \in \mathbb{Q}, ac \neq 0 \right\}$ and $G_1 = \left\{ \begin{pmatrix} 1 & a \\ 0 & 1 \end{pmatrix} | a \in \mathbb{Q} \right\}$.
Prove that $G_1 \lhd G$.

13. If a_1, a_2, a_3, and a_4 are elements of a group G, show that
$a_1^{-1}a_2^{-1}a_1a_2a_3^{-1}a_4^{-1}a_3a_4 \in G'$.

14. If $G = \left\{ \begin{pmatrix} a & b \\ 0 & c \end{pmatrix} | a, b, c \in \mathbb{Q}, ac \neq 0 \right\}$, prove that
$G' = \left\{ \begin{pmatrix} 1 & d \\ 0 & 1 \end{pmatrix} | \, d \in \mathbb{Q} \right\}$.

15. Suppose $N \lhd G$. Prove that G/N is Abelian if and only if $G' < N$.

16. Let's consider a multiplicative group $G = \{1, -1, i, -i\}$

   (a) Prove that $G_1 = \{1, -1\} \lhd G$.

   (b) Find the elements of G/G_1 and construct its Cayley table.

17. If $L \lhd G$ and $H < G$, prove that $L \cap H \lhd H$ and $L \lhd LH$.

Chapter 5

Homomorphisms and Isomorphisms

Since there is a bijection from the additive group $2\mathbb{Z}$ to $\mathbb{Z}$, can we say that they are the same groups in some sense? The answer is yes (in some sense) and we'd like to confirm this intuition using algebraic structures.

When we say that two groups are the same, they don't have to be the same physically or element-wise. We will say that two groups are the same when there exists a bijection between them and this bijection preserves their group structure. In this case, we don't use the word "same" because they don't have to be exactly same.

5.1 Homomorphisms

First, we define a function that preserves group structures.

Definition 5.1.1. *Let θ be a map from a group $(G_1, *_1)$ to a group $(G_2, *_2)$. This θ is called* **a homomorphism** *if $\theta(g_1 *_1 g_2) = \theta(g_1) *_2 \theta(g_2)$ for any $g_1, g_2 \in G_1$.*

If θ is a homomorphism from a group to itself, it is called **an endomorphism.** and the following two examples are obvious but still useful ones.

Example 5.1.1. *If we define a map θ from a group G_1 to another group G_2 by $\theta(g) = e_2$ for any $g \in G_1$ where e_2 is the identity of G_2, then θ is an homomorphism.*

Solution. For $g_1, g_2 \in G_1$, we have that $\theta(g_1 g_2) = e_2$. On the other hand, $\theta(g_1)\theta(g_2) = e_2 e_2 = e_2$. So $\theta(g_1 g_2) = \theta(g_1)\theta(g_2)$ and this implies that θ is a group homomorphism. □

Example 5.1.2. *If we define a map θ from G to itself by $\iota(g) = g$ for all $g \in G$, ι is a group endomorphism of G.*

Solution. For $g_1, g_2 \in G$, we have that $\iota(g_1 g_2) = g_1 g_2$. On the other hand, $\iota(g_1)\iota(g_2) = g_1 g_2$. So $\iota(g_1 g_2) = \iota(g_1)\theta(g_2)$ and this implies that ι is a group homomorphism. □

The map ι is called the identity map on G.

Definition 5.1.2. *Suppose θ be a group homomorphism. Then, we define the following.*

1. *θ is called a* **monomorphism** *if it is injective (one to one).*
2. *θ is called an* **epimorphism** *if it is surjective (onto).*
3. *θ is called an* **isomorphism** *if it is bijective (one to one & onto).*

4. θ is called an **automorphism** *if it is an isomorphism from a group to itself.*

If there exists an isomorphism from a group G_1 to another group G_2, we call them **isomorphic** and denote $G_1 \cong G_2$.

Example 5.1.3. $(\mathbb{Z}_4, + \text{ } mod \text{ } 4) \cong (M_4 = \{1, -1, i, -i\}, \times)$

Solution. Let's define θ from $\mathbb{Z}_4$ to M_4 as $\theta(\overline{0}) = 1$, $\theta(\overline{1}) = i$, $\theta(\overline{2}) = -1$, and $\theta(\overline{3}) = -i$. Then, θ is an isomorphism. □

Definition 5.1.3. *Let θ be a homomorphism from a group G_1 to a group G_2. Then, the set of all the elements in G_1 whose image is the identity in G_2 is called* **the kernel of** θ *and denoted by $ker(\theta)$*

This means that $ker(\theta) = \{g \in G_1 | \theta(g) = e_2\}$ where e_2 is the identity of G_2. We also use $im(\theta)$ for the set of all the images under θ, i.e., $im(\theta) = \{\theta(g) | g \in G_1\}$ and the following is an important property of them.

Theorem 5.1.1. *Let θ be a (group) homomorphism from a group G_1 to a group G_2. Let e_1 and e_2 be the identities of G_1 and G_2 respectively. Then*

1. $ker(\theta) \lhd G_1$

2. $im(\theta) < G_2$

Proof. 1. For $g_1, g_2 \in ker(\theta)$, we have that $\theta(g_1 g_2) = \theta(g_1)\theta(g_2) = e_2 e_2 = e_2$, i.e., $g_1 g_2 \in ker(\theta)$. This implies that $ker(\theta)$ is closed under the operation. $\theta(g_1 g_1^{-1}) = \theta(g_1)\theta(g_1^{-1}) = e_2\theta(g_1^{-1}) = e_2$. This implies that $\theta(g_1^{-1}) = e_2$ so $g_1^{-1} \in$

ker(θ) this makes $ker(\theta) < G_1$. For all $g \in G_1$ and $g_1 \in ker(\theta)$, $\theta(g^{-1}g_1g) = \theta(g^{-1})\theta(g_1)\theta(g) = \theta(g^{-1})e_2\theta(g) = \theta(g^{-1}g) = e_2$. This implies that $g^{-1}g_1g \in ker(\theta)$ and this makes $ker(\theta) \lhd G_1$.

2. Since $\theta(e_1) = e_2 \in im(\theta)$, $im(\theta)$ is not the empty set. For any $g_1', g_2' \in im(\theta)$, there are $g_1, g_2 \in G_1$ such that $\theta(g_1) = g_1'$ and $\theta(g_2) = g_2'$. This implies that $\theta(g_1g_2) = g_1'g_2'$, i.e., $g_1'g_2' \in im(\theta)$ so we derive that $im(\theta)$ is closed under operation. By $\theta(g_1g_1^{-1}) = e_2$, we have that $\theta(g_1)\theta(g_1^{-1}) = e_2$. This implies that $g_1'\theta(g_1^{-1}) = e_2$ and this makes $(g_1')^{-1} = \theta(g_1^{-1})$, i.e., $(g_1')^{-1} \in im(\theta)$. Thus, $im(\theta) < G_2$. □

Example 5.1.4. *Let θ be an isomorphism from a group G_1 to a group G_2. Then $ker(\theta) = \{e\}$ and $im(\theta) = G_2$.*

Example 5.1.5. *If θ be a map from $GL(n, \mathbb{R})$ to $\mathbb{R}^*$ such that $\theta(A)$ is the determinant of A, show that θ is an epimorphism and find its kernel.*

Solution. For any $A, B \in GL(n, \mathbb{R})$, $\theta(AB) = |AB| = |A||B| = \theta(A)\theta(B)$. This implies that θ is a homomorphism. For $A \in ker(\theta)$, we have that $\theta(A) = |A| = 1$. Conversely, for $A \in SL(n, \mathbb{R})$, we have that $\theta(A) = 1$. This implies that $ker(\theta) = SL(n, \mathbb{R})$. Since for any $a \in \mathbb{R}^*$, $\theta(\begin{pmatrix} a & 0 & 0 & \cdots & 0 & 0 \\ 0 & 1 & 0 & \cdots & 0 & 0 \\ & & & \cdots & & \\ 0 & 0 & 0 & \cdots & 1 & 0 \\ 0 & 0 & 0 & \cdots & 0 & 1 \end{pmatrix}) = a$.

This implies that θ is onto, i.e., $im(\theta) = R^*$. □

We know that a composition of injections is injective and we can state a similar property for a composition of surjections. This type of transitivity is true for homomorphisms.

Lemma 5.1.1. *Let θ_1 be a homomorphism from a group G_1 to a group G_2 and θ_2 be a homomorphism from a group G_2 to a group G_3. Then $\theta_2\theta_1$ is a homomorphism from G_1 to G_3.*

Proof. For any $g_1, g_2 \in G_1$, we have that

$$(\theta_2\theta_1)(g_1g_2) = \theta_2(\theta_1(g_1g_2)) = \theta_2(\theta_1(g_1)\theta_1(g_2)) =$$
$$\theta_2(\theta_1(g_1))\theta_2(\theta_1(g_2)) = (\theta_2\theta_1)(g_1)(\theta_2\theta_1)(g_2).$$

This implies that $\theta_2\theta_1$ is a homomorphism from G_1 to G_3. $\square$

The set of all homomorphisms from G_1 to G_2 is denoted by $Hom(G_1, G_2)$ and the set of all endomorphisms of G is denoted by $End(G)$, i.e.,

$$Hom(G_1, G_2) = \{\theta | \theta : G_1 \to G_2, \theta \text{ is a homomorphism}\}$$

and $End(G)$ is defined similarly, and we can show that these are groups under function composition using Lemma 5.1.1.

The following is an important property of the kernel of a homomorphism.

Lemma 5.1.2. *Let θ be a homomorphism from a group G_1 to a group G_2. θ is one to one if and only if $ker(\theta)$ is the trivial group.*

Proof. Assume that θ is one to one. This implies that since $e \in ker(\theta)$, we have that $ker(\theta) = \{e\}$. Conversely let us assume that $ker(\theta)$ is the trivial group. Let us assume that $\theta(g_1) = \theta(g_2)$ for $g_1, g_2 \in G_1$. This implies that

$\theta(g_1g_2^{-1}) = e_2$ where e_2 is the identity of G_2 and this makes $g_1g_2^{-1} = e_1$ where e_1 is the identity of G_1. So, $g_1 = g_2$ and this implies that θ is one to one. □

We combine these two lemmas as a theorem for compositions of isomorphisms. Its proof is obvious.

Theorem 5.1.2. *If θ_1 and θ_2 are isomorphisms, then so is $\theta_2\theta_1$.*

5.2 Isomorphism Theorems

In this section, we look at homomorphisms and their kernels. First, we'll check that a kernel is a normal subgroup so we can construct a quotient group. In this case, we can build an isomorphism naturally.

Theorem 5.2.1. *Suppose θ is a homomorphism from a a group G_1 to a group G_2.*

1. *If $N_1 \lhd G_1$, then $\theta(N_1) \lhd \theta(G_1)$.*

2. *If $N_2 \lhd \theta(G_1)$, then $\theta^{-1}(N_2) \lhd G_1$.*

Proof. 1. By Theorem 5.1.1, $\theta(N_1)$ is a subgroup of $\theta(G_1)$. Suppose $g_1' \in \theta(G_1)$. Then, there is $g_1 \in G_1$ such that $\theta(g_1) = g_1'$. This implies that $(g_1')^{-1}\theta(N_1)g_1' = \theta(g_1)^{-1}\theta(N_1)\theta(g_1) = \theta(g_1^{-1}N_1g_1) = \theta(N_1)$ and this makes $\theta(N_1)$ a normal subgroup of $\theta(G_1)$.
2. Since $e_1 \in \theta^{-1}(N_2)$, $\theta^{-1}(N_2) \neq \phi$. Suppose $g_1, g_2 \in \theta^{-1}(N_2)$. Then, $\theta(g_1), \theta(g_2) \in N_2$ so $\theta(g_1)\theta(g_2) = \theta(g_1g_2) \in N_2$ and this implies that $g_1g_2 \in \theta^{-1}(N_2)$. Also since $\theta(g_2^{-1}) = (\theta(g_2))^{-1} \in N_2$, $g_2^{-1} \in \theta^{-1}(N_2)$. Thus $\theta^{-1}(N_2)$ is a subgroup of G_1 by

the two-step subgroup test. Now, suppose $g_1 \in G_1$. Then, $\theta(g_1^{-1}\theta^{-1}(N_2)g_1) = \theta(g_1^{-1})\theta(\theta^{-1}(N_2))\theta(g_1) = N_2$ and this implies that $g_1^{-1}\theta^{-1}(N_2)g_1 = \theta^{-1}(N_2)$. Therefore, $\theta^{-1}(N_2)$ is a normal subgroup of G_1. □

Lemma 5.2.1. *Suppose N is a normal subgroup of a group G. If the map $\mathfrak{p} : G \to G/N$ is defined by $\mathfrak{p}(g) = Ng$ for $g \in G$, $\mathfrak{p}$ is a group homomorphism with N as its kernel.*

Proof. For all $g_1, g_2 \in G$, $\mathfrak{p}(g_1g_2) = Ng_1g_2 = Ng_1Ng_2 = \mathfrak{p}(g_1)\mathfrak{p}(g_2)$. So $\mathfrak{p}$ is a group homomorphism. If $n \in N$, $\mathfrak{p}(n) = Nn = N$ so $N \subseteq ker(\mathfrak{p})$. If $g \in ker(\mathfrak{p})$, $\mathfrak{p}(g) = Ng = N$ and this implies that $g \in N$. Thus the kernel $ker(\mathfrak{p})$ is N and this proves the lemma. □

We call the map $\mathfrak{p}$ of Lemma 5.2.1 the **canonical homomorphism** from a group G to G/N.

Theorem 5.2.2. (First Isomorphism Theorem)
If θ is a homomorphism from G, then $G/ker(\theta) \cong im(\theta)$.

Proof. Let $N = ker(\theta)$ and $\overline{\theta}$ be a map from G/N to $im(\theta)$ defined by

$$\overline{\theta}(Ng) = \theta(g)$$

for $Ng \in G/N$. We will show that $\overline{\theta}$ is an isomorphism using the following commutative diagram:

Figure 5.2.1.

$$\begin{array}{ccc} G & \longrightarrow & im(\theta) \\ & \searrow \quad \nearrow & \\ & G/N & \end{array}$$

1. $\overline{\theta}$ is well-defined and one to one.

$$\begin{aligned} Ng_1 = Ng_2 &\iff Ng_1g_2^{-1} = N \\ &\iff g_1g_2^{-1} \in N \\ &\iff \theta(g_1)\theta(g_2)^{-1} = \theta(g_1)\theta(g_2^{-1}) = \theta(g_1g_2^{-1}) = e \\ &\iff \theta(g_1) = \theta(g_2) \\ &\iff \overline{\theta}(Ng_1) = \overline{\theta}(Ng_2) \end{aligned}$$

2. $\overline{\theta}$ is a homomorphism.

   For $Ng_1, Ng_2 \in G/N$, $\overline{\theta}(Ng_1Ng_2) = \overline{\theta}(Ng_1g_2) = \theta(g_1g_2) = \theta(g_1)\theta(g_2) = \overline{\theta}(Ng_1)\overline{\theta}(Ng_2)$ and this implies that $\overline{\theta}$ is a homomorphism.

3. $\overline{\theta}$ is onto.

   For any $g_2 \in \theta(G)$, there is $g_1 \in G$ such that $\theta(g_1) = g_2$. This implies that $\overline{\theta}(Ng_1) = \theta(g_1) = g_2$ and this makes the function surjective. □

Example 5.2.1. *Let θ be a group homomorphism from $\mathbb{Z}_{24}$ to $\mathbb{Z}_{12}$ defined by $\theta(\overline{1}) = \overline{2}$. Apply the First Isomorphism Theorem to θ.*

Solution. Note that $ker(\theta) = \{\overline{0}, \overline{6}, \overline{12}, \overline{18}\}$ and $im(\theta) = \{\overline{0}, \overline{2}, \overline{4}, \overline{6}, \overline{8}, \overline{10}\}$ since $\theta(\overline{n}) = \overline{2n}\ mod\ 12$. Then, the canonical isomorphism $\overline{\theta}$ from $\mathbb{Z}_{24}/ker(\theta)$ to $im(\theta)$ should be defined as

$$\overline{\theta}(\overline{i} + ker(\theta)) = \theta(\overline{i})$$

for $\overline{i} + ker(\theta) \in \mathbb{Z}_{24}/ker(\theta)$. □

Corollary 5.2.1. *If θ is an epimorphism from G_1 to G_2, then $G_1/ker(\theta) \cong G_2$.*

Proof. Since θ is surjective, $im(\theta) = G_2$ so $G_1/ker(\theta) \cong G_2$ by Theorem (5.2.2). □

Corollary 5.2.2. *If N is a normal subgroup of a group G, then N can be a kernel of a homomorphism from G to another group.*

Proof. Let's define a map θ from G to G/N as follows:

$$\theta(g) = Ng$$

for $g \in G$. Then, θ is a homomorphism from G to G/N and if $g \in ker(\theta)$, $\theta(g) = Ng = N$. This implies that $g \in N$ and this makes $ker(\theta) \subset N$. Conversely if $g \in N$, $\theta(g) = Ng = N$ and it implies that $N \subset ker(\theta)$. Therefore, N is the kernel of θ. □

Now, we will state the 2nd and 3rd theorems by looking at the following lemma.

Lemma 5.2.2. *1. If $H < G$ and $N \lhd G$, then $HN < G$.*

2. If $H, N \lhd G$, then $HN \lhd G$.

Proof.

1. Since the identity of G is an element of HN, HN is not empty. For $h_1n_1, h_2n_2 \in HN$, $h_1n_1(h_2n_2)^{-1} = h_1(n_1n_2^{-1})h_2^{-1} = h_1h_2^{-1}n_3 \in HN$ for some $n_3 \in N$. This implies that HN is a subgroup of G.

2. Since H and N are normal subgroups of G, $g^{-1}HNg = g^{-1}Hgg^{-1}Ng = HN$ for any $g \in G$ and this implies that HN is a normal subgroup of G. $\square$

Theorem 5.2.3. (Second and Third Isomorphism Theorems) *Suppose H and L are normal subgroups of a group G.*

1. *If L is a subgroup of H, then G/H and $(G/L)/(H/L)$ are isomorphic.*

2. *LH/L and $H/(L \cap H)$ are isomorphic.*

Proof.

1. Let us define a map θ from G/L to G/H as follows:

$$\theta(gL) = gH$$

for $g \in G$. If $g_1L, g_2L \in G/L$, $\theta(g_1Lg_2L) = \theta(g_1g_2L) = g_1g_2H = g_1Hg_2H = \theta(g_1L)\theta(g_2L)$ and this implies that θ is a homomorphism. Obviously θ is surjective. Now, let's take a look at its kernel. Note that

$$\begin{aligned} gL \in ker(\theta) &\iff \theta(gL) = gH = H \\ &\iff g \in H \\ &\iff gL \in H/L \end{aligned}$$

and this implies that the kernel of θ is H/L. Thus G/H and $(G/L)/(H/L)$ are isomorphic by the 1st Isomorphism Theorem.

2. LH is a normal subgroup of G and $L \cap H$ is a normal subgroup of H by

Lemma 5.2.2. Let's define a map θ from LH to $H/H \cap L$ as follows:

$$\theta(lh) = h(H \cap L)$$

for $lh \in LH$ and we'll show that θ is a well-defined epimorphism.

(a) Assume that $l_1 h_1 = l_2 h_2 \in LH$ where $l_1, l_2 \in L$ and $h_1, h_2 \in H$. Then, $l_2^{-1} l_1 = h_2 h_1^{-1} \in H \cap L$ and this implies that $h_1(H \cap L) = h_2(H \cap L)$. So $\theta(l_1 h_1) = \theta(l_2 h_2)$ and it shows that θ is well-defined.

(b) Now we'll show that θ is a homomorphism using the following commutative diagram:

Figure 5.2.2.

$$\begin{array}{ccc} LH & \longrightarrow & H/H \cap L \\ & \searrow \quad \swarrow & \\ & LH/L & \end{array}$$

If $l_1 h_1, l_2 h_2 \in LH$, since $H \cap L$ is a normal subgroup of G, we have that $\theta(l_1 h_1 l_2 h_2) = \theta(l_1 l_2' h_1 h_2) = h_1 h_2 (H \cap L) = \theta(l_1 h_1)\theta(l_2 h_2)$ for some $l_2' \in L$. This implies that θ is a homomorphism.

(c) θ is surjective since $\theta(eh) = h(H \cap L)$ for any $h \in H$.

(d) If $lh \in ker(\theta)$, $\theta(lh) = h(H \cap L) = H \cap L$ and this implies that $h \in L$. So $lh \in L$ and this makes $ker(\theta) \subset L$. Conversely, if $h \in L$, since $\theta(he) = e(H \cap L) = H \cap L$, $L \subset ker(\theta)$ and we derive that $ker(\theta) = L$.

Therefore, LH/L and $H/(L \cap H)$ are isomorphic by the 1st Isomorphism Theorem. $\square$

Example 5.2.2. *Check the 2nd Isomorphism Theorem for $G = \mathbb{Z}_{24}$ and its subgroups $L =< \overline{2} >$ and $H =< \overline{6} >$.*

Solution. Since G is Abelian, L and H are normal subgroups of G. Then, $G/L = \{L, \overline{1} + L\}$, $G/H = \{H, \overline{1} + H, \overline{2} + H, \overline{3} + H, \overline{4} + H, \overline{5} + H\}$, and $L/H = \{H, \overline{2} + H, \overline{4} + H\}$. So, $G/L \cong \mathbb{Z}_2$ and $(G/H)/(L/H) = \{H + L/H, \overline{1} + H + L/H\} \cong \mathbb{Z}_2$. Therefore, $G/L \cong (G/H)/(L/H)$. □

5.3 Isomorphic Infinite Groups

Recall that $\mathbb{Z}/ < n >= \mathbb{Z}/n\mathbb{Z} \cong \mathbb{Z}_n$ and we can see this through the modulo homomorphism from $\mathbb{Z}$ to $\mathbb{Z}_n$ with its kernel $< n >= n\mathbb{Z}$. We'd like to look at number groups including $\mathbb{Z}$ and $\mathbb{Z}_n$ using homomorphisms between them. We also look at a much bigger additive group $\mathbb{R}$ and a multiplicative group $\mathbb{R}^*$, and find a homomorphism between them.

We can see that two groups are not isomorphic if two groups are finite with different orders. If two groups are infinite, it is not always true.

Example 5.3.1. *Show that $< 2 >= 2\mathbb{Z} \cong \mathbb{Z}$.*

Solution. Let θ be a map from $2\mathbb{Z}$ to $\mathbb{Z}$ defined by

$$\theta(2a) = a$$

for $2a \in 2\mathbb{Z}$. Then for $2a, 2b \in 2\mathbb{Z}$, $\theta(2a+2b) = \theta(2(a+b)) = a+b = \theta(2a)+\theta(2b)$ and if $\theta(2a) = 0$, then $a = 0$ so θ is a monomorphism. For any $b \in \mathbb{Z}$, we have that $\theta(2b) = b$ and this shows that θ is onto. Therefore, θ is an isomorphism from $2\mathbb{Z}$ to $\mathbb{Z}$. $\square$

By Example 5.3.1, we can not distinguish the additive group $\mathbb{Z}$ with its subgroup $2\mathbb{Z}$ since they are isomorphic. The following is another example of a group and its isomorphic subset but with different operations.

Example 5.3.2. *Show that* $(\mathbb{R}, +)$ *and* $(\mathbb{R}^+, *)$ *are isomorphic.*[1]

Solution. Let's define a map θ from $\mathbb{R}$ to $\mathbb{R}^+$ as follows:

$$\theta(a) = e^a$$

where e is the base of the natural logarithm. Then $\theta(a + b) = e^{a+b} = e^a e^b = \theta(a)\theta(b)$ shows that θ is a homomorphism. If $\theta(a) = 1$, then $e^a = 1$. This implies that $a = 0$ and θ is one to one since the kernel is trivial. If $x \in \mathbb{R}^+$, $\theta(ln(x)) = x$ since $x = e^{lnx}$ so θ is onto. Therefore θ is a desired isomorphism between $\mathbb{R}$ and $\mathbb{R}^+$ $\square$

As we see from these examples, cardinalities of infinite groups are quite different from ones of finite groups and they could be studies in a course of Set Theory.

Chapter 5 Exercises.

1. Suppose θ is a group endomorphism of $\mathbb{Z}_8$ with $\theta(\overline{1}) = \overline{3}$.

   (a) Find $\theta(\overline{4})$ and $\theta(\overline{5})$.

[1] $\mathbb{R}^+$ is the set of all positive real numbers.

(b) Is θ one to one? Onto? Prove your claim.

(c) Find $ker(\theta)$.

(d) Find a non-trivial non-isomorphic endomorphism of $\mathbb{Z}_8$.

2. Let θ be a group endomorphism of $\mathbb{Z}_9$ with $\theta(\overline{1}) = \overline{4}$.

   (a) Find $\theta(\overline{4})$ and $\theta(\overline{5})$.

   (b) Is θ one to one and onto? Prove your claim.

   (c) Find $ker(\theta)$.

   (d) Find a non-trivial non-isomorphic group endomorphism of $\mathbb{Z}_9$.

   (e) Find a non-identity automorphism of $\mathbb{Z}_{12}$.

3. (a) Find a non-trivial group endomorphism of $\mathbb{Z}_3$.

   (b) Identify all the group endomorphisms of $\mathbb{Z}_3$.

   (c) Find a non-trivial group endomorphism of $\mathbb{Z}_p$ where p is prime.

   (d) Identify all the group homomorphisms from $\mathbb{Z}_p$ to itself.

4. Let θ be a group homomorphism from $\mathbb{Z}_{12}$ to $\mathbb{Z}_6$ with $\theta(\overline{1}) = \overline{2}$.

   (a) Find $\theta(\overline{2})$, $\theta(\overline{3})$, $\theta(\overline{5})$, and $\theta(\overline{6})$.

   (b) Find $ker(\theta)$.

   (c) Find the order of 4 in $\mathbb{Z}_{12}$ and $\mathbb{Z}_6$.

   (d) Find $|\theta(4)|$.

5. Let $\theta : \mathbb{Z}_6 \to \mathbb{Z}_{12}$ be a map defined by $\theta(a) = 2a$ for $a \in \mathbb{Z}_6$.

   (a) Prove that θ is a homomorphism.

   (b) Find $\theta(\overline{0})$, $\theta(\overline{1})$, $\theta(\overline{2})$, and $\theta(\overline{3})$.

(c) Find $ker(\theta)$.

(d) Is θ one to one? Onto? Prove your claim.

6. Let $\theta : \mathbb{Z} \to \mathbb{Z}_8$ be a map such that $\phi(a) = \overline{a}$ for $a \in \mathbb{Z}$.

   (a) Prove that θ is a homomorphism.

   (b) Find $ker(\theta)$.

   (c) Is θ one to one? Onto? Prove your claim.

7. (a) Find an isomorphism between $\mathbb{Z}$ and $3\mathbb{Z}$.

   (b) Find an isomorphism between $\mathbb{Z}$ and $m\mathbb{Z}$ for any natural number m.

   (c) Find an isomorphism between $m\mathbb{Z}$ and $n\mathbb{Z}$ for any natural numbers m and n.

8. Let θ be a group isomorphism from a group G_1 to another group G_2.

   (a) Prove that if H_1 is an Abelian subgroup of G_1, then $\theta(H_1)$ is an Abelian subgroup of G_2.

   (b) Prove that if H_2 is an Abelian subgroup of G_2, then $\theta^{-1}(H_2)$ is an Abelian subgroup of G_1.

9. Let $\theta : \mathbb{R}^* \to \mathbb{R}^*$ be the following map where $\mathbb{R}^*$ is the group of all non-zero elements of $\mathbb{R}$ under the multiplication.

   (a) If $\theta(a) = |a|$, then show that θ is a group homomorphism and find $ker(\theta)$.

   (b) If $\theta(a) = a^2$, then show that θ is a group homomorphism and find $ker(\theta)$.

10. Suppose $\theta_i : G_i \to G_{i+1}$ is a homomorphism where G_i is a group for $i \in \mathbb{N}$.

(a) Prove that $\theta_{k+2} \circ \theta_{k+1} \circ \theta_k$ is a homomorphism from G_k to G_{k+3} for $k \in \mathbb{N}$ where $\circ$ is the composition of maps.

(b) Prove that $\theta_k \circ \theta_{k-1} \circ \cdots \circ \theta_{k-p}$ is a homomorphism from G_{k-p} to G_{k+1}.

11. Suppose $\theta : G_1 \longrightarrow G_2$ is a group homomorphism.

(a) If $H_2 < G_2$, prove that $\theta(\theta^{-1}(H_2)) < G_2$. Does the equality $\theta(\theta^{-1}(H_2)) = H_2$ hold? Prove your claim.

(b) If $H_1 < G_1$, prove that $\theta^{-1}(\theta(H_1)) < G_1$. Does the equality $\theta^{-1}(\theta(H_1)) = H_1$ hold? Prove your claim.

12. If $\theta : \mathbb{Z}_n \longrightarrow \mathbb{Z}_n$ is defined by $\theta(x) = -x$ for $x \in \mathbb{Z}_n$, prove that θ is a homomorphism.

13. Suppose $\theta_1 : G_1 \longrightarrow G_2$ and $\theta_2 : G_2 \longrightarrow G_3$ are group homomorphisms.

(a) If $N_1 \lhd G_1$, prove that $\theta_2\theta_1(N_1) \lhd G_3$.

(b) If $N_3 \lhd G_3$, prove that $\theta_1^{-1}(\theta_2^{-1}(N_3)) \lhd G_1$.

14. Prove that a group G is Abelian if and only if $\theta : G \longrightarrow G$ is an endomorphism where $\theta(a) = a^{-1}$ for $a \in G$.

15. Prove that a group G is Abelian if and only if $\theta : G \longrightarrow G$ is an endomorphism where $\theta(a) = a^2$ for any $a \in G$.

16. Recall that $Hom(G, H)$ is the set of all homomorphisms from G to H.

(a) Identify $Hom(\mathbb{Z}_4, \mathbb{Z}_7)$.

(b) Identify $Hom(\mathbb{Z}_m, \mathbb{Z}_n)$ if $m \neq n$.

17. Recall that $End(G) = Hom(G, G)$.

(a) Identify $End(\mathbb{Z}_4)$. Which one is an automorphism? Is there $\theta \in End(\mathbb{Z}_4)$ such that $\theta(\overline{2}) = \overline{3}$?

(b) Identify $End(\mathbb{Z}_m)$.

18. (a) Find a monomorphism from $3\mathbb{Z}$ to $\mathbb{Z}$. Is it an isomorphism?

(b) Find a epimorphism from $3\mathbb{Z}$ to $\mathbb{Z}$. Is it an isomorphism?

(c) Identify all the isomorphisms from $3\mathbb{Z}$ to $\mathbb{Z}$.

19. Suppose θ is a map from $\mathbb{Z}_{48}$ to $\mathbb{Z}_{12}$ with $\theta(\overline{1}) = \overline{4}$.

(a) Extend θ to a homomorphism from $\mathbb{Z}_{48}$ to $\mathbb{Z}_{12}$.

(b) Find $ker(\theta)$.

(c) Check the First Isomorphism Theorem using this homomorphism.

20. Define a group isomorphism θ from $\mathbb{Z}/<n>$ to $\mathbb{Z}_n$.

21. Check the Third Isomorphism Theorem if $G = \mathbb{Z}_{24}$, $L =<\overline{4}>$, and $H =<\overline{6}>$.

Chapter 6

Cyclic Groups

If g is an element of a group G, we define the cyclic subgroup generated by g in G (Chapter 2). It is denoted by $< g >$ and $< g >= \{g^i | i \in \mathbb{Z}\}$ if G is multiplicative. If G is additive, $< g >= \{ng | n \in \mathbb{Z}\}$.

6.1 Definition and Examples

Since a subgroup is a group, we'd like to look at a cyclic subgroup as a group.

Definition 6.1.1. *A group G is* **cyclic** *if there is an element g of G such that $< g >= G$. This g is called a* **generator of** G.

Example 6.1.1. *The additive group $\mathbb{Z}$ is a cyclic group since $\mathbb{Z} =< 1 >$. Note that $\mathbb{Z} =< -1 >$ as well so there are only two generators for the cyclic group $\mathbb{Z}$.*

Example 6.1.2. *$\mathbb{Z}_n$ is an additive cyclic group since $< \bar{1} >= \mathbb{Z}_n$. $\bar{1}$ is a generator of the group.*

Example 6.1.3. *Show that the multiplicative subgroup $\{1, -1, i, -i\}$ is cyclic.*

Solution. Since $< i >= \{1, -1, i, -i\}$, it is a cyclic group and i is an generator of the group. We can also see that $< i >=< -i >$ so there are two generators of the group. $\square$

Note that 1 and -1 are not the generators of the cyclic group in Example 6.1.3.

Example 6.1.4. *The Klein 4-group $K_4 = \{e, a, b, c\}$ is a cyclic group since $a^2 = b^2 = c^2 = e$, i.e., there is no element $g \in K_4$ such that $< g >= K_4$.*

6.2 Properties of Cyclic Groups

From Example 6.1.4, we can derive the following proposition.

Proposition 6.2.1. *Suppose G is cyclic with $|G| = n$. Then there exists $g \in G$ with $|g| = n$.*

Corollary 6.2.1. *For each element g of a group, $|g| = | < g > |$.*

Proposition 6.2.2. *A cyclic group is Abelian.*

Proof. Suppose $G =< g >$ and $x, y \in G$. Then, $x = g^m$ and $y = g^n$ for some integers m and n, and we derive that

$$xy = g^m g^n = g^{m+n} = g^{n+m} = g^n g^m = yx.$$

Thus G is Abelian. □

Theorem 6.2.1. *Suppose G is cyclic group and $G =< g >$.*

1. *$g^i = g^j$ for some two distinct integers, i and j if and only if G is a finite group.*

2. *If G is infinite and $g \in G$ is not the identity, then $g^i \neq g^j$ for any two distinct integers, i and j.*

Proof.

1. Assume that $G =< g >$ is a finite group and $|G| = |g| = n.$, then $g^n = e$ and this implies that $g^i = g^{n+i}$ for any integer i. Conversely, assume that $g^i = g^j$ with $i > j$. Then $g^{i-j} = e$ and this means that there exists a positive integer n such that $g^n = e$. If we say that n is the smallest such an integer (the Well Ordering Principle), $\{e, g, \cdots, g^{n-1}\}$ is the group G. Therefore, G is finite.

2. Suppose that G is an infinite group. If we assume that $g^i = g^j$ for some integers $i > j$, then $g^{i-j} = e$ and this implies that G is a finite group by the 1st part of the theorem, a contradiction. Thus we have proven the theorem. □

Theorem 6.2.2. *Let G_1 be a cyclic group and $\theta : G_1 \longrightarrow G_2$ is a group homomorphism. Then, $im(\theta) = \theta(G_1)$ is a cyclic subgroup of G_2.*

Proof. Suppose $< g >= G_1$. Since for any $g_2 \in im(\theta)$, there is $g_1 \in G_1$ such that $\theta(g_1) = g_2$. Then, this implies that $g_1 = g^k$ for some $k \in \mathbb{Z}$. So, we derive that $\theta(g_1) = \theta(g^k) = (\theta(g))^k$ and this implies that $im(\theta) \subseteq =< \theta(g) >$. Since the other direction is obvious, we conclude that $im(\theta) =< \theta(g) >$, a cyclic subgroup of G_2. □

Let's check some properties of cyclic groups related to isomorphisms.

Theorem 6.2.3. *Suppose $\theta : G_1 \longrightarrow G_2$ is a group isomorphism.*

1. *$ker(\theta) = \{e\}$*
2. *If G_1 is cyclic, then G_2 is cyclic.*
3. *$|g| = |\theta(g)|$ for any $g \in G_1$.*
4. *If G_1 is a finite group, then G_2 is a finite group.*

Proof.

1. Since θ is an isomorphism, it is one to one. Thus by Lemma 5.1.2, $ker(\theta)$ is the trivial group.
2. Since G_1 is cyclic, there exists $g \in G_1$ such that $G_1 =< g >$. Then, since $G_2 = \theta(G_1)$, for any $g' \in G_2$, there exists $g^m \in G_1$ so that $g' = \theta(g^m)$ and this implies that $(\theta(g_1))^m = g'$. Thus, $G_2 =< \theta(g) >$.
3. Let $|g| = n$ and $|\theta(g)| = m$. Since $(\theta(g))^n = \theta(g^n) = e'$ where e' is the identity of G_2, we derive that $m|n$. On the other hand, since $\theta(g_1^m) =$

$(\theta(g))^m = e'$ and since θ is an isomorphism, we derive that $g^m = e$ where e is the identity of G_1. This implies that $n|m$. Thus $m = n$.

4. Since G is finite, let's say that $G_1 = \{g_1, \cdots, g_n\}$. Then, since θ is an isomorphism, we see that $G_2 = \{\theta(g_1), \cdots, \theta(g_n)\}$ and this makes G_2 finite.

□

We saw that a cyclic group is Abelian in Proposition 6.2.2 and this commutativity of a cyclic group is based on exponents. Since all the exponents are integers, we can extend this intuition to the following theorem.

Theorem 6.2.4. *Let G be a cyclic group.*

1. *If G is infinite, then G is isomorphic to $\mathbb{Z}$.*
2. *If G is finite, then G is isomorphic to $\mathbb{Z}_n$ and $|G| = n$.*

Proof.

1. Suppose that G is an infinite cyclic group and $G =< g >$. Let's define a map $\theta : G \longrightarrow \mathbb{Z}$ by $\theta(g^n) = n$ for $g \in G$. We'll show that θ is an isomorphism. Obviously, it is well-defined and from

$$\theta(g^i g^j) = \theta(g^{i+j}) = i + j = \theta(g^i) + \theta(g^j)$$

we derive that θ is a group homomorphism. If $\theta(g^i) = \theta(g^j)$, then $i = j$ by Theorem 6.2.1 since G is infinite, and this implies that θ is injective. If $k \in \mathbb{Z}$, since $\theta(g^k) = k$, we can say that θ is surjective. Therefore, θ is a desired isomorphism.

2. Suppose that $G =< g >$ is a finite group. Since $\{g^i | i \in \mathbb{Z}\}$ is finite, there are $i, j \in \mathbb{N}$ such that $g^i = g^j$ for some integers $i > j$ by Theorem 6.2.1. This implies that $g^{i-j} = e$ and there is the minimal positive integer n such that $g^n = e$ by the Well Ordering Principle. This makes $G = \{e, g, \cdots, g^{n-1}\}$ and $g^i \neq g^j$, $1 \leq i \neq j \leq n-1$ and if we define a map θ from G to $\mathbb{Z}_n$ by $\theta(g^i) = \bar{i}$ for $g^i \in G$, we can prove that θ is a desired isomorphism. The remaining proof is similar to the 1st part of the theorem.

□

The following theorem is useful to check if a group is Abelian by looking at its center.

Theorem 6.2.5. *Let G be a group and $\mathfrak{Z}$ be the center of G. If $G/\mathfrak{Z}$ is cyclic, then G is Abelian.*

Proof. Since $G/\mathfrak{Z}$ is cyclic, there is an element $x \in G$ such that $G/\mathfrak{Z} = < x\mathfrak{Z} >$. If $a\mathfrak{Z}, b\mathfrak{Z} \in G/\mathfrak{Z}$, we have that $a\mathfrak{Z} = (x\mathfrak{Z})^n = x^n\mathfrak{Z}$ and $b\mathfrak{Z} = (x\mathfrak{Z})^m = x^m\mathfrak{Z}$ where $n, m \in \mathbb{N}$. This implies that $a = x^n z_1$ and $b = x^m z_2$ for some $z_1, z_2 \in \mathfrak{Z}$ and this makes $ab = (x^n z_1)(x^m z_2) = x^m z_2 x^n z_1 = ba$. Therefore, G is Abelian.

□

6.3 Fundamental Theorem of Cyclic Groups

Lagrange Theorem says that the order of a subgroup divides the order of the group but it doesn't guarantee the existence of a subgroup of order n when n is a divisor of the group order. In fact, it is not true in general but it is true for cyclic groups. In this section, we start with a property of a subgroup of a cyclic group.

Lemma 6.3.1. *A subgroup of a cyclic group is cyclic.*

Proof. Suppose that $G =< g >$ and $H < G$. If $H = \{e\}$, then there is nothing to prove, so assume that $H \neq \{e\}$. For any $e \neq g_1 \in H \subset G$, there exists an integer m such that $g_1 = g^m \in H$. Since $g^{-m} \in H$, we can assume that m is a positive integer. Then, by the Well Ordering Principle, there is the minimal positive integer p such that $g^p \in H$. For any $g^s \in H$, we have that $s = pq + r$ such that $0 \leq r < p$. If $r \neq 0$, then $g^{s-pq} = g^r \in H$. This contradicts the minimality of p, so $r = 0$, i.e., $s = pq$. This implies that $H << g^p >$ and since $< g^p >$ is a subgroup of H, we conclude that $H =< g^p >$. □

The following theorem is the other direction of Lagrange Theorem and it is called the Fundamental Theorem of Cyclic Groups with Lemma 6.3.1.

Theorem 6.3.1. (Fundamental Theorem of Cyclic Groups)
If G is a finite cyclic group and k divides $|G|$, then there is a unique subgroup H of G such that $|H| = k$.

Proof. Let us assume that $G =< g >$ for $g \in G$. If $|G| = n$, $< g^{\frac{n}{k}} >$ is a subgroup of order k. □

The uniqueness of a subgroup of each order is not proved rigorously even though it is not too difficult. Instead of prove it, we'd like to see it through examples.

First, we'd like to see how to find a generator of a cyclic group. There can be several generator for a cyclic group and the following theorem explains how to find them.

Theorem 6.3.2. *For a finite cyclic group* $G =< g >$, $G =< g^i >$ *if and only if* $(i, n) = 1$ *where* $|G| = n$.

Proof. If $(i, n) = 1$, then there are $s, t \in \mathbb{Z}$ such that $is + nt = 1$ and this makes

$$g = g^{is+nt} = g^{is}g^{nt} = g^{is}(g^n)^t = g^{is}e^t = g^{is} \in < g^i >$$

This implies that $< g > \subset < g^i >$ and since $< g^i > \subset < g >$, we have that $< g >=< g^i >$. Conversely, assume that $G =< g >=< g^i >$. Then, $g \in < g^i >$, i.e., $g = g^{is}$ for some $s \in \mathbb{N}$. This implies that $g^{1-is} = e$ and this makes $1 - is = nk$ for some $k \in \mathbb{Z}$, i.e., $1 = is + nk$. Therefore, $(i, n) = 1$ and this proves the theorem. □

Example 6.3.1. *For the additive group* $\mathbb{Z}_7$, *since* $(1, 5) = 1$, *we have that* $< \overline{1} >=< \overline{5} >$. *Similarly we have that* $< \overline{1} >=< \overline{5} >=< \overline{2} >=< \overline{3} >$.

Theorem 6.3.3. *If* $|G =< g > | = n$, $< g^i >=< g^{(n,i)} >$. *Furthermore,* $|g^i| = \dfrac{n}{(n, i)}$.

Proof. Let $(n, i) = k$. Then, $i = kd$ for some $d \in \mathbb{N}$. We need to prove that $< g^{kd} >=< g^k >$. Obviously, $< g^k > \supset < g^{kd} >=< g^i >$. Since $(n, i) = k$, there are $s, t \in \mathbb{Z}$ such that

$$g^k = g^{ns+it} = g^{ns}g^{it} = (g^n)^s g^{it} = g^{it} = (g^i)^t$$

and this implies that $g^k \in < g^i >$. So $< g^k > \subset < g^i >$ and this makes $< g^k >=< g^i >$ and this proves the 1st part of the theorem. It also means that $|g^i| = |g^{(n,i)}|$.

If $d|n$, $(g^d)^{\frac{n}{d}} = g^n = e$. If $|g^d| < \frac{n}{d}$, then $(g^d)^{|g^d|} \neq e$, a contradiction. So, $|g^d| = \frac{n}{d}$ and this implies that $|g^i| = |g^{(n,i)}| = \frac{n}{(n,i)}$. □

Example 6.3.2. *For the additive group $\mathbb{Z}_{24}$, we have that $<\overline{4}>=<\overline{(4,24)}>=<\overline{(20,24)}>=<\overline{20}>$. Also, the order of $\overline{18}$ is $\frac{24}{(24,18)} = \frac{24}{6} = 4$.*

Example 6.3.3. *Let g, h be elements of a group G such that $g^3 \in \mathfrak{Z}(h)$, $|g| = 2n$, and $(3, n) = 1$ where n is a positive integer. Show that $gh = hg$.*[1]

Solution. Since g^3 is an element of $\mathfrak{Z}(h)$, $< g^3 >$ is a subgroup of $\mathfrak{Z}(h)$. Since $(3, 2n) = 1$, we have that $< g^3 >=< g^{(3,2n)} >=< g >$ and this implies that $< g >$ is a subgroup of $\mathfrak{Z}(h)$. Thus, $gh = hg$. □

Example 6.3.4. *Among the subgroups of $\mathbb{Z}_{16}$, find all the ones of order 4.*

Solution. If $\frac{16}{(16,n)} = 4$, $n = 4$ or 12. So, by Theorem 6.3.3, $<\overline{4}>$ and $<\overline{12}>$ have order 4. Note that

$$
\begin{aligned}
<\overline{4}> &= \{\overline{4}, \overline{8}, \overline{12}, \overline{0}\} \\
<\overline{12}> &= \{\overline{12}, \overline{8}, \overline{4}, \overline{0}\}
\end{aligned}
$$

So there is only one subgroup of order 4. □

Theorem 6.3.3 can be stated as following and its proof is obvious.

Theorem 6.3.4. *For a finite cyclic group $G =< g >$, $< g^i >=< g^j >$ if and only if $(i, |G|) = (j, |G|)$ where $g \in G$.*

[1] $\mathfrak{Z}(h)$ is the centralizer of h in G.

Example 6.3.5. *Let* $G = \mathbb{Z}_{24}$*. Then, since* $(3, |G|) = 3 = (9, |G|) = (15, |G|) = (21, |G|)$, $<\overline{3}>=<\overline{9}>=<\overline{15}>=<\overline{21}>$ *. In fact,*

$$\begin{aligned} <\overline{3}> &= \{\overline{3}, \overline{6}, \overline{9}, \overline{12}, \overline{15}, \overline{18}, \overline{21}, \overline{0}\} \\ <\overline{9}> &= \{\overline{9}, \overline{18}, \overline{3}, \overline{12}, \overline{21}, \overline{6}, \overline{15}, \overline{0}\} \\ <\overline{15}> &= \{\overline{15}, \overline{6}, \overline{21}, \overline{12}, \overline{3}, \overline{18}, \overline{9}, \overline{0}\} \\ <\overline{21}> &= \{\overline{21}, \overline{18}, \overline{15}, \overline{12}, \overline{9}, \overline{6}, \overline{3}, \overline{0}\} \end{aligned}$$

Once again, we can see that there is only one subgroup of order $\frac{24}{(24,n)} = 8$. Here $n = 3, 9, 15$, or 21. If we want to change the order 8 to $|\mathbb{Z}_{24}| = 24 = \frac{24}{(24,n)}$, n must be relatively prime to 24. So, we state the following corollary.

Corollary 6.3.1. *The set of all the generators of* $\mathbb{Z}_n$ *is the set* $U(n) = \{\overline{a} \in \mathbb{Z}_n | (a, n) = 1\}$.

Proof. The proof of the corollary is straightforward by Theorem 6.3.4. □

Example 6.3.6.

$$\begin{aligned} U(8) &= \{\overline{1}, \overline{3}, \overline{5}, \overline{7}\} \\ U(20) &= \{\overline{1}, \overline{3}, \overline{7}, \overline{9}, \overline{11}, \overline{13}, \overline{17}, \overline{19}\} \\ U(13) &= \{\overline{1}, \overline{2}, \overline{3}, \overline{4}, \overline{5}, \overline{6}, \overline{7}, \overline{8}, \overline{9}, \overline{10}, \overline{11}, \overline{12}\} \end{aligned}$$

If we look at the following Caylay Table for K_4, the Klein-4 group, we can see that $K_4 \cong U(8)$.

Figure 6.3.1.

$\cdot$	$\overline{1}$	$\overline{3}$	$\overline{5}$	$\overline{7}$
$\overline{1}$	$\overline{1}$	$\overline{3}$	$\overline{5}$	$\overline{7}$
$\overline{3}$	$\overline{3}$	$\overline{1}$	$\overline{7}$	$\overline{5}$
$\overline{5}$	$\overline{5}$	$\overline{7}$	$\overline{1}$	$\overline{3}$
$\overline{7}$	$\overline{7}$	$\overline{5}$	$\overline{3}$	$\overline{1}$

We can also notice that $U(n)$ is a group under multiplication modulo n. Another one we can notice is the following.

Corollary 6.3.2. *The set of all the generators of* $\mathbb{Z}_p$ *is the set* $\{\overline{1}, \cdots, \overline{p-1}\}$ *if* p *is a prime number.*

Proof. The proof of the corollary is straightforward by Corollary 6.3.1. □

The **Euler** ϕ **function** is a map from all the positive integers to itself defined as follows: $\phi(1) = 1$ and if n is a positive integer and $n \neq 1$, then $\phi(n)$ is the number of positive integers which are relatively prime to n. Using this ϕ, we state the following proposition. The proof is straightforward.

Proposition 6.3.1. *$U(n)$ is a multiplicative group and $|U(n)| = \phi(n)$.*

Example 6.3.7. $|U(20)| = |\{\overline{1}, \overline{3}, \overline{7}, \overline{9}, \overline{11}, \overline{13}, \overline{17}, \overline{19}\}| = 8 = \phi(20)$.

Chapter 6 Exercises.

1. Prove that there are only two generators for the additive cyclic group $\mathbb{Z}$.

2. Find each of the following. Here, p is a prime number and n is a positive integer.

   (a) $\phi(4)$ (b) $\phi(12)$ (c) $\phi(p)$ (d) $\phi(p^2)$ (e) $\phi(p^n)$

3. Let $G = \mathbb{Z}_n$ and solve the following questions for each given n.

   (a) Find all possible orders of subgroups of G.

   (b) Find all the subgroups of G.

   (c) Draw the subgroup lattice of G.

   (d) Find all the generators of G.

   (e) Find the orders of $\overline{8}$ and $\overline{80}$ in G?.

   (f) Is $< \overline{3} >=< \overline{7} >$ in G?

   (g) Is $< \overline{14} >=< \overline{18} >$ in G.?

   (h) Find all the $a's \in G$ such that $< \overline{6} >=< a >$.

   (i) Find $U(n)$.

      i. $n = 13$ ii. $n = 24$ iii. $n = 25$ iv. $n = 40$ v. $n = 60$

4. Is $\mathbb{Q}$ a cyclic group? What about $\mathbb{Q}[\sqrt{2}]$? Explain.

5. **(Generalized Euler ϕ-function)** Suppose $n, k \in \mathbb{N}$ such that $k|n$. Let's define

$$U_k(n) = \{\overline{x} \in U(n) | (x, k) = 1\}$$ [2]

   for each of the following questions.

   (a) Find $U_7(35)$ (b) $U_2(24)$ (c) $U_5(55)$

   (d) Prove that $U_k(n)$ is a multiplicative group.

[2]We call $\phi_k(n) = |U_k(n)|$ a **generalized Euler ϕ-function.**

(e) Prove that $U_5(105) \cong U(21)$.

(f) Prove that then $U_n(n) = U(n)$ if $k = n$.

6. Solve the following questions for each given n.

(a) Find all the elements of $U(n)$.

(b) Compute the orders of the elements of $U(n)$.

(c) Is $U(n)$ cyclic? If it is, find the generators.

i. $n = 5$ ii. $n = 20$ iii. $n = 36$

7. Let g, h be elements of a group G and $n, m \in \mathbb{N}$.

(a) If $g^5 \in \mathfrak{Z}(h)$ and $|g| = 33$, prove that $gh = hg$.

(b) If $g^5 \in \mathfrak{Z}(h)$, $|g| = 2n$, and $(5, n) = 1$, prove that $gh = hg$.

(c) If $g^m \in \mathfrak{Z}(h)$, $|g| = n$, and $(m, n) = 1$, prove that $gh = hg$.

8. Find an isomorphism from $U(8)$ to the Klein 4-group $K_4 = \{e, a, b, c\}$.

9. If $|G| = p^2$, then prove that there is an element $g \in G$ such that $|g| = p$ where p is prime.

10. If $|G| = p^2$, prove that every normal subgroup N of G with $|N| = p$ is a subgroup of the center G.

Chapter 7

Permutation Groups

A group theory has a long history and permutation groups were very early examples of groups, even before groups were defined. Another reason for studying permutation groups is that every group is isomorphic to some permutation group.

7.1 Definition and Notations

Definition 7.1.1. *A* **permutation** σ *on the set* $I = \{1, \cdots, n\}$ *is a one to one and onto function from* I *to itself. A notation for* σ *is*

$$\sigma = \begin{pmatrix} 1 & 2 & \cdots & n \\ \sigma(1) & \sigma(2) & \cdots & \sigma(n) \end{pmatrix}$$

The set of all permutations on I *is called* **the symmetric group of order** n *and denoted by* $\mathfrak{S}_n$.

A subgroup of a symmetric group is called **permutation group**.

Example 7.1.1. $\sigma = \begin{pmatrix} 1 & 2 & 3 & 4 \\ 2 & 3 & 1 & 4 \end{pmatrix}$ *means that* σ *is a function from* $I = \{1, 2, 3, 4\}$ *to itself such that* $\sigma(1) = 2$, $\sigma(2) = 3$, $\sigma(3) = 1$, *and* $\sigma(4) = 4$.

Example 7.1.2. $\sigma = \begin{pmatrix} 1 & 2 & 3 & 4 \\ 1 & 2 & 3 & 4 \end{pmatrix}$ *means that* σ *is the identity function* ι *from* $I = \{1, 2, 3, 4\}$ *to itself.*

Since a permutation on I is a map from I to itself, we can define the multiplication of two permutations as the composition of them. So for $\sigma_1, \sigma_2 \in \mathfrak{S}_n$ and for $k \in I$, we have that $\sigma_1\sigma_2(k) = \sigma_1(\sigma_2(k))$. Since σ_1 and σ_2 are one to one and onto, so is $\sigma_1\sigma_2$, i.e., it is an element of $\mathfrak{S}_n$. For $\sigma \in \mathfrak{S}_n$, since σ is one to one and onto, we can find the inverse σ^{-1} that is one to one and onto as well. Thus we derive the following proposition.

Proposition 7.1.1. $\mathfrak{S}_n$, *the symmetric group of order* n *is a group under the function composition.*

The following example shows that $\mathfrak{S}_n$ is not Abelian.

Example 7.1.3. *Let* $\sigma_1 = \begin{pmatrix} 1 & 2 & 3 \\ 3 & 2 & 1 \end{pmatrix}, \sigma_2 = \begin{pmatrix} 1 & 2 & 3 \\ 2 & 1 & 3 \end{pmatrix} \in \mathfrak{S}_3$. *Then,* $\sigma_2\sigma_1 \neq \sigma_1\sigma_2$ *since* $(\sigma_2\sigma_1)(1) = \sigma_2(\sigma_1(1)) = \sigma_2(3) = 3$ *and* $(\sigma_1\sigma_2)(2) = \sigma_1(\sigma_2(2)) = \sigma_1(2) = 2$.

Note that from Example (7.1.3) we see that $\sigma_1\sigma_2 = \begin{pmatrix} 1 & 2 & 3 \\ 2 & 1 & 3 \end{pmatrix} \begin{pmatrix} 1 & 2 & 3 \\ 3 & 2 & 1 \end{pmatrix} = \begin{pmatrix} 1 & 2 & 3 \\ 3 & 1 & 2 \end{pmatrix}$ and $\sigma_2\sigma_1 = \begin{pmatrix} 1 & 2 & 3 \\ 3 & 2 & 1 \end{pmatrix} \begin{pmatrix} 1 & 2 & 3 \\ 2 & 1 & 3 \end{pmatrix} = \begin{pmatrix} 1 & 2 & 3 \\ 2 & 3 & 1 \end{pmatrix}$.

Example 7.1.4. *Find all the elements of* $\mathfrak{S}_3$.

Solution. If $\sigma \in \mathfrak{S}_3$, there are 3 choices for $\sigma(1)$. Once $\sigma(1)$ is picked, there are 2 choices for $\sigma(2)$ and after $\sigma(1)$ and $\sigma(2)$ are decided, there is only one choice for $\sigma(3)$ left since σ is a bijection from $I = \{1, 2, 3\}$ to itself. In fact, $\mathfrak{S}_3$ is

$$\left\{ \begin{pmatrix} 1 & 2 & 3 \\ 1 & 2 & 3 \end{pmatrix}, \begin{pmatrix} 1 & 2 & 3 \\ 1 & 3 & 2 \end{pmatrix}, \begin{pmatrix} 1 & 2 & 3 \\ 2 & 1 & 3 \end{pmatrix}, \begin{pmatrix} 1 & 2 & 3 \\ 2 & 3 & 1 \end{pmatrix}, \begin{pmatrix} 1 & 2 & 3 \\ 3 & 2 & 1 \end{pmatrix}, \begin{pmatrix} 1 & 2 & 3 \\ 3 & 1 & 2 \end{pmatrix} \right\}$$

□

Using the argument used in Example 7.1.4, we can derive the following property of $\mathfrak{S}_n$.

Proposition 7.1.2. $|\mathfrak{S}_n| = n!$

Note that under $\sigma = \begin{pmatrix} 1 & 2 & 3 & 4 \\ 2 & 3 & 1 & 4 \end{pmatrix}$, 4 is fixed but not $1, 2$, and 3. We denote $\sigma = (123)$ and this means that $\sigma(1) = 2$, $\sigma(2) = 3$, $\sigma(3) = 1$, and $\sigma(4) = 4$. This notation is called **a cycle notation of** σ. The first notation we used in Definition 7.1.1 is called **an array notation of** σ.

Example 7.1.5. *1.* $\iota = \begin{pmatrix} 1 & 2 & 3 & 4 \\ 1 & 2 & 3 & 4 \end{pmatrix}$ *is denoted by* (1)

2. $\begin{pmatrix} 1 & 2 & 3 & 4 & 5 & 6 & 7 \\ 2 & 3 & 1 & 4 & 6 & 7 & 5 \end{pmatrix} = (123)(567)$

3. $\begin{pmatrix} 1 & 2 & 3 & 4 & 5 & 6 \\ 2 & 3 & 4 & 5 & 6 & 1 \end{pmatrix} = (123456)$

We call (123456) a 6-cycle and (23) a 2-cycle. A 2-cycle is also called **a transposition.**

Example 7.1.6. *1.* $(14523) = (45231) = (52314) = (23145) = (31452)$

2. The inverse of $\begin{pmatrix} 1 & 2 & 3 & 4 & 5 \\ 3 & 5 & 2 & 1 & 4 \end{pmatrix}$ *is* $\begin{pmatrix} 1 & 2 & 3 & 4 & 5 \\ 4 & 3 & 1 & 5 & 2 \end{pmatrix}$

3. The inverse of (13254) *is* (14523)

4. The inverse of $(i_1 i_2 \cdots i_n) = \begin{pmatrix} i_1 & \cdots & i_{n-1} & i_n \\ i_2 & \cdots & i_n & i_1 \end{pmatrix}$ *is* $\begin{pmatrix} i_2 & \cdots & i_n & i_1 \\ i_1 & \cdots & i_{n-1} & i_n \end{pmatrix} = (i_n i_{n-1} \cdots i_2 i_1)$.

7.2 Properties of Permutations

Note that (12)(23) can be written as (123) but (12)(34) can not be written as a 3-cycle. We call (12)(34) a product of **disjoint cycles**. (12) and (23) are not disjoint.

Definition 7.2.1. $\sigma = (a_1 \cdots a_r)$ *and* $\tau = (b_1 \cdots b_s)$ *are called* **disjoint** *if* $\{a_1, \cdots, a_r\} \cap \{b_1, \cdots, b_s\} = \emptyset$.

Theorem 7.2.1. *Every permutation of a symmetric group can be written as a product of disjoint cycles.*

Proof. Suppose $\sigma \in \mathfrak{S}_n$. Using σ, a bijection from $I = \{1, \cdots, n\}$ to itself, let's define $I_a = \{\sigma^n(a) | n \in \mathbb{N}\}$ for $a \in I$. Since I is a finite set, there are $i, j \in \mathbb{N}$, such that $\sigma^i(a) = \sigma^j(a)$ where $i > j$. This implies that $\sigma^{i-j}(a) = a$

and by the Well Ordering Principle, there exists the minimal positive integer n_1 such that $\sigma^{n_1}(a) = a$. If we let $\sigma^u(a) = a_u$, $1 \leq u \leq n_1$, then we have a cycle $(aa_1a_2 \cdots a_{n_1})$. If $I_1 = I$, then $\sigma = (aa_1a_2 \cdots a_{n_1})$, a single cycle and we have thus proven the theorem. If there is an element $b \in I - I_1$, then we can derive that $(bb_1b_2 \cdots b_{n_2})$ using a similar argument. If $x \in \{a_1, a_2, \cdots, a_{n_1}\} \cap \{b_1, b_2, \cdots, b_{n_2}\}$, then $x = \sigma^k(a) = \sigma^t(b)$ for some k and $t \in \mathbb{W}$. This implies that $\sigma^{k-t}(a) = b \in S_1$, a contradiction since $b \in I - I_1$,. Thus $\{a_1, a_2, \cdots, a_{n_1}\}$ and $\{b_1, b_2, \cdots, b_{n_2}\}$ are disjoint. By continuing these processes repeatedly, we can write σ as $(a_1a_2 \cdots a_{n_1})(b_1b_2 \cdots b_{n_2})(c_1c_2 \cdots c_{n_k}) \cdots$, a product of disjoint cycles. □

Theorem 7.2.2. *If σ and τ are disjoint cycles, $\sigma\tau = \tau\sigma$.*

Proof. Suppose $\sigma = (a_1a_2 \cdots a_{n_1}), \tau = (b_1b_2 \cdots b_{n_2}) \in \mathfrak{S}_n$ and let's pick a number $c \in I = \{1, \cdots, n\}$. Then, we have three cases:

Case I: $c \in \{a_1, a_2, \cdots, a_{n_1}\}$

Case II: $c \in \{b_1, b_2, \cdots, b_{n_2}\}$

Case III: $c \notin \{a_1, a_2, \cdots, a_{n_1}\} \cup \{b_1, b_2, \cdots, b_{n_2}\}$

Case I. There must be i such that $c = a_i$. Then, since $\sigma\tau(a_i) = \sigma(a_i) = a_{i+1}$ and $\tau(\sigma(a_i)) = \tau(a_{i+1}) = a_{i+1}$, $\sigma\tau = \tau\sigma$ for this type of c's.

Case II. This case is similar to Case I.

Case III. It is clear that $\sigma\tau(c) = c = \tau\sigma(c)$. □

Example 7.2.1. *1.* $\begin{pmatrix}1 & 2 & 3 & 4\end{pmatrix}\begin{pmatrix}5 & 6 & 7 & 8\end{pmatrix} = \begin{pmatrix}5 & 6 & 7 & 8\end{pmatrix}\begin{pmatrix}1 & 2 & 3 & 4\end{pmatrix}$

2. $\left(\begin{pmatrix}1 & 2 & 3 & 4\end{pmatrix}\begin{pmatrix}5 & 6 & 7 & 8\end{pmatrix}\right)^3 = \begin{pmatrix}1 & 2 & 3 & 4\end{pmatrix}^3\begin{pmatrix}5 & 6 & 7 & 8\end{pmatrix}^3$

3. $\begin{pmatrix}1 & 2 & 3 & 4\end{pmatrix}^2 = \begin{pmatrix}1 & 2 & 3 & 4\end{pmatrix}\begin{pmatrix}1 & 2 & 3 & 4\end{pmatrix} = \begin{pmatrix}1 & 3\end{pmatrix}\begin{pmatrix}2 & 4\end{pmatrix}$

4. $\begin{pmatrix}1 & 2 & 3 & 4\end{pmatrix}^3 = \begin{pmatrix}1 & 4 & 3 & 2\end{pmatrix}$

5. $\begin{pmatrix}1 & 2 & 3 & 4\end{pmatrix}^4 = \iota$

6. $\begin{pmatrix}1 & 2 & 3 & \cdots & n \\ 2 & 3 & 4 & \cdots & 1\end{pmatrix}^n = \iota$

Note that $(132) = (12)(13)$ and $(13254) = (14)(15)(12)(13)$. Therefore, we can derive that $(a_1 a_2 \cdots a_n) = (a_1 a_n) \cdots (a_1 a_2)$.

Lemma 7.2.1. *1.* $(a_1 a_2 \cdots a_n) = (a_1 a_n) \cdots (a_1 a_2)$

2. Every cycle is a product of 2*-cycles.*

3. Every permutation is a product of 2*-cycles.*

4. The order of an n-cycle is n.

Lemma 7.2.2. *If $\sigma = (a_1 \cdots a_k)$, then σ^t is a product of disjoint cycles on the set $\{a_1, \cdots, a_k\}$ where $t \in \mathbb{N}$. Also, $\sigma^t(c) = c$ if $c \in S - \{a_1, \cdots, a_k\}$.*

Note that $|(12)(345)| = 6$ since $(12)(345) = \begin{pmatrix}1 & 2 & 3 & 4 & 5 \\ 2 & 1 & 4 & 5 & 3\end{pmatrix}$ In fact, $|(12)(345)| = |(12)||(345)| = 2 \cdot 3$ since it is a product of disjoint cycles, and this can be extended to a general case.

Theorem 7.2.3. (Ruffini's Theorem)

If $\sigma_1, \cdots, \sigma_k$ are disjoint cycles, then $|\sigma_1 \cdots \sigma_k| = lcm(|\sigma_1|, \cdots, |\sigma_k|)$.

Proof. Since it is obviously true if $k = 1$, let's assume that $k > 1$. Let $\sigma_1 = (a_1 \cdots a_n), \cdots, \sigma_k = (b_1 \cdots b_m)$, t be the least common multiple of $|\sigma_1|, \cdots, |\sigma_k|$, and $\sigma_1 \cdots \sigma_k = d$. Then, since $(\sigma_1 \cdots \sigma_k)^t = \sigma_1^t \cdots \sigma_k^t = \iota$, $d|t$. We also have that $(\sigma_1 \cdots \sigma_k)^d = \iota$ and this implies that $\sigma_1^d \cdots \sigma_k^d = \iota$. Then, by Lemma 7.2.2, $\sigma_1^d = \iota, \cdots, \sigma_k^d = \iota$ and this makes $n|d, \cdots, m|d$. So $t|d$ that implies $t = d$ and it completes the proof. □

7.3 Alternating Groups

One important subgroup of a symmetric group is an alternating group. In fact, this type of subgroup is more important than the group itself. We start with elements in the subgroup.

Definition 7.3.1. *A permutation is called* **even** *if it is a product of an even number of transpositions: otherwise it is called* **odd**.

Example 7.3.1. $(123) = (13)(12)$ *is an even permutation but* (12) *is an odd permutation.*

Let's define a polynomial

$$f(x_1, \cdots, x_n) = \prod_{1 \leq i < j \leq n} (x_i - x_j)$$

For example,

$$\begin{aligned} f(x_1, x_2) &= x_1 - x_2 \\ f(x_1, x_2, x_3) &= (x_1 - x_2)(x_1 - x_3)(x_2 - x_3) \\ f(x_1, x_2, x_3, x_4) &= (x_1 - x_2)(x_1 - x_3)(x_1 - x_4)(x_2 - x_3)(x_2 - x_4)(x_3 - x_4) \end{aligned}$$

If $\sigma = (12)$, $\sigma\big(f(x_1, \cdots, x_n)\big) = -f(x_1, \cdots, x_n)$. For, if $x_u - x_v$ is a factor of f then we have three cases:

Case I: $u, v \notin \{1, 2\}$

Case II: $u, v \in \{1, 2\}$

Case III: only one of u and $v \in \{1, 2\}$

Case I. This means that u and v are fixed by σ so $\sigma(x_u - x_v) = x_u - x_v$.

Case II. Since u and v are not fixed by σ, $\sigma(x_u - x_v) = x_v - x_u = -(x_u - x_v)$.

Case III. For any $w \notin \{1, 2\}$, there are factors, $(x_1 - x_w)$ and $(x_2 - x_w)$. If we apply σ on the product of these factors, we have that $\sigma\big((x_1 - x_w)(x_2 - x_w)\big) = \big((x_2 - x_w)(x_1 - x_w)\big)$.

Since any factor of f is one of the three cases and σ fixes Case I and II, we have showed the claim and we can derive the following lemma as well.

Lemma 7.3.1. *If σ is a 2-cycle, $\sigma\big(f(x_1, \cdots, x_n)\big) = -f(x_1, \cdots, x_n)$*

Proof. This proof is almost identical to the case for $\sigma = (12)$ except Case III. If $\sigma = (ab)$, then Case II becomes that only one of u and $v \in \{a, b\}$. Then, for any $w \notin \{a, b\}$, there are factors, $\pm(x_a - x_w)$ and $\pm(x_b - x_w)$. For example, if $a < w$, it is $x_a - x_w$ but if $a > w$, it becomes $x_w - x_a = -(x_a - x_w)$ If we apply σ on the product of these factors, we have that $\sigma\big(\pm(x_a - x_w)(x_b - x_w)\big) = \big(\pm(x_3 - x_w)(x_2 - x_w)\big)$ and the order of $\pm$ is the same. This means that σ fixes $(x_a - x_w)(x_b - x_w)$ and this proves the lemma. □

Theorem 7.3.1. *A permutation is either even or odd, not both.*

Proof. Suppose σ is a permutation and let's write it as

$$\sigma = \sigma_1 \cdots \sigma_r = \tau_1 \cdots \tau_s$$

where each σ_i and τ_j are a 2-cycle. Then,

$$\begin{aligned} \sigma\big(f(x_1, \cdots, x_n)\big) &= \sigma\left(\prod_{1 \le i < j \le n} (x_i - x_j)\right) \\ &= \sigma_1 \cdots \sigma_r\left(\prod_{1 \le i < j \le n} (x_i - x_j)\right) \\ &= (-1)^r \prod_{1 \le i < j \le n} (x_i - x_j) \end{aligned}$$

On the other hand,

$$\begin{aligned} \sigma\big(f(x_1, \cdots, x_n)\big) &= \sigma\left(\prod_{1 \le i < j \le n} (x_i - x_j)\right) \\ &= \tau_1 \cdots \tau_s\left(\prod_{1 \le i < j \le n} (x_i - x_j)\right) \\ &= (-1)^s \prod_{1 \le i < j \le n} (x_i - x_j) \end{aligned}$$

Thus, we derive that

$$(-1)^{t-s} \prod_{1\le i<j\le n} (x_i - x_j) = \prod_{1\le i<j\le n} (x_i - x_j)$$

Therefore, we conclude that t and s are both even or both odd. □

Since every permutation is a product of transpositions, we have the following properties.

Proposition 7.3.1. *1. The product of two even permutations is even.*

2. The product of an even permutation and an odd permutation is odd.

3. The product of two odd permutations is even.

4. The identity permutation is even.

5. For any permutation σ, both of σ and σ^{-1} are either even or odd together.

Let A_n be the set of all even permutations of $\mathfrak{S}_n$. This A_n is called **the alternating group of order** n.

Example 7.3.2.

$$A_1 = \{(1)\}$$

$$A_2 = \{(1)\}$$

$$A_3 = \{(1),(123),(132)\}$$

$$A_4 = \{(1),(123),(132),(124),(142),(134),(143),(234),(243),$$
$$(12)(34),(13)(24),(14)(23)\}$$

Theorem 7.3.2. $A_n \lhd \mathfrak{S}_n$.

Proof. A composition of even permutations is even since it is a product of two even numbers of 2-cycles. Note that ι is an even permutation and the inverse of an even permutation is even by Proposition 7.3.1. Since the associativity of function compositions is inherited from $\mathfrak{S}_n$, $A_n < \mathfrak{S}_n$. Let O_n be the set of all odd permutations of $\mathfrak{S}_n$. If we define a map θ from A_n to O_n as $\theta(\sigma) = (12)\sigma$ for $\sigma \in A_n$, θ is a bijection.[1] Thus $|A_n| = |O_n| = \frac{n!}{2}$ since $|A_n| + |O_n| = |\mathfrak{S}_n| = n!$ and this implies that $[\mathfrak{S}_n : A_n] = 2$. Therefore, A_n is a normal subgroup of $\mathfrak{S}_n$ because A_n has index 2. □

Example 7.3.3. *A_n is generated by 3-cycles for $n \geq 3$.*

Solution. Since A_n is the set of all the even permutations, it is generated by $(ik)(ij)$ and $(ij)(kl)$ where $i, j, k, l \in \{1, \cdots, n\}$. Note that $(ik)(ij) = (ijk)$ and $(ij)(kl) = (ij)(jk)(jk)(kl) = (ji)(jk)(kj)(kl) = (jki)(klj)$. Also note that $\iota = (1) = (123)^3$. Thus A_n is generated by 3-cycles. □

7.4 Cayley's Theorem and Group Actions

Recall that a subgroups of a symmetric group is called a permutation group.

Theorem 7.4.1. (Cayley's Theorem)
Every group is isomorphic to a permutation group.

[1] This θ sends an even permutation to an odd permutation vice versa.

Proof. Let $G = \{e = g_1, g_2, \cdots, g_n\}$ and $\mathfrak{S}_n$ be the symmetric group on the set $\{e, g_2, \cdots, g_n\}$. For a given $g \in G$, we define a map $l_g : G \longrightarrow G$ as follows:

$$l_g(x) = gx$$

for $x \in G$. Suppose that $l_g(g_i) = l_g(g_j)$. Then, $gg_i = gg_j$ and this implies that $g_i = g_j$, i.e., l_g is one to one. l_g is also onto since $l_g(g^{-1}g_1') = g_1'$ for any $g_1' \in G$. This means that $l_g \in \mathfrak{S}_n$. Note that $l_{g_1'g_2'} = l_{g_1'}l_{g_2'}$ since $l_{g_1'g_2'}(g_3) = g_1'g_2'g_3 = l_{g_1'}(g_2'g_3) = l_{g_1'}l_{g_2'}(g_3)$ for any $g_3 \in G$. Now, let's define a map $\theta : G \longrightarrow \mathfrak{S}_n$ as follows:

$$\theta(g) = l_g$$

for $g \in G$. Then, θ is a homomorphism since $\theta(g_1''g_2'') = l_{g_1''g_2''} = l_{g_1''}l_{g_2''} = \theta(g_1'')\theta(g_2'')$ for any $g_1'', g_2'' \in G$. θ is also one to one since $\theta(g_1) = \theta(g_2) \Longrightarrow l_{g_1} = l_{g_2} \Longrightarrow g_1g = g_2g$ for any $g \in G \Longrightarrow g_1 = g_2$. Therefore G is isomorphic to a subgroup of the symmetric group $\mathfrak{S}_n$. □

Since we can assume that every group is a permutation group by Theorem 7.4.1, it's time to consider a group action.

Definition 7.4.1. *Let G be a group and S be a set. We say that G* **acts on** *S if $g * s$ is defined and*

1. $g * s \in S$ *and* $e * s = s$

2. $(g_1g_2) * (s) = g_1 * (g_2 * s)$

for any $g, g_1, g_2 \in G$ and $s \in S$. Note that e is the identity of G.

Example 7.4.1. *A group G acts on itself as follows:*

$$g * g_1 \in G$$

where $g, g_1 \in G$. For any $g \in G$, the stabilizer $stab_G(g) = \{e\}$. The orbit $orb_G(g) = G$ for any $g \in G$. □

There are two important subsets, one for the given group and one for the given set when a group action is defined.

Definition 7.4.2. 1. **The stabilizer of** *s is the subset $\{g \in G | g * s = s\}$ of G and denoted by $stab_G(s)$.*

2. **The orbit of** *s is the subset $\{g * s | g \in G\}$ of S and denoted by $orb_G(s)$*

Example 7.4.2. *$\mathfrak{S}_n$ acts on the set $\{1, \cdots, n\}$ as*

$$\sigma * x = \sigma(x)$$

for $\sigma \in \mathfrak{S}_n$ and $x \in \{1, \cdots, n\}$.

From now on, if there is no ambiguity, then we will omit $*$ of a group action.

Example 7.4.3. *Suppose a group G acts on a set S. Then, let's define a relation $\sim$ on x, y in S as follows: $x \sim y$ if and only if $x = gy$ for some $g \in G$. Then,*

1. *Prove that $\sim$ is an equivalence relation.*

2. *Prove that for $x \in G$, the equivalence class of x is $orb_G(x)$.*

Solution.

1. Suppose $x, y, z \in S$. Then, $x \sim x$ since $ex = x$. If $x \sim y$, then $x = gy$ for some $g \in G$. This implies that $g^{-1}x = y$, so $y \sim x$. If $x \sim y$ and $y \sim z$, then there are $g_1, g_2 \in G$ such that $x = g_1 y$ and $x_1 = g_2 z$. This implies that $x = g_1(g_2 z) = (g_1 g_2)z$, so $x \sim z$ and this shows that $\sim$ is an equivalence relation.

2. The proof is straightforward by the definition of the orbit. □

Theorem 7.4.2. (Orbit-Stabilizer Theorem)
Let a group G act on a set S. Then,

1. $stab_G(s) < G$ *for each* $s \in S$

2. $[G : stab_G(s)] = |orb_G(s)|$

Proof.

1. Since $es = s$, we have that $e \in stab_G(s) \neq \emptyset$. If $g_1, g_2 \in stab_G(s)$, $g_1 s = s$ and $g_2 s = s$, and then $g_1 g_2(s) = g_1(g_2 s) = g_1 s = s$. This implies that $g_1 g_2 \in stab_G(s)$. For $g \in stab_G(s)$, $gs = s$ implies that $s = g^{-1}g_1 s = g^{-1}s$, so that $g^{-1} \in stab_G(s)$. Since the associativity is obvious, we conclude that $stab_G(s)$ is a subgroup of G.

2. Let's define a map $\theta : G/stab_G(s) \longrightarrow orb_G(s)$ as

$$\theta(stab_G(s)g) = gs$$

for $stab_G(s)g \in G/stab_G(s)$. If $\theta(stab_G(s)g_1) = \theta(stab_G(s)g_2)$, then

$$\begin{aligned} g_1 s = g_2 s &\implies g_1^{-1} g_2 s = s \\ &\implies g_1^{-1} g_2 \in stab_G(s) \\ &\implies stab_G(s)g_1 = stab_G(s)g_2 \end{aligned}$$

So, θ is one to one. Suppose $s_1 \in orb_G(s)$. Then, there is $g_3 \in G$ such that $g_3 s = s_1$. This derives that $\theta(stab_G(s)g_3) = g_3 s = s_1$ and this implies that θ is onto. Since θ is obviously a homomorphism, it is an isomorphism. Therefore,

$$|G/stab_G(s)| = \frac{|G|}{|stab_G(s)|} = [G : stab_G(s)] = |orb_G(s)| \quad \square$$

Example 7.4.4. *Let's consider a group action of $\mathfrak{S}_3$ on the set $I = \{1, 2, 3\}$.*

1. $stab_{\mathfrak{S}_3}(1) = \left\{ \begin{pmatrix} 1 & 2 & 3 \\ 1 & 2 & 3 \end{pmatrix}, \begin{pmatrix} 1 & 2 & 3 \\ 1 & 3 & 2 \end{pmatrix} \right\}$

2. $\begin{pmatrix} 1 & 2 & 3 \\ 1 & 3 & 2 \end{pmatrix}(1) = 1$, $\begin{pmatrix} 1 & 2 & 3 \\ 2 & 1 & 3 \end{pmatrix}(1) = 2$, *and* $\begin{pmatrix} 1 & 2 & 3 \\ 3 & 1 & 2 \end{pmatrix}(1) = 3$. *This implies that* $orb_{\mathfrak{S}_3}(1) = \{1, 2, 3\}$

3. *We can see that* $3 = |orb_{\mathfrak{S}_3}(1)| = [\mathfrak{S}_3 : stab_{\mathfrak{S}_3}(1)] = \dfrac{|\mathfrak{S}_3|}{|stab_{\mathfrak{S}_3}(1)|} = \dfrac{6}{2}$

Chapter 7 Exercises.

1. Calculate each of the following. Write your answer as a cycle or a product of disjoint cycles.

(a) $\begin{pmatrix} 1 & 2 & 3 \\ 3 & 2 & 1 \end{pmatrix} \begin{pmatrix} 1 & 2 & 3 \\ 2 & 3 & 1 \end{pmatrix}$

(b) $\begin{pmatrix} 1 & 2 & 3 & 4 \\ 3 & 2 & 1 & 4 \end{pmatrix} \begin{pmatrix} 1 & 2 & 3 & 4 \\ 1 & 2 & 3 & 4 \end{pmatrix}$

(c) $\begin{pmatrix} 1 & 2 & 3 & 4 \\ 3 & 1 & 2 & 4 \end{pmatrix} \begin{pmatrix} 1 & 2 & 3 & 4 \\ 2 & 4 & 3 & 1 \end{pmatrix}$

(d) $\begin{pmatrix} 1 & 2 & 3 & 4 & 5 & 6 \\ 5 & 6 & 3 & 1 & 2 & 4 \end{pmatrix} \begin{pmatrix} 1 & 2 & 3 & 4 & 5 & 6 \\ 4 & 6 & 2 & 1 & 3 & 5 \end{pmatrix}$

(e) $\begin{pmatrix} 1 & 2 & 3 & 4 & 5 & 6 & 7 \\ 5 & 6 & 7 & 3 & 1 & 2 & 4 \end{pmatrix} \begin{pmatrix} 1 & 2 & 3 & 4 & 5 & 6 & 7 \\ 5 & 7 & 6 & 3 & 4 & 2 & 1 \end{pmatrix}$

(f) $\begin{pmatrix} 1 & 2 & 3 \\ 3 & 2 & 1 \end{pmatrix} \begin{pmatrix} 1 & 2 & 3 \\ 2 & 1 & 3 \end{pmatrix} \begin{pmatrix} 1 & 2 & 3 \\ 1 & 3 & 2 \end{pmatrix}$

2. Find the orders of the following.

(a) $\begin{pmatrix} 1 & 2 & 3 & 4 & 5 \end{pmatrix}$

(b) $\begin{pmatrix} 1 & 2 & 3 & 4 \end{pmatrix} \begin{pmatrix} 5 & 6 & 7 \end{pmatrix}$

(c) $\begin{pmatrix} 1 & 2 & 3 & 4 & 5 & 6 & 7 & 8 \\ 5 & 4 & 3 & 2 & 1 & 7 & 6 & 8 \end{pmatrix}$

3. Find the inverses of the following element. Write your answer as a cycle or a product of disjoint cycles.

(a) $\begin{pmatrix} 1 & 2 & 3 \\ 3 & 1 & 2 \end{pmatrix}$ (b) (134) (c) (13457) (d) $(a_k a_{k-1} \cdots a_1)$

(e) Find the inverse of $\begin{pmatrix} 1 & 2 & \cdots & n \\ i_1 & i_2 & \cdots & i_n \end{pmatrix}$

4. (a) Find A_3.

   (b) Find A_4.

5. (a) Find two permutations σ_1 and σ_2 of $\mathfrak{S}_7$ such that $\sigma_1\sigma_2 = \sigma_2\sigma_1$.

   (b) Find two permutations σ_1 and σ_2 of $\mathfrak{S}_7$ such that $\sigma_1\sigma_2 \neq \sigma_2\sigma_1$.

   (c) Write $\begin{pmatrix} 1 & 2 & 3 & 4 & 5 & 6 & 7 \\ 6 & 5 & 4 & 1 & 2 & 3 & 7 \end{pmatrix}$ as a product of disjoint cycles. What is the order?

   (d) Write $\begin{pmatrix} 1 & 2 & 3 & 4 & 5 & 6 & 7 \\ 6 & 5 & 4 & 1 & 2 & 3 & 7 \end{pmatrix}$ as a product of 2-cycles. Is it odd or even?

   (e) Find the inverse of $\begin{pmatrix} 1 & 2 & 3 & 4 & 5 & 6 & 7 \\ 6 & 5 & 4 & 1 & 2 & 3 & 7 \end{pmatrix}$

6. Is there an element of $\mathfrak{S}_4$ whose order is 6? What about order 8 or 10? What is the maximum order? Answer the same questions for $\mathfrak{S}_5$

7. If possible, find an element of $\mathfrak{S}_7$ whose order is

   (a) 8 (b) 12 (c) 14 (d) 36 (e) of the maximum order.

8. Find a permutation group that is isomorphic to the Klein 4-group.

9. Find a permutation group that is isomorphic to the the dihedral group D_3. What is the center of D_3?

10. Let

$$A = \left\{ \begin{pmatrix} 1 & 2 & 3 & 4 \\ 3 & 2 & 1 & 4 \end{pmatrix}, \begin{pmatrix} 1 & 2 & 3 & 4 \\ 1 & 3 & 2 & 4 \end{pmatrix} \begin{pmatrix} 1 & 2 & 3 & 4 \\ 2 & 1 & 3 & 4 \end{pmatrix}, \begin{pmatrix} 1 & 2 & 3 & 4 \\ 2 & 3 & 1 & 4 \end{pmatrix}, \begin{pmatrix} 1 & 2 & 3 & 4 \\ 3 & 2 & 1 & 4 \end{pmatrix}, \begin{pmatrix} 1 & 2 & 3 & 4 \\ 3 & 1 & 2 & 4 \end{pmatrix} \right\}$$

(a) Prove that A is a permutation group.

(b) Find the order of every element of A.

(c) Compare A with $\mathfrak{S}_3$. Are they isomorphic?

(d) Find the center of A.

(e) Find the maximal normal subgroup of A.

11. Prove that m-cycle is an even permutation if m is odd and an odd permutation if m is even.

12. (a) Find all the normal subgroups of $\mathfrak{S}_3$.

    (b) Find all normal subgroups of $\mathfrak{S}_4$.

    (c) Find all normal subgroups of $\mathfrak{S}_5$.

13. (a) Find thc commutator subgroup of $\mathfrak{S}_3$.

    (b) Find the commutator subgroup of $\mathfrak{S}_4$.

    Prove that A_n is a normal subgroup of $\mathfrak{S}_n$ without using Theorem 7.3.2.

14. For $f(x_1, x_2, x_3, x_4) = (x_1 - x_2)(x_1 - x_3)(x_1 - x_4)(x_2 - x_3)(x_2 - x_4)(x_3 - x_4)$, find $\sigma(f(x_1, x_2, x_3, x_4))$ for the following.

    (a) $\sigma = (23)$ (b) $\sigma = (14)$ (c) $\sigma = (14)(23)$ (d) $\sigma = (123)$

15. 24. Let H be a proper subgroup of a group G. If H acts on G as a left multiplication, find $stab_G(g)$ and $org_G(g)$ for $g \in G$. Is it possible that G acts on the subgroup H.

16. Let's define $\sigma * a = \sigma(a)$ for $\sigma \in \mathfrak{S}_3$ and $a \in I = \{1, 2, 3\}$.

    (a) Prove that $\mathfrak{S}_3$ acts on I.

    (b) Find $orb_{\mathfrak{S}_3}(2)$ and $stab_{\mathfrak{S}_3}(2)$.

(c) Confirm that $|orb_{\mathfrak{S}_3}(2)| = |\mathfrak{S}_3 : stab_{\mathfrak{S}_3}(2)|$.

17. Prove that $\mathfrak{S}_n$ acts on $\{1, \cdots, n\}$ by

$$\sigma * x = \sigma(x)$$

for $\sigma \in \mathfrak{S}_n$ and $x \in \{1, \cdots, n\}$.

18. If H is a subgroup of G, then prove that H acts on G by (left) translation:

$$h * x = hx$$

for $h \in H$ and $g \in G$.

19. If $K \lhd G$ and S is the set of all left cosets of K in G, then prove that G acts on S by translation:

$$g * xK = gxK$$

20. If $H < G$, prove then that H acts on G by conjugation:

$$h * x = hxh^{-1}$$

21. If $H < G$, and S is the set of all subgroups of G, then prove that H acts on S by conjugation:

$$h * K = hKh^{-1}$$

where K is a subgroup of G and h is an element of H.

22. If G acts on a set S, prove that G acts on $P(S)$ by

$$g * A = gA$$

where $P(S)$ is the power set[2] of S, $A \in P(S)$, $g \in G$, and $gA = \{ga|a \in A\}$. Note that if A is empty, then so is gA.

23. Prove that if a group G acts on a set X, then for $x \in X$, the stabilizer $G_x = \{g \in G|gx = x\}$ of x is a subgroup of the group G.

[2]The set of all subsets is called the power set.

Chapter 8

Automorphisms

Recall that an automorphism of a group G is an isomorphism from G to itself. Therefore, a group automorphism is a way of mapping a group to itself that still preserves the group structure. Naturally, finding all the automorphisms of a group is essential. We start this chapter with automorphisms of cyclic groups.

8.1 Automorphisms of Cyclic Groups

Theorem 8.1.1. *Let $G =< g >$ and $\theta \in Aut(G)$. Then,*

1. *$\theta(g)$ is a generator of G.*
2. *If H is an Abelian subgroup of G, then so is $\theta(H)$.*

Proof.

1. For all $g_2 \in G$, there is $g_1 \in G$ such that $\theta(g_1) = g_2$. This implies that $g_1 = g^k$ for some $k \in \mathbb{N}$. So we have that $\theta(g_1) = \theta(g^k) = (\theta(g))^k = g_2$ and this shows that $\theta(g)$ is a generator of G.

2. For $g_1', g_2' \in \theta(H)$, there are g_1 and g_2 such that $\theta(g_i) = g_i'$, $1 \leq i \leq 2$. This implies that $g_1'g_2' = \theta(g_1)\theta(g_2) = \theta(g_1g_2) = \theta(g_2g_1) = \theta(g_2)\theta(g_1) = g_2'g_1'$. Thus $\theta(H)$ is an Abelian subgroup of G. $\square$

Example 8.1.1. *Show that* $Aut(\mathbb{Z}_6) \cong U(6)$.

Solution. All the generators of $\mathbb{Z}_6$ are $\overline{1}$ and $\overline{5}$, so $U(6) = \{\overline{1}, \overline{5}\}$ and it is isomorphic to $\mathbb{Z}_2$. Suppose $\theta \in Aut(\mathbb{Z}_6)$. Then $\theta(\overline{1}) = \overline{1}$ or $\theta(\overline{1}) = \overline{5}$ by Theorem 8.1.1. If $\theta(\overline{1}) = \overline{1}$, then θ is the identity map, i.e., the identity automorphism. If $\theta(\overline{1}) = \overline{5}$, then the order of θ is 2 since

$$\begin{aligned}
\theta^2(\overline{1}) &= \theta(\overline{5}) \\
&= \theta(\overline{1} + \overline{1} + \overline{1} + \overline{1} + \overline{1}) \\
&= \theta(\overline{1}) + \cdots + \theta(\overline{1}) \\
&= \overline{5} + \overline{5} + \overline{5} + \overline{5} + \overline{5} = \overline{25} = \overline{1}.
\end{aligned}$$

So $Aut(\mathbb{Z}_6)$ has 2 elements and is isomorphic to $\mathbb{Z}_6$. This proves that $Aut(\mathbb{Z}_6) \cong U(6)$. $\square$

From Example 8.1.1, we can see that the automorphism θ is determined by $\theta(\overline{1})$ and the following a generalization.

Theorem 8.1.2. $Aut(\mathbb{Z}_n) \cong U(n)$ *for* $n \in \mathbb{N}$.

Proof. Since the set of all the generators of $\mathbb{Z}_n$ is $U(n) = \{\overline{x} \in \mathbb{Z}_n | (x, n) = 1\}$, let's define a map $\psi : Aut(\mathbb{Z}_n) \longrightarrow U(n)$ as follows:

$$\psi(\theta) = \theta(\overline{1}) = \overline{k} = k\overline{1}$$

for $\theta \in Aut(\mathbb{Z}_n)$ where $\overline{k} \in U(n)$. If $\theta_1, \theta_2 \in Aut(\mathbb{Z}_n)$, we have that $\psi(\theta_i) = k_i\overline{1}$, $i = 1, 2$, where $\overline{k_i} \in U(n)$. Then,

$$\psi(\theta_2\theta_1) = (\theta_2\theta_1)(\overline{1}) = (\theta_2(\theta_1(\overline{1})) = \theta_2(k_1\overline{1}) = k_2k_1\overline{1} = k_2\overline{1}k_1\overline{1} = \psi(\theta_2)\psi(\theta_1)$$

and this implies that ψ is a group homomorphism. Suppose that $\psi(\theta_1) = \psi(\theta_2)$. Then, $k_1\overline{1} = k_2\overline{1}$, i.e., $\overline{k_1} = \overline{k_2}$ where $\overline{k_1} = \overline{k_2} \in U(n)$. We then have that $\theta_1 = \theta_2$, i.e., ψ is one to one. Since there is an automorphism $\theta \in Aut(\mathbb{Z}_n)$ with $\theta(\overline{1}) = \overline{r}$ for each $\overline{r} \in U(n)$, $\psi(\theta) = \overline{r}$ and this implies that ψ is onto. So ψ is an isomorphism and therefore We have therefore proven the theorem. □

Corollary 8.1.1. $|Aut(\mathbb{Z}_n)| = \phi(n)$ *for* $n \in \mathbb{N}$.

Proof. It is obvious since $\phi(n) = |U(n)|$. □

8.2 Inner Automorphisms

Now, we'd like to look at special automorphisms.

Theorem 8.2.1. *Let G be a group and $g \in G$. If we define a map θ_g from G to itself as*

$$\theta_g(g_1) = gg_1g^{-1} \tag{8.1}$$

for $g_1 \in G$, then θ_g is an automorphism of G.

Proof. If $g_1, g_2 \in G$, we have that

$$\theta_g(g_1g_2) = gg_1g_2g^{-1} = gg_1g^{-1}gg_2g^{-1} = \theta_g(g_1)\theta_g(g_2)$$

and this implies that θ_g is a homomorphism. Suppose that $\theta_g(g_1) = \theta_g(g_2)$. Then, we derive that $gg_1g^{-1} = gg_2g^{-1}$ and this implies that $g_1 = g_2$, i.e., θ_g is one to one. For any $g_3 \in G$, we have that $\theta_g(g^{-1}g_3g) = g_3$. i.e., θ_g is onto. Thus θ_g is an automorphism. $\square$

This type of automorphisms in Theorem 8.2.1 are called **inner automorphisms** of G.

Definition 8.2.1. *For a given group G, we denote* **Aut(G)** *for the set of all the automorphisms of G and* **Inn(G)** *for the set of all the inner automorphisms of G.*

Theorem 8.2.2. *For a group G, $Aut(G)$ is a group under the function composition and $Inn(G)$ is a subgroup of $Aut(G)$.*

Proof. Since the identity map I from G to itself is an automorphism, I is an element of $Aut(G)$. This implies that $Aut(G)$ is not the empty set. For any $\theta_1, \theta_2 \in Aut(G)$, $\theta_1\theta_2$ is a homomorphism from G to itself. Since it is one to one and onto, $\theta_1\theta_2$ is an element of $Aut(G)$. Obviously θ has the inverse θ^{-1} which is an automorphism of G. So $Aut(G)$ is a group. By (8.1), for any $g \in G$, we have that

$$\theta_{g_1}\theta_{g_2}(g) = \theta_{g_1}(g_2gg_2^{-1}) = g_1g_2gg_2^{-1}g_1^{-1} = g_1g_2g(g_1g_2)^{-1} = \theta_{g_1g_2}(g),$$

i.e., $\theta_{g_1}\theta_{g_2} = \theta_{g_1g_2}$. This implies that $Inn(G)$ is closed under the operation. This also implies that $\theta_{g^{-1}}$ is the inverse of θ_g. Therefore, $Inn(G)$ is a subgroup of $Aut(G)$ by the Two-Step Subgroup Test. $\square$

Theorem 8.2.3. *If G is an Abelian group, then $Inn(G)$ is a trivial group.*

Proof. For any $g \in G$ and $\theta \in Inn(G)$, $\theta(g) = hgh^{-1} = g$ for some $h \in G$. This implies that $\theta = I$. $\square$

Theorem 8.2.4. *For any group G, $G/\mathfrak{Z}$ is isomorphic to $Inn(G)$.*

Proof. For any $g \in G$, let us define a map θ from $G/\mathfrak{Z}$ to $Inn(G)$ as follows:

$$\theta(g\mathfrak{Z}) = \psi_g$$

where $\psi_g(x) = gxg^{-1}$ for any $x \in G$. For any $g_1\mathfrak{Z}, g_2\mathfrak{Z} \in G/\mathfrak{Z}$, we have that $\theta(g_1\mathfrak{Z}g_2\mathfrak{Z}) = \theta(g_1g_2\mathfrak{Z}) = \psi_{g_1g_2}$. On the other hand, we have that $\theta(g_1\mathfrak{Z})\theta(g_2\mathfrak{Z}) = \psi_{g_1}\psi_{g_2}$. So for $x \in G$, we have that $\psi_{g_1g_2}(x) = g_1g_2x(g_1g_2)^{-1} = g_1g_2xg_2^{-1}g_1^{-1}$. We also have that $\psi_{g_1}\psi_{g_2}(x) = \psi_{g_1}(g_2xg_2^{-1})) = g_1g_2xg_2^{-1}g_1^{-1}$. This implies that

$$\psi_{g_1g_2}(x) = \psi_{g_1}\psi_{g_2}(x), \tag{8.2}$$

i.e., θ is a group homomorphism. Next, suppose $\theta(g_1\mathfrak{Z}) = \theta(g_2\mathfrak{Z})$ for $g_1\mathfrak{Z}, g_2\mathfrak{Z} \in G/\mathfrak{Z}$. Then,

$$\begin{aligned}
\theta(g_1\mathfrak{Z}) = \theta(g_2\mathfrak{Z}) &\implies \psi_{g_1} = \psi_{g_2} \\
&\implies \psi_{g_1}(x) = \psi_{g_2}(x) \text{ for } x \in G \\
&\implies g_1xg_1^{-1} = g_2xg_2^{-1} \\
&\implies g_2^{-1}g_1x = xg_2^{-1}g_1 \\
&\implies g_2^{-1}g_1 \in \mathfrak{Z} \\
&\implies g_1\mathfrak{Z} = g_2\mathfrak{Z}
\end{aligned}$$

This implies that θ is one to one. For any $\psi_g \in Inn(G)$, we have that $\theta(g\mathfrak{Z}) = \psi_g$.

This implies that θ is onto. Therefore, $G/\mathfrak{Z} \cong Inn(G)$. □

Lemma 8.2.1. *If $H < G$, the centralizer $\mathfrak{Z}(H)$ is a normal subgroup of the normalizer $N(H)$.*

Proof. For any $h \in H$, $n \in N(H)$, and $c \in \mathfrak{Z}(H)$, we need to show that $ncn^{-1}h = hncn^{-1}$. Since $n^{-1}hn \in H$, we have that

$$nc(n^{-1}hn)c^{-1}n^{-1} = ncc^{-1}n^{-1}hnn^{-1} = h,$$

i.e., $ncn^{-1}h = hncn^{-1}$. This implies that $ncn^{-1} \in \mathfrak{Z}(H)$ so $n\mathfrak{Z}(H)n^{-1} \subseteq \mathfrak{Z}(H)$ for any $n \in N(H)$. Therefore, $\mathfrak{Z}(H) \lhd N(H)$. □

The following is simply called the N/C-**Theorem**.

Theorem 8.2.5. (Normalizer-Centralizer Theorem)
If $H < G$, $N(H)/\mathfrak{Z}(H)$ is isomorphic to a subgroup of $Aut(H)$.

Proof. Since $\mathfrak{Z}(H)$ is a normal subgroup of $N(H)$, $N(H)/\mathfrak{Z}(H)$ is a group. For $g \in N(H)$, let us define a map θ from $N(H)$ to $Aut(H)$ as follows:

$$\theta(g) = \psi_g \tag{8.3}$$

where $\psi_g(x) = gxg^{-1}$ for any $x \in H$. Then, by (8.2) θ is a homomorphism, i.e., for $g, h \in N(H)$, $\theta(gh) = \theta(g)\theta(h)$. Then, $h \in ker(\theta)$ if and only if $\psi_h(x) = hxh^{-1} = x$, i.e., $xh = hx$. This implies that h is an element of the centralizer $\mathfrak{Z}(H)$ of H and this makes $ker(\theta) = \mathfrak{Z}(H)$. Since ψ_g is an automorphism of H, we can conclude that $N(H)/\mathfrak{Z}(H)$ is isomorphic to a subgroup of $Aut(H)$ by Theorem 5.2.2. □

Another way to state the N/C-Theorem is that $N(H)/\mathfrak{Z}(H)$ is embedded into $Aut(H)$ and denoted by $N(H)/\mathfrak{Z}(H) \hookrightarrow Aut(H)$. Here, $\hookrightarrow$ means that there exists a monomorphism.

The following is a useful equivalent condition to an Abelian group using an automorphism.

Example 8.2.1. *A group G is Abelian if and only if the map θ from G to itself is an automorphism where θ is defined as follows:*

$$\theta(a) = a^{-1}$$

for $a \in G$.[1]

Proof. Let us assume that G is Abelian. For any $a, b \in G$, we have that

$$\theta(ab) = (ab)^{-1} = b^{-1}a^{-1} = a^{-1}b^{-1} = \theta(a)\theta(b).$$

Thus θ is a homomorphism. If $\theta(a) = e$, then $a^{-1} = e$ where e is the identity of G. This implies that θ is one to one. For any $a \in G$, we have that $a = (a^{-1})^{-1}$ holds. This implies that $\theta(a^{-1}) = a$. So θ is onto and it makes θ an automorphism. Conversely, let us assume that θ is an automorphism. For any $a, b \in G$, we have that $\theta(ab) = (ab)^{-1} = b^{-1}a^{-1}$. On the other hand, we have that $\theta(a)\theta(b) = a^{-1}b^{-1}$. This implies that $b^{-1}a^{-1} = a^{-1}b^{-1}$. This also implies that $ab = ba$. Therefore, G is Abelian. $\square$

Chapter 8 Exercises.

1. (a) Find $Aut(\mathbb{Z})$.

[1]Compare this example with Exercise 14 in Chapter 5.

(b) Find $Aut(2\mathbb{Z})$.

(c) Find $Aut(p\mathbb{Z})$ where p is a prime number.

2. (a) Find $Aut(\mathbb{Z}_7)$.

   (b) Find $Aut(\mathbb{Z}_p)$ if p is prime.

3. (a) Find $Aut(\mathbb{Z}_8)$.

   (b) Find $Aut(\mathbb{Z}_m)$ if m is not prime.

4. (a) Find $Aut(U(3))$.

   (b) Find $Aut(U(6))$.

   (c) Find $Aut(U(p))$ where p is a prime number.

5. Find $Inn(\mathbb{Z}_4)$.

6. Find $Inn(G)$ if G is Abelian.

7. (a) Find $Inn(D_3)$. Here D_3 is the dihedral group on the equilateral triangle.

   (b) Find $Inn(D_4)$.

   (c) Find $Inn(D_n)$ where $n > 4$.

8. (a) Find $Inn(\mathfrak{S}_2)$ and $Aut(\mathfrak{S}_2)$.

   (b) Find $Inn(\mathfrak{S}_3)$ and $Aut(\mathfrak{S}_3)$.

9. If θ is an automorphism of G, prove that $\theta(G') = G'$ where G' is the set of all commutators in a group G. The commutator of a and b is $a^{-1}b^{-1}ab$ and denoted by $[a, b]$.

10. (a) Find a non-identity automorphism on $\mathbb{Q}$.

(b) Find a non-identity automorphism on $\mathbb{Q}[\sqrt{2}]$.

(c) Find a non-identity automorphism on $\mathbb{Q}[\sqrt{6}]$.

11. Suppose $\theta \in Aut(G)$. If $\mathfrak{C}_G(g)$ is the centralizer of g in G, prove that $\theta(\mathfrak{C}_G(g))$ is the centralizer of $\theta(g)$ in G.

12. Prove that θ of (8.3) is a group homomorphism.

Chapter 9

Abelian Groups and Solvable Groups

In this chapter, we investigate relations between solvability and Abelian property of a group. Abelian groups themselves have a very important role in the Galois Theory.

9.1 Abelian Groups with Prime Orders

One important example of Abelian groups is a cyclic group and the following is another one.

Lemma 9.1.1. *For a finite group G with the identity e and $G \neq \{e\}$, G has only two subgroups G and $\{e\}$ if and only if $|G|$ is a prime p. In this case, $G \cong \mathbb{Z}_p$.*

Proof. If $|G| = p$, G has only trivial subgroups, G itself and $\{e\}$ by the Lagrange Theorem. Conversely, suppose that G has only trivial subgroups and

$G \neq \{e\}$. Assume that $|G| \neq p$. Then, $|G| = mn$ for some $m, n \in \mathbb{N}$ with $(m, n) = 1$ and $m, n > 1$. For any non-identity element $g \in G$, $g^{|G|} = g^{mn} = (g^m)^n = e$. If $g^m = e$, then $< g >$ is a proper subgroup of G. This leads a contradiction. If $g^m \neq e$, then $< g^m >$ is a proper subgroup of G. This also leads another contradiction. Since $|G| \neq p$ leads a contradiction always, $|G|$ is therefore a prime. □

Not every groups are Abelian but every groups always have some Abelian subgroup. The following theorem proves it.

Theorem 9.1.1. (Cauchy's Theorem)
If G is a finite group whose order is divisible by a prime p, G contains an element of order p.

Proof. We prove the theorem by induction on the order of G. If $|G|$ is 1 or prime, then by Lemma 9.1.1, the statement is true so let's say that $|G|$ is not 1 or a prime. For induction, let's assume that the theorem is true for any subgroup of G whose order is less than $|G|$. Let g be a non-identity element of G such that $|g| = qn$ for some prime p and $n \in \mathbb{N}$. This implies that $|g^n| = q$. If q is p, then we have proven the theorem. If q is not p, then p divides $|G/< g >|$. Since $|G/< g >|$ is less than $|G|$, by induction, there is a non identity element $g_1 < g >$ whose order is p, i.e., $g_1^p < g >=< g >$. This means that $g_1^p \in< g >$ so there exists $h \in \mathbb{N}$ such that $g_1^p = g^h$. If $g^h = e$, then $|g_1| = p$ so we have proven the theorem. If $g^h \neq e$, then $g_1^{pk} = e$ for some $k \in \mathbb{N}$. Since the order of g_1 is pk, $g_1^k \neq e$. So we have that $|g_1^k| = p$. This implies that g_1^k is an element whose order is p and this competes the proof. □

Corollary 9.1.1. *If G is a finite group, G contains an Abelian subgroup.*

Proof. If $|G| = 1$, there is nothing to prove. If $|G| \neq 1$, there must be a prime number that divides $|G|$ since G is finite. Then, G contains an element g of order p by Theorem 9.1.1 and $< g >$ is an Abelian subgroup of G by Lemma 9.1.1. □

9.2 Solvable Groups

For a given group G, we have defined **the commutator subgroup of** G as $G' = [G, G] = \{g^{-1}h^{-1}gh || g, h \in G\}$. Recall that G' is a normal subgroup of G.

If we let $G^{(0)} = G$ and $G^{(2)} = [G', G']$, we can inductively define

$$G^{(n)} = [G^{(n-1)}, G^{(n-1)}]. \tag{9.1}$$

Note that $G^{(n-1)} \lhd G^{(n)}$ for each $n \in \mathbb{N}$ and the following sequence of subgroups of G is called **the derived series of** G:

$$\cdots \subseteq G^{(k)} \subseteq G^{(k-1)} \subseteq \cdots \subseteq G^{(1)} \subseteq G^{(0)} = G$$

Definition 9.2.1. 1. *A group G is called* **solvable (or soluble)** *if there exists $n \in \mathbb{N}$ such that $G^{(n)} = \{e\}$.*

2. *The smallest integer n satisfying $G^{(n)} = \{e\}$ is called* **the solvability class of** G.

Example 9.2.1. 1. *Abelian groups are solvable since*

$$g^{-1}h^{-1}gh = g^{-1}gh^{-1}h = e.$$

The commutator subgroups are trivial.

2. *A non-trivial group is Abelian if and only if the solvability class is* 1.

Example 9.2.2. $\mathbb{Z}_n$ *is solvable of class* 1 *or* 0 *since it is Abelian for all* $n \in \mathbb{N}$.

Example 9.2.3. *Since* $\mathfrak{S}_3^{(1)} = \{(1), (13)(12), (12)(13)\}$ *and* $\mathfrak{S}_3^{(2)} = \{\iota\}$, $\mathfrak{S}_3$ *is solvable of class* 2.

We have more (obvious) solvable groups.

Example 9.2.4. 1. *A subgroup of a solvable group is solvable.*

2. *A homomorphic image of a solvable group is solvable.*

3. *A quotient group of a solvable group is solvable.*

4. *If two groups are solvable, so is their direct product.*[1]

Lemma 9.2.1. *For a group* G, G/G' *is Abelian.*

Proof. By Proposition 4.2.1, we know that G' is a normal subgroup of G. For $gG', hG' \in G/G'$,

$$(gG')^{-1}(hG')^{-1}gG'hG' = g^{-1}h^{-1}ghG' = G'.$$

This implies that $ghG' = hgG'$. Therefore, G/G' is Abelian. □

[1]The direct product of groups will be defined in the next chapter.

Lemma 9.2.2. *If G is a solvable group, then G has a series of subgroups such that*

$$\{e\} = G_k \subseteq G_{k-1} \subseteq \cdots \subseteq G_0 = G \tag{9.2}$$

where $G_i \lhd G_{i-1}$ and G_{i-1}/G_i is Abelian for each $i = 1, \ldots, k$.

Proof. If G is solvable, its derive series can be (9.2). □

Lemma 9.2.3. *If a group G has the following series of subgroups*

$$\{e\} = G_k \subset G_{k-1} \subset \cdots \subset G_1 \subset G$$

where $G_i \lhd G_{i-1}$ and G_{i-1}/G_i is Abelian for each $i = 1, \ldots, k$, then G is solvable.

Proof. This implies that $G' = G^{(1)} = [G, G] \subset G_1$, $G^{(2)} = [G^{(1)}, G^{(1)}] \subset [G_1, G_1] \subset G_2$, $\cdots$, $G^{(n)} \subset [G_{n-1}, G_{n-1}] \subset G_n$, $G^{(n+1)} \subset [G_n, G_n] \subset \{e\}$. This implies that $G^{(n+1)} = \{e\}$. Thus G is solvable.[2] □

We combine Lemma 9.2.2 and Lemma 9.2.3 as one theorem, and this can be a definition of a solvable group.

Theorem 9.2.1. *A group G is solvable if and only if G has the following series of subgroups*

$$\{e\} = G_k \subset G_{k-1} \subset \cdots \subset G_1 \subset G$$

where $G_i \lhd G_{i-1}$ and G_{i-1}/G_i is Abelian for each $i = 1, \ldots, k$.

Since a homomorphic image and a quotient group of a solvable group are solvable,[3] we derive another equivalent condition to a solvable group.

[2]Check Chapter 9 Exercise 12.
[3]Check Chapter 9 Exercises 9 and 10.

Theorem 9.2.2. *Suppose $N \lhd G$. Then, N and G/N are solvable if and only if G is solvable.*

Proof. The sufficient condition is satisfied by Chapter 9 Exercises 9 and 10. Let's suppose N and G/N are solvable. Since N is solvable, we have the following solvable series of N

$$N = N_0 \supset \cdots \supset N_{k-1} \supset N_k = \{e\}.$$

Since G/N is solvable, we have the following solvable series

$$G/N = G_0/N \supset G_1/N \supset \cdots \supset G_{p-1}/N \supset G_p/N = \{e\}.$$

Note that $(G_i/N)/(G_{i+1}/N)$ is Abelian and $(G_i/N)/(G_{i+1}/N) \cong G_i/G_{i+1}$, $p \geq i \geq 0$ by the 3rd Isomorphism Theorem. This makes G_i/G_{i+1} Abelian and by Lemma 9.2.3, we derive that

$$G \supset G_1 \supset \cdots \supset G_{p-1} \supset G_p = N \supset N_1 \supset \cdots \supset N_{k-1} \supset N_k = \{e\}$$

and this makes G solvable. □

The following famous theorem is stated here without proof because it is more than 200 pages and beyond the scope of this book.

Theorem 9.2.3. (Feit-Thompson Theorem, 1963)
Every finite group of odd order is solvable.

A series of its subgroups can be useful including the derived series. We state several common series without stating their usages.

Definition 9.2.2. *If G is a group, there might exist the following series.*

1. **A subnormal series of G:**

$$\{e\} = G_n \subseteq G_{n-1} \subseteq \cdots \subseteq G_1 \subseteq G_0 = G$$

if $G_i \lhd G_{i-1}$ for each $i = 1 \cdots n$.

2. **A normal series of G:**

$$\{e\} = G_n \subseteq G_{n-1} \subseteq \cdots \subseteq G_1 \subseteq G_0 = G$$

if $G_i \lhd G$ for each $i = 1 \cdots n$.

3. **A composition series of G:**

$$\{e\} = G_n \subseteq G_{n-1} \subseteq \cdots \subseteq G_1 \subseteq G_0 = G$$

if G_i is a maximal normal subgroup of G_{i-1} for each $i = 1 \cdots n$.

4. **The derived series of G:**

$$\cdots \subseteq G^{(n)} \subseteq G^{(n-1)} \subseteq \cdots \subseteq G^{(1)} \subseteq G^{(0)} = G$$

5. **The solvable series of G:**

$$\{e\} = G^{(n)} \subseteq G^{(n-1)} \subseteq \cdots \subseteq G^{(1)} \subseteq G^{(0)} = G$$

For each series, **the length of the series** is a number of distinct subgroups, i.e. $G_i \subsetneqq G_{i-1}$. Note that the length of a subnormal series of a group is maximal if an only if the subnormal series is a composition series.

Chapter 9 Exercises.

1. If $g^2 = e$ for all g in a group G, then prove that G is Abelian.

2. Suppose G is a finite group such that $G' = \{e\}$. Prove that $G \cong \mathbb{Z}_p$ for some prime number p.

3. (a) Find all the elements of order 9 in $\mathbb{Z}_9$.

   (b) Find all the elements of order p^2 in $\mathbb{Z}_{p^2}$.

4. A group G is called **perfect group** if $G = G'$. Show that if a group is perfect, it is not solvable.

5. Prove that a subgroup of a solvable group is solvable.

6. Prove that $\mathfrak{S}_4$ is solvable.

7. Prove that the dihedral group D_n is solvable. (Hint: D_n has a cyclic subgroup of index 2.)

8. Prove that a group of order p^n solvable where p is prime.

9. Prove that a homomorphic image of a solvable group is solvable.

10. Prove that a quotient group of a solvable group is solvable.

11. Show that S_5 is not solvable.

12. Prove that G/N is Abelian if and only if $G' \subseteq N$.[4]

13. If $\{e\} = G_n \subseteq G_{n-1} \subseteq \cdots \subseteq G_1 \subseteq G_0 = G$ is a composition series of a group G, prove that G_i/G_{i-1} doesn't have a non-trivial normal subgroup

[4]This means that if G/N is Abelian, $G' = N$ is the smallest.

for each $i = 1 \cdots n$. Recall that trivial subgroups are $\{e\}$ and the group itself.[5]

[5]If a group doesn't have a non-trivial normal subgroup, it is called **a simple group** and will be defined in Chapter 11.

Chapter 10

Products

When we see the notation $\mathbb{R}^2$, we know that it describes the Euclidean plan and it uses two $\mathbb{R}$'s, the Euclidean lines. In other words, using two (number) groups, we can create a new group, $\mathbb{R}^2$ and each element of this group has two coordinates. We'd like to use other groups to create a new group with multiple coordinates and check what kind of group structures are inherited to a new group.

10.1 External Direct Products

First, we create multiple coordinates.

Definition 10.1.1. *Suppose $G_1, \cdots, G_n$ are groups (with their own binary operations). Then,*

1. **The external direct product of** $G_1, \cdots, G_n$ *is the set*

$$G_1 \oplus \cdots \oplus G_n = \{(g_1, \cdots, g_n) | g_i \in G_i, 1 \leq i \leq n\}$$

and it is a group under the following binary operation:

$$(g_1, \cdots, g_n)(h_1 \cdots, h_n) = (g_1h_1, \cdots, g_nh_n)$$

for any $(g_1, \cdots, g_n), (h_1 \cdots, h_n) \in G_1 \oplus \cdots \oplus G_n$.

2. *It is possible to make an external direct product of infinite number of groups and it can be denoted by*

$$\oplus_{i\in J} G_i$$

where J *is a countable infinite index set. We can use* $\oplus_{i\in J} g_i$ *for an element of the direct product.*

3. *There is another direct product of infinitely many groups and it is called* **the weak direct product of infinitely many groups** *denoted by*

$$\oplus^w_{i\in J} G_i$$

Each element of the weak direct product has all e *but a finite number of coordinates.*

The external direct product is sometimes called simply the direct product and it is easy to check that $\oplus^w_{i\in J} G_i$ is a subgroup of the direct product $\oplus_{i\in J} G_i$.

Theorem 10.1.1. *For each* $\oplus_{i\in J} g_i \in G = \oplus_{i\in J} G_i$*, if* $p_j : \oplus_{i\in J} G_i \longrightarrow G_j$ *is defined by*

$$p_j(\oplus_{i\in J} g_i) = g_j,$$

then each p_j *is a group epimorphism where* J *is a given index set.*

Proof. Suppose $(g_1, \cdots, g_n)$ and $(h_1, \cdots, h_n) \in \oplus_{i\in J} G_i$. Then,

$$\begin{aligned} p_j\big((g_1, \cdots, g_n)(h_1, \cdots, h_n)\big) &= p_j\big((g_1h_1, \cdots, g_nh_n)\big) \\ &= g_jh_j \\ &= p_j\big((g_1, \cdots, g_n)\big)p_j\big((h_1, \cdots, h_n)\big) \end{aligned}$$

Thus, p_j is a homomorphism and onto (obviously). □

Theorem 10.1.2. *1. $\oplus^w_{i\in J} G_i \lhd \oplus_{i\in J} G_i$*

2. If $N_i \lhd G_i$ for each $i \in J$, then $\oplus_{i\in J} N_i \lhd \oplus_{i\in J} G_i$.

Proof.

1. Suppose $\oplus_{i\in J} g_i \in \oplus_{i\in J} G_i$ and $\oplus^w_{i\in J} n_i \in \oplus^w_{i\in J} G_i$. Then,

$$\begin{aligned} (\oplus_{i\in J} g_i)^{-1}(\oplus^w_{i\in J} n_i)(\oplus_{i\in J} g_i) &= (\oplus_{i\in J} g_i^{-1})(\oplus^w_{i\in J} n_i)(\oplus_{i\in J} g_i) \\ &= \oplus_{i\in J} g_i^{-1} n_i g_i \in \oplus^w_{i\in J} G_i \end{aligned}$$

2. This proof is similar to the 1st one. □

10.2 Internal Direct Products

Even though $\mathbb{R} \oplus \mathbb{R}$ is made of $\mathbb{R}$ and $\mathbb{R}$, $\mathbb{R} \oplus \mathbb{R}$ doesn't include $\mathbb{R}$. $\mathbb{R}$ is not even a subset of $\mathbb{R} \oplus \mathbb{R}$ but $\mathbb{R} \oplus \mathbb{R}$ includes an isomorphic copy of $\mathbb{R}$: $\mathbb{R} \oplus \{e\}$ or $\{e\} \oplus \mathbb{R}$. This is why $\mathbb{R} \oplus \mathbb{R}$ is called the external direct product because it doesn't include each component group physically. In this section, we define a product of groups and it contains each component group as a subgroup.

Definition 10.2.1. *Suppose G_i is a subgroup of G for $1 \leq i \leq n$. We call G* **the internal direct product** of $G_1, \ldots, G_n$ *if*

1. *for any element $g \in G$, there exists $g_i \in G_i$ for $1 \leq i \leq n$ such that $g = g_1 \cdots g_n$ and*

2. *if $g_i, h_i \in G_i$ such that $g_1 \cdots g_n = h_1 \cdots h_n$, then $g_i = h_i$ for $1 \leq i \leq n$.*

In this case we denote G as $G_1 G_2 \cdots G_n$, $\prod_{i=1}^{n} G_i$, or $G_1 \times \cdots \times G_n$.

Lemma 10.2.1. *Let G be the internal direct product of normal subgroups N_1, $\cdots$, N_n of G. If $N_i \cap N_j = \{e\}$ for $i \neq j$, then for $a_i \in N_i$ and $b_j \in N_j$, $a_i b_j = b_j a_i$, i.e., $a_i b_j a_i^{-1} b_j^{-1} = e$.*

Proof. For any $a_i \in N_i$ and $a_j \in N_j$, we have that $a_i^{-1} a_j^{-1} a_i a_j = (a_i^{-1} a_j^{-1} a_i) a_j \in N_j$ since $N_j \lhd G$ and $a_i^{-1} a_j^{-1} a_i a_j = a_i^{-1} (a_j^{-1} a_i a_j) \in N_i$ since $N_i \lhd G$. This implies that $a_i^{-1} a_j^{-1} a_i a_j \in N_i \cap N_j = \{e\}$. So we have that $a_i^{-1} a_j^{-1} a_i a_j = e$ and we derive that $a_i a_j = a_j a_i$. $\square$

Theorem 10.2.1. *Let G is the internal direct product of normal subgroups $N_1, \cdots, N_n$ of G. If $N_i \cap N_j = \{e\}$ for $1 \leq i \neq j \leq n$, then G is isomorphic to the external direct product $N_1 \oplus \cdots \oplus N_n$.*

Proof. If we define a map θ from $G = N_1 N_2 \cdots N_n$ to $\bigoplus_{1 \leq i \leq n} N_i$ as

$$\theta(a_1 \cdots a_n) = (a_1, \cdots, a_n)$$

for $a_1 \cdots a_n \in G$, since $a_1 \cdots a_n$ should be written uniquely, the mapping is well defined and one to one. Since $N_i \cap N_j = \{e\}$, $a_i a_j = a_j a_i$ for any $a_i \in N_i$ and $a_j \in N_j$ by Lemma 10.2.1. Thus for $a_1 \cdots a_n$ and $b_1 \cdots b_n \in G$, $(a_1 \cdots a_n)(b_1 \cdots b_n) =$

$a_1b_1 \cdots a_nb_n$ and this makes

$$\begin{aligned}\theta\big((a_1 \cdots a_n)(b_1 \cdots b_n)\big) &= \theta\big((a_1b_1 \cdots a_nb_n)\big)\\ &= (a_1b_1, \cdots, a_nb_n)\\ &= (a_1, \cdots, a_n)(b_1, \cdots, b_n)\\ &= \theta\big((a_1 \cdots a_n)\big)\theta\big((a_1 \cdots b_n)\big)\end{aligned}$$

Thus θ is a group homomorphism. For any $(c_1, \cdots, c_n) \in N_1 \oplus \cdots \oplus N_n$, $\theta((c_1 \cdots c_n)) = (c_1, \cdots, c_n)$ and this makes θ onto. Therefore, θ is an isomorphism from the internal direct product to the external direct product. $\square$

Corollary 10.2.1. *Let N_1 and N_2 be normal subgroups of a group G. If $N_1 \cap N_2 = \{e\}$ and $N_1N_2 = G$, then $G \cong N_1 \bigoplus N_2$.*

Example 10.2.1. *Find two subgroups of $\mathbb{Z}_6$ such that $\mathbb{Z}_6$ is the internal direct product of them.*

Solution. Let $G_1 = \{\overline{0}, \overline{2}, \overline{4}\}$ and $G_2 = \{\overline{0}, \overline{3}\}$. Then, they are subgroups of $\mathbb{Z}_6$ such that $G = G_1 \times G_2$. Moreover, since $G_1 \cap G_2 = \{\overline{0}\}$, we can say that $G \cong G_1 \oplus G_2$. $\square$

Example 10.2.2. *Since A_3 and $K = \{(1), (12)\}$ are subgroups of $\mathfrak{S}_3$ and they satisfy $\mathfrak{S}_3 = A_3 \times K$ and $A_3 \cap K = \{(1)\}$, we can say that $\mathfrak{S}_3 \cong A_3 \oplus K$.*

Example 10.2.3. *$M_2(\mathbb{R}) \bigoplus M_3(\mathbb{R})$ is isomorphic to the group generated by* *matrices*

$$\begin{pmatrix} a_{11} & a_{12} & 0 & 0 & 0 \\ a_{21} & a_{22} & 0 & 0 & 0 \\ 0 & 0 & a_{33} & a_{34} & a_{35} \\ 0 & 0 & a_{43} & a_{44} & a_{45} \\ 0 & 0 & a_{53} & a_{54} & a_{55} \end{pmatrix}$$

where $a_{11}, \cdots, a_{55} \in \mathbb{R}$.

Example 10.2.4. *If a group is of order 4, then it is isomorphic to either* $\mathbb{Z}_4$ *or* $\mathbb{Z}_2 \bigoplus \mathbb{Z}_2$.

Solution. Suppose $|G| = 4$. If $g \in G$, $|g| = 1, 2$, or 4. If $|g| = 4$, then G is a cyclic group of order 4 and it is isomorphic to $\mathbb{Z}_4$. If g is not the identity, then $|g|$ must be 2 and $| < g > | = 2$ as well. Since G has four elements, there exists $g_1 \in G$ but not in $< g >$. Note that $| < g_1 > | = 2$ and $|gg_1| = 2$. Thus G is isomorphic to the Klein 4-group that is isomorphic to $\mathbb{Z}_2 \bigoplus \mathbb{Z}_2$. $\square$

Chapter 10 Exercises

1. Prove that the external direct product of several groups is actually a group.

2. Prove that $\oplus^w_{i \in J} G_i < \oplus_{i \in J} G_i$ and $\oplus^w_{i \in J} G_i \lhd \oplus_{i \in J} G_i$.

3. Prove that $\mathbb{Z}_p \times \mathbb{Z}_p$ is not isomorphic to $\mathbb{Z}_{p^2}$ but isomorphic to $\mathbb{Z}_p \oplus \mathbb{Z}_p$ where p is prime.

4. For groups G_1 and G_2, prove that $G_1 \oplus G_2 \cong G_2 \oplus G_1$.

5. Find two groups H and K such that $\mathbb{Z}_{20}$ is isomorphic to $H \bigoplus K$.

6. Find all the subgroups of $\mathbb{Z}_2 \oplus \mathbb{Z}_2 \oplus \mathbb{Z}_2$.

7. Prove that G and H are Abelian if and only if $G \oplus H$ is Abelian.

8. Prove that $\mathbb{Z}_8 \oplus \mathbb{Z}_2$ is not isomorphic to $\mathbb{Z}_4 \oplus \mathbb{Z}_4$.

9. Prove that $\mathbb{Z} \oplus \mathbb{Z}$ is not cyclic.

10. Let N_1 and N_2 be normal subgroups of a group G. Prove that $G = N_1 \times N_2$ if and only if for any $a \in G$, a can be written as the product of the elements $n_1 \in N_1$ and $n_2 \in N_2$ uniquely.

11. Find two subgroups G_1 and G_2 of $\mathbb{Z}_{pq}$, such that $\mathbb{Z}_{pq} = G_1 \times G_2$ where p and q are distinct prime numbers.

12. (a) Prove that $\mathbb{Z}_{30}$ is isomorphic to an internal and external direct product of three groups.

    (b) Prove that $\mathbb{Z}_{pqr}$ is isomorphic to an internal and external direct product of three subgroups of $\mathbb{Z}_{pqr}$ where p, q, and r are distinct prime numbers.

13. (a) For groups $G_1 = M_m(\mathbb{R})$ and $G_2 = M_m(\mathbb{R})$, find a subgroup of $M_{2m}(\mathbb{R})$ which is isomorphic to $M_m(\mathbb{R}) \bigoplus M_m(\mathbb{R})$.

    (b) For groups $G_1 = M_m(\mathbb{R})$ and $G_2 = M_n(\mathbb{R})$, find a subgroup of $M_{m+n}(\mathbb{R})$ which is isomorphic to $M_m(\mathbb{R}) \bigoplus M_n(\mathbb{R})$.

14. Prove that an internal and external direct product of solvable groups is solvable.

Chapter 11

Sylow Theorems

When we study finite groups, we can say many things just using its order. For example, if $|G| = p$, a prime number, then it is isomorphic to $\mathbb{Z}_p$. If $|G| = 6$, it doesn't have an element of order 4. We'd like to investigate properties of a group based on its order so we consider only finite groups in this chapter.

Suppose $H < G$. Then $|H|$ is a divisor of $|G|$ by Lagrange theorem. If m is a positive integer which divides $|G|$, does $|G|$ have a subgroup of order $|m|$? This is the converse of Lagrange's theorem and it is true for Abelian or cyclic groups, however it may be false for arbitrary groups. For example, A_4 is a group of order 12. But it has no subgroup of order 6. The Sylow Theorems are the first basic step in understanding the structure of non-Abelian finite groups.

11.1 First Sylow Theorem

To prove the 1st Sylow Theorem, we need to recall group actions we have studied in Chapter 7. We start this chapter with the following example.

Example 11.1.1. *If a group G acts on itself as follows*

$$g * x = gxg^{-1} \text{ for } g \in G \text{ and } x \in G,$$

then $stab_G(x) = \{g \in G | gxg^{-1} = x\}$ *and* $orb_G(x) = \{g * x = gxg^{-1} | g \in G\}$. *Note that* $stab_G(x)$ *is the centralizer* $C_G(g)$ *of* x *in* G *so we have*

$$|orb_G(x)| = [G : stab_G(x)] = [G : C_G(x)].$$

Example 11.1.2. *A subgroup H of G acts on the set S of all the subgroups of G as follows:*

$$g * K = gKg^{-1} \text{ for } g \in H \text{ and } K \in S.$$

Then $stab_H(K) = \{g \in H | gKg^{-1} = K\}$ *and* $orb_H(K) = \{gKg^{-1} | g \in H\}$. *Note that* $stab_H(K)$ *is* $N_H(K)$, *the normalizer of* K *in* H, *and we have*

$$|orb_H(K)| = [H : stab_H(K)] = [H : N_H(K)].$$

We start to prove the first Sylow Theorem using the following lemma.

Lemma 11.1.1. *Let a group G acts on itself as Example 11.1.1.*

1. *If you define a relation $\sim$ on G as follows: $g_1 \sim g_2$ if there is g in G such that $g_1 = gg_2g^{-1}$, then $\sim$ is an equivalence relation and each equivalence class is the orbit $orb_G(g)$ for g in G. This class is called* **the conjugate class of** *g and the set of all conjugate classes, therefore makes a partition of G.*

2. *For any $g \in G$, the stabilizer $stab_G(g)$ is the centralizer $C_G(g)$ of g in G and $|orb_G(g)| = [G : C_G(g)]$.*

3. *For a finite group G, we have that*

$$|G| = \sum_g [G : C_G(g)] = |\mathfrak{Z}| + \sum_{g \notin \mathfrak{Z}} [G : C_G(g)]. \tag{11.1}$$

Proof.

1. Since $g = ege^{-1}$, we have that $g \sim g$ where $g \in G$. If $g_1 \sim g_2$, then there is $g \in G$ such that $g_1 = gg_2g^{-1}$. This implies that $g_2 = g^{-1}g_1g = g^{-1}g_1g = ((g^{-1})^{-1})^{-1}g_1(g^{-1})^{-1}$. We have that $g_2 \sim g_1$. If $g_1 \sim g_2$ and $g_2 \sim g_3$, then there g and h of G such that $g_1 = gg_2g^{-1}$ and $g_2 = hg_3h^{-1}$. This implies that $g_1 = ghg_3h^{-1}g^{-1} = ghg_3(gh)^{-1}$ and this makes $g_1 \sim g_3$. So $\sim$ is an equivalence relation.

2. The first part of the proof is straightforward by Example 11.1.1 and the second part is straightforward by Theorem 7.4.2.

3. Since the group G acts on itself by conjugation, G can be partitioned as its orbits, $|G| = \sum_g orb_G(g)$, i.e., $|G| = \sum_g [G : stab_G(g)]$. If $g \in \mathfrak{Z}$, then $[G : stab_G(g)] = 1$ and $stab_G(g) = G$. This implies that $|G| = \sum_g [G : C_G(g)] = |\mathfrak{Z}| + \sum_{g \notin \mathfrak{Z}} [G : C_G(g)]$. □

The equation (11.1) is called **the class equation of the group** G.

Definition 11.1.1. *For a finite group G, a subgroup H of G is called* p-**group** *if $|H| = p^r$ where p is prime and $r \in \mathbb{N}$. If p^n is the largest power of p that divides $|G|$, then a subgroup of G of order p^n is called* **a Sylow** p-**subgroup.**

Theorem 11.1.1. (First Sylow Theorem)

Let G be a finite group. If p is a prime and p^k divides $|G|$, then G has a subgroup of order p^k for $k \in \mathbb{N}$.

Proof. We prove the theorem by induction on the order $|G| = n$. If $n = 1$, then there is nothing to prove. Let us assume that the theorem is true for all groups whose orders are less than n where $p||G|$. Let $\mathfrak{Z}$ be the center of G. Then, we have two cases:

1. If p divides $|\mathfrak{Z}|$, then there is an element $g \in \mathfrak{Z}$ such that $| < g > | = p$ by Cauchy Theorem. Since $< g >$ is a subgroup of $\mathfrak{Z}$, it is a normal subgroup of G. Since $\frac{n}{p} = |G|/| < g > | < n$, by induction there is a subgroup H of G such that $|H/< g >| = p^{k-1}$. This implies that $|H| = p^k$. So we have proven the theorem.

2. If p does not divide $|\mathfrak{Z}|$, we use the conjugation group action by G on G. The class equation is $n = |\mathfrak{Z}| + \sum_{g \notin \mathfrak{Z}} [G : C_G(g)]$ with $[G : C_G(g)] > 1$. Since p divides $|G|$ and p does not divides $|\mathfrak{Z}|$, we have that p does not divides $[G : C_G(g)]$ for some g. This implies that p^k divides $|C_G(g)|$ and $C_G(g)$ is a subgroup of G. By induction, $C_G(g)$ has a subgroup H of order p^k. Let H act on H by conjugation. Since $|H| = p^k$, the class equation of H is $p^k = |\mathfrak{Z}| + \sum_{g \notin \mathfrak{Z}} [H : C_H(g)]$. So we have that $\overline{0} \equiv \overline{|\mathfrak{Z}|} \bmod p$. This implies that $|\mathfrak{Z}| \geq p$ since $|\mathfrak{Z}| \geq 1$ so we derive that there is a normal subgroup K of H such that $|K| = p^r$ for $r \leq k$. By induction on k of p^k, there is a subgroup L of H such that $|L| = p^s$ for $s \leq k$.

Thus we have proven the theorem. $\square$

Example 11.1.3. *$\mathfrak{S}_4$ has order* 24 *and the divisors of* 24 *are* $1, 2, 3, 4, 6, 8, 12,$ *and 24. Then, the 1st Sylow Theorem guarantees that* $\mathfrak{S}_4$ *has a subgroup of order* 2, 3, $2^2 = 4$, $2^3 = 8$.

Using Example 11.1.3, we can derive that $\mathfrak{S}_4$ has subgroups of order 6 and 12 as well.[1]

Corollary 11.1.1. (Generalized Cauchy Theorem of a finite group) *Let G be a finite group and p is a prime number. If $p||G|$, then there is an element g of G such that $| < g > | = p$.*

Proof. This is a special case of the 1st Sylow Theorem with $k = 1$ in Theorem 11.1.1. □

Corollary 11.1.2. *Let G be a finite group and $|G| = p^k m$ where p is a prime number. If $(p, m) = 1$, then there is a subgroup H of G such that $|H| = p^k$.*

Proof. This is another special case of the 1st Sylow Theorem with the maximal power of p. □

This subgroup H in Example 11.1.3 is a Sylow p-subgroup of G.

Corollary 11.1.3. *If G is a p-group with $|G| = p^m$, then G has subgroups of order p^k for $k = 0, \cdots, m$.*

---

[1]See Chapter 11 Exercise 1.

11.2 Second Sylow Theorem and Simple Groups

Theorem 11.2.1. (Second Sylow Theorem)

If H is a Sylow p-subgroup of a finite group G, then H is conjugate to another Sylow p-subgroup of G.

Proof. Suppose P be a Sylow p-subgroup of G and consider the group action by H on the set $S = \{xP | x \in G\}$ given by

$$h * xP = hxP$$

for $h \in H$ and $xP \in S$. Then, since $|S|$ is $[G : P]$, S can be decomposed as

$$\begin{aligned} [G : P] &= |S| \\ &= |S_0| + |orb_H(x_1P)| + \cdots + |orb_H(x_rP)| \\ &= |S_0| + [H : stab_H(x_1P)] + \cdots + [H : stab_H(x_rP)] \end{aligned} \quad (11.2)$$

where $S_0 = \{xP | hxP = xP \text{ for all } h \in H\}$. This implies that if $gP \in S_0$, $orb_H(gP) = \{gP\}$ and $stab_H(gP) = H$, so we derive that $1 = |\{gP\}| = |orb_H(gP)| = [H : stab_H(gP)]$. If $x_iP \notin S_0$, then $|orb_H(x_iP)| \neq 1$, i.e., $stab_H(x_iP) \neq H$, $1 \leq i \leq r$. Since p does not divide $|S| = [G : P]$ and $[H : stab_H(x_iP)] > p$ for $1 \leq i \leq r$, we have that $|S_0| \not\equiv 0 \text{ mod } p$, i.e., $|S_0| > 1$. Therefore S_0 is not the empty set, which implies that by the definition of S_0, there is an element $xP \in S_0$ such that for any $h \in H$, $hxP = xP$. This also implies that for all $x^{-1}hxP = P$ and we derive that $x^{-1}hx \in P$ for all $h \in H$, i.e., $x^{-1}Hx \subset P$. This makes H a subset of xPx^{-1}. Since H and xPx^{-1} have the same order, they are the same group. Therefore, $H = xPx^{-1}$. □

Example 11.2.1. $\mathfrak{S}_4$ *has three Sylow 2-subgroups:*

$$\begin{aligned} H_1 &= \{(1),(1234),(13)(24),(1432),(24),(12)(34),(13),(14)(32)\} \\ H_2 &= \{(1),(1243),(14)(23),(1342),(14),(23),(12)(34),(13)(24)\} \\ H_3 &= \{(1),(1324),(12)(34),(1423),(12),(34),(13)(24),(14)(23)\} \end{aligned}$$

All of them are conjugate to each other so isomorphic.

Corollary 11.2.1. *If there is a unique Sylow p-subgroup, then the Sylow p-subgroup is normal.*

Proof. Let P be a Sylow p-subgroup of a group G. The 2nd Sylow Theorem says that $x^{-1}Px$ is another Sylow p-subgroup of G for any $x \in G$ but there exists only one Sylow p-subgroup of G. This implies that $P = x^{-1}Px$ for any $x \in G$ and this is an exact definition of a normal subgroup. □

Definition 11.2.1. *A group G is called* **simple** *if G does not have a non-trivial normal subgroup, i.e., the only normal subgroups of G are $\{e\}$ and G.*

The following is a re-statement of Corollary 11.2.1.

Proposition 11.2.1. *If a group G has a unique Sylow subgroup, then it is not simple.*

Proof. If P be the unique Sylow p-subgroup of G, P is a non-trivial normal subgroup of G by Corollary 11.2.1. Therefore, G is not simple. □

Example 11.2.2. $\mathbb{Z}_p$ *is simple if p is prime and $\mathbb{Z}_n$ is not simple if n is not prime.*

Example 11.2.3. *For $n \geq 3$, $\mathfrak{S}_n$ is not simple since $A_n \lhd \mathfrak{S}_n$.*

Lemma 11.2.1. *Let P and Q be Sylow p-subgroups of a group G. If P is a subgroup of the normalizer $N_G(Q)$ of Q, then $P = Q$.*

Proof. Let us assume that there is an element $g \in P$ such that $g \notin Q$. Since Q is a normal subgroup of $N_G(Q)$, $< g > Q$ is subgroup of $N_G(Q)$ and $| < g > Q| > |Q|$. Since the order $| < g > Q|$ of $< g > Q$ is a power of p. This contradicts the fact that Q is a Sylow p-subgroup of G. So every elements in P are in Q, i.e., P is a subgroup of Q. Since P and Q have the same order, we have that $P = Q$. □

11.3 Third Sylow Theorem and Examples

Let $Syl_p(G)$ denote the set of all the Sylow p-subgroups of a group G.

Theorem 11.3.1. (Third Sylow Theorem)
The number of Sylow p-subgroups of a finite group G divides $|G|$ and is of the form $1 + pk$ for some non-negative integer k.

Proof. Let's consider the group action by G on $Syl_p(G)$ by conjugation and let P be a Sylow p-subgroup of G. By the 2nd Sylow Theorem, every Sylow p-subgroup of G is conjugate to P. This implies that $orb_G(P) = \{gPg^{-1} | g \in G\} = Syl_p(G)$. Since $stab_G(P) = N_G(P)$, $|Syl_p(G)| = [G : N_G(P)]$ and it divides $|G|$.

Let Q be a Sylow p-subgroup of G and let P act on $S = Syl_p(G)$ by conjugation. Then, $|S|$ can be written as

$$|S| = |S_0| + \sum_{P_i \in S} [P : stab_P(P_i)].$$

By the definition of S_0, $Q \in S_0$ if and only if $gQg^{-1} = Q$ for all $g \in P$. This condition holds if and only if P is a subgroup of $N_G(Q)$. Since Q is normal in $N_G(Q)$ and $|Q| = |P|$, $Q = P$ by Lemma 11.2.1. This means that $S_0 = \{P\}$. Since $|S| \equiv |S_0| \equiv 1 \bmod p$, therefore $|S| = kp+1$ for some non-negative integer k. □

Example 11.3.1. *Since $\mathfrak{S}_3$ has order 6, every subgroup of order 2 is a Sylow 2-subgroup. Since $2k+1$ divides 6, we have that $k = 0$ or $k = 1$. In fact, $\mathfrak{S}_3$ has three Sylow 2-subgroups:*

$$H_1 = \{(1), (12)\}, \quad H_2 = \{(1), (13)\}, \quad H_3 = \{(1), (23)\}$$

so $k = 1$.

Also $\mathfrak{S}_3$ has one Sylow 3-group, $H_4 = \{(1), (123), (132)\}$ so $H_4 \lhd \mathfrak{S}_3$. For example, $(13)H_1(13)^{-1} = H_3$ and it is easy to check that $gH_4g^{-1} = H_4$ for any $g \in \mathfrak{S}_3$.

Example 11.3.2. *Since $\mathfrak{S}_4$ has order 24, every subgroup of order 8 is a Sylow 2-subgroup. In fact, $\mathfrak{S}_4$ has three Sylow 2-subgroups:*

$$\begin{aligned} H_1 &= \{(1), (1234), (13)(24), (1432), (24), (12)(34), (13), (14)(32)\} \\ H_2 &= \{(1), (1243), (14)(23), (1342), (14), (23), (12)(34), (13)(24)\} \\ H_3 &= \{(1), (1324), (12)(34), (1423), (12), (34), (13)(24), (14)(23)\} \end{aligned}$$

Also it has four Sylow 3-subgroups:

$$\begin{aligned} H_4 &= \{(1), (123), (132)\} \\ H_5 &= \{(1), (124), (142)\} \\ H_6 &= \{(1), (134), (143)\} \\ H_7 &= \{(1), (234), (243)\} \end{aligned}$$

Example 11.3.3. *The order of A_4 is $12 = 2^2 \cdot 3$. Since $2k+1$ is a divisor of 12, A_4 has either one Sylow 2-subgroup or three Sylow 2-subgroups. In fact, A_4 has one Sylow 2-subgroup that is isomorphic to $\mathbb{Z}_2 \oplus \mathbb{Z}_2$:*

$$H_1 = \{(1), (12)(34), (13)(24), (14)(32)\}$$

Also, it has four Sylow 3-subgroups.

$$\begin{aligned} H_2 &= \{(1), (123), (132)\} \\ H_3 &= \{(1), (124), (142)\} \\ H_4 &= \{(1), (134), (143)\} \\ H_5 &= \{(1), (234), (243)\} \end{aligned}$$

We can check that these Sylow 3-subgroups are conjugate to each other. For example, $(12)H_4(12)^{-1} = \{(1), (234), (243)\} = H_5$.

Example 11.3.4. *Show that there is no simple group whose order is 63.*

Solution. Suppose G is a group of order $63 = 3^2 \cdot 7$. Then, the number of Sylow 7-subgroups is a divisor of 63 of the form $1 + 7k$, $k \geq 0$. This implies that

$k = 0$ and it implies that there is only one Sylow 7-subgroup that is a normal subgroup of G. Therefore, G is not simple. $\square$

Example 11.3.5. *Show that there is no simple group whose order is* 350.

Solution. Suppose G is a group of order $350 = 2 \cdot 5^2 \cdot 7$. Then, the number of Sylow 5-subgroups is a divisor of 350 and of the form $1 + 5k$. But $k = 0$ is the the only number satisfying these conditions. So G has a normal subgroup of order 25 and G is not simple. $\square$

Example 11.3.6. *Show that G is cyclic if* $|G| = 35$.

Solution. Since G has only one Sylow 5-subgroup H and only one Sylow 7-subgroup K, they are normal and $H \cap K = \{e\}$. If we let $H =< a >$ and $K =< b >$, then $G =< ab >$. Thus G is a cyclic group. $\square$

We have checked only simple cases using Sylow Theorems and here is a summary.

Suppose

$$|G| = p_1^{i_1} \cdots p_s^{i_s}$$

where $p_1, \cdots, p_s$ are distinct prime numbers and $i_1, \cdots, i_n \in \mathbb{W}$. Then, by the 1st Sylow Theorem, there is a subgroup $H < G$ with $|H| = p_u^{j_u}$ where $1 \leq u \leq s$ and $0 \leq j_u \leq i_u$. Let $H_{v,1}, \cdots, H_{v,m}$ be Sylow p_v-subgroups of G where $1 \leq v \leq s$. Then by the 2nd Sylow Theorem, there is $g \in G$ such that $gH_{v,u}g^{-1} = H_{v,w}$, $1 \leq u, w \leq m$. Finially, by the 3rd Theorem, $m = kp_v + 1$ which divides $|G|$ where $k = 0, 1, \cdots$, i.e.,

$$(1 + kp_v)|(p_1^{i_1} \cdots \widehat{p_v^{i_v}} p_{v+1}^{i_{v+1}} \cdots p_s^{i_s})$$

where $\widehat{p_v^{i_v}}$ means that the factor is omitted.

See other references for more complicated examples.

Chapter 11 Exercises.

1. Show that $\mathfrak{S}_4$ has subgroups of order 6 and 12 using the 1st Sylow Theorem. You don't have to find them.

2. Prove that groups of the following orders are not simple.

   (a) 56 (b) 66 (c) 88 (d) 91 (e) 165 (f) 455 (g) 595

   (h) 805 (i) $117 = 3^2 \cdot 13$ (j) $195 = 3 \cdot 5 \cdot 13$ (k) $3 \cdot 5 \cdot 17$

   (l) $3^4 \cdot 5^2 \cdot 17$ (m) $3^5 \cdot 5^2 \cdot 17$ (n) $3^3 \cdot 5^2 \cdot 23$

3. (a) Find all the 2-Sylow subgroups of $\mathfrak{S}_4$. Pick 3 distinct ones and call them H_1, H_2, and H_3. Your answers will vary.

   (b) Find $g \in \mathfrak{S}_4$ such that $gH_1g^{-1} = H_1$.

   (c) Find $g \in \mathfrak{S}_4$ such that $gH_1g^{-1} = H_2$.

   (d) Find $g \in \mathfrak{S}_4$ such that $gH_2g^{-1} = H_3$.

4. Prove that any non-trivial group homomorphism from a simple group is one to one.

5. Find a non-simple group whose order is of the following form:

   (a) pq where p and q are prime.

   (b) pn where p is prime and n is not prime.

   (c) pqn where p and q are prime, and n is not prime.

6. (a) If $|G| = 15$, prove that G is not simple and it is cyclic.

   (b) If $|G| = pq$ for prime numbers p and q, prove that G is not simple and it is cyclic.

   (c) If $|G| = pq$ for prime numbers p and q, draw the subgroup lattice. (Since G is cyclic, they are all normal.)

7. Let G be a group where $|G| = pqr$ where p, q, and r are distinct primes.

   (a) Find primes p, q, and r such that G is not simple.

   (b) If possible, find primes p, q, and r such that G is simple.

8. Let P be a Sylow p-subgroup of a group G. Let $N_G(P)$ be the normalizer of P in the group G. Prove that P is the only Sylow p-subgroup of $N_G(P)$. (This implies that $P \triangleleft N_G(P)$.)

9. Let G be a p-group where p is prime.

   (a) Prove that the order of the center of G is greater than 1. (Hint: Use the class equation.)

   (b) If the order of G is p^2, then prove that G is Abelian.

10. Suppose $G_1 < G$ such that

$$|G| = p_1^{i_1} \cdots p_{n-1}^{i_{n-1}} p_n^{i_n} \quad \text{and} \quad |G_1| = p_1^{i_1} \cdots p_{n-1}^{i_{n-1}} p_n^{j_n}$$

where $p_1 < \cdots < p_n$ are primes, $i_1, \cdots, i_n \in \mathbb{W}$, $i_n > 1$, and $j_n \leq i_n$ Prove that if G_1 has a normal Sylow p_n-subgroup, then G is not simple [2].

Chapter 12

Basic Ring Theory

When we say that $\mathbb{Z}$ is a group, we mean that it is a group under the addition but the addition is not the only binary operation defined in $\mathbb{Z}$. There are several operations including the subtraction, the multiplication, and the division defined (but sometimes not well-defined) in $\mathbb{Z}$.

Starting this chapter, we'd like to study algebraic structures with multiple operations, for example $\mathbb{R}$ under the addition and the multiplication so now we have two tools. Other examples are $\mathbb{Z}$ and $2\mathbb{Z}$. They are isomorphic groups so there is no difference as groups but they are totally different as rings.

12.1 Definition and Examples of Rings

Definition 12.1.1. *A* **ring** R *is a non-empty set with two binary operations, typically* $+$ *and* $\cdot$ *satisfying the following axioms:*

1. *R is an additive Abelian group.*

2. *$r_1 \cdot r_2 \in R$ for any $r_1, r_2 \in R$.*

3. $(r_1 \cdot r_2) \cdot r_3 = r_1 \cdot (r_2 \cdot r_3)$ *for any* $r_1, r_2, r_3 \in R$.

4. $r_1 \cdot (r_2 + r_3) = r_1 \cdot r_2 + r_1 \cdot r_3$ *and* $(r_1 + r_2) \cdot r_3 = r_1 \cdot r_3 + r_2 \cdot r_3$ *for any* $r_1, r_2, r_3 \in R$.

Note 12.1.1. *1. A ring is an additive Abelian group under the 1st operation but it is not even a group under the 2nd operation.*

2. The 1st and the 2nd operations are typically $+$ *and* $\cdot$ *but not necessarily. Any set is a ring as far as defined two operations satisfy the 5 axioms of the definition. Also, one set could have different ring structures if different pairs of operations are used.*

3. A ring R *is called* **a commutative ring** *if if* $r_1 \cdot r_2 = r_2 \cdot r_1$ *for any* $r_1, r_2 \in R$. *So, if a ring* R *is non-commutative, there are two elements* $r_1, r_2 \in R$ *such that* $r_1 \cdot r_2 \neq r_2 \cdot r_1$.

4. 0 *is the additive identity of* R.

5. A ring may not have a multiplicative identity but if it has one, we call it 1 *and it satisfies* $1 \cdot r = r \cdot 1 = r$ *for* $r \in R$.

6. Sometime this "1" is called **unity** *and if a ring has a unity, the ring is called* **a unitary ring**.

7. If it is obvious, we will omit the multiplication $\cdot$*, but not the addition (or the 1st operation).*

Example 12.1.1. $\mathbb{Z}$ *is a ring with the addition* $+$ *and the multiplication* $\cdot$. *We denote* $(\mathbb{Z}, +, \cdot)$.

It makes sense to denote $(\mathbb{Q}, +, \cdot)$, $(\mathbb{R}, +, \cdot)$ and $(\mathbb{C}, +, \cdot)$.

Example 12.1.2. *The set* $\mathbb{Z}^-$ *of all negative integers is not a ring with the addition* $+$ *and the multiplication* $\cdot$. $\mathbb{Q}^-$ *is not a ring as well.*

Example 12.1.3. $M_2(\mathbb{Z})$*, the set of all* 2×2 *matrices over* $\mathbb{Z}$ *is a non-commutative unitary ring under the usual matrix addition and multiplication.*

Note. $\begin{pmatrix} 0 & 0 \\ 0 & 0 \end{pmatrix}$ is the additive identity and $I_2 = \begin{pmatrix} 1 & 0 \\ 0 & 1 \end{pmatrix}$ is the multiplicative identity of $M_2(\mathbb{Z})$ since

$$\begin{pmatrix} 1 & 0 \\ 0 & 1 \end{pmatrix}\begin{pmatrix} a_{11} & a_{12} \\ a_{21} & a_{22} \end{pmatrix} = \begin{pmatrix} a_{11} & a_{12} \\ a_{21} & a_{22} \end{pmatrix}\begin{pmatrix} 1 & 0 \\ 0 & 1 \end{pmatrix} = \begin{pmatrix} a_{11} & a_{12} \\ a_{21} & a_{22} \end{pmatrix}.$$

Also, it is non-commutative because

$$\begin{pmatrix} 1 & 2 \\ 2 & 1 \end{pmatrix}\begin{pmatrix} 2 & 1 \\ 3 & 1 \end{pmatrix} \neq \begin{pmatrix} 2 & 1 \\ 3 & 1 \end{pmatrix}\begin{pmatrix} 1 & 2 \\ 2 & 1 \end{pmatrix}$$

Similarly, we can see that I_n is the multiplicative identity for $\mathbb{Z}_n$.

Example 12.1.4. $M_n(\mathbb{Z})$, $M_n(\mathbb{Q})$, $M_n(\mathbb{R})$, *and* $M_n(\mathbb{C})$ *are all non-commutative unitary rings under the usual matrix addition and multiplication.*

Example 12.1.5. $\mathbb{Z}_n$ *is a commutative unitary ring under the addition modulo* n *and the multiplication modulo* n.

So we can say that $(\mathbb{Z}_n, +/n, \cdot/n)$ is a ring. Note that $\mathbb{Z}_0 = \mathbb{Z}$ and $\mathbb{Z}_1 = \{0\}$.

Example 12.1.6. $\mathbb{Z}[i] = \{a + bi | a, b \in \mathbb{Z}\}$ *is a commutative unitary ring with the usual addition and the multiplication.*

Recall that $i = \sqrt{-1}$ and $\mathbb{Z}[i]$ is called **Gaussian Ring of Integers.**

Example 12.1.7. $\mathbb{Q}[\sqrt{7}] = \{a + b\sqrt{7} | a, b \in \mathbb{Q}\}$ *is a commutative unitary ring with the usual addition and the multiplication.*

12.2 Polynomial Rings: New Ring From A Ring

Let's consider $\mathbb{R}[x] = \{a_n x^n + \cdots + a_1 x + a_0 | a_i \in \mathbb{R}, 0 \leq i \leq n, n \in \mathbb{N}\}$.

1. x is called **an indeterminate** or **a variable.**

2. An element $a_i x^i$ is called **a monomial.**

3. A sum of monomials is called **a polynomial.**

4. **The degree of a monomial** $a_i x^i$ is i and denoted by $deg(a_i x^i)$.

5. If $f(x) = a_n x^n + \cdots + a_1 x + a_0$ in the decreasing order of a degree of each monomial, a_i is called **the coefficient of** x^i.

6. For a polynomial, the monomial with the highest degree is called **the leading term** of the polynomial and the coefficient of the leading term is called **the leading coefficient** of the polynomial. Note that a leading coefficient can not be 0.

7. **The degree of a polynomial** $f(x)$ is the degree of the leading term of $f(x)$ and denoted by $deg(f(x)$. This means that $deg(f(x)$ is the highest

degree of the monomial of $f(x)$ whose coefficient is not zero so if $f(x) = a_nx^n + \cdots + a_1x + a_0$ with $a_n \neq 0$, then $deg(f(x) = n$.

8. If $deg(f(x)) = 0$, $f(x)$ is called **a constant polynomial**. If $deg(f(x)) = 1$, $f(x)$ is called **a linear polynomial** and a degree 2 polynomial is called **a quadratic polynomial.**[1]

9. Two polynomials are the same if and only if they have the same degrees with the same corresponding coefficients, i.e., if

$$\begin{aligned} f(x) &= a_nx^n + \cdots + a_1x + a_0 \\ g(x) &= b_mx^m + \cdots + b_1x + b_0 \end{aligned}$$

then $f(x) = g(x)$ if and only if $n = m$ and $a_n = b_n(= b_m), \cdots, a_0 = b_0$.

10. Suppose $f(x) = a_nx^n + \cdots + a_1x + a_0$ and $g(x) = b_nx^n + \cdots + b_1x + b_0 \in \mathbb{R}[x]$. Note that we don't care about the degrees so a_n could be 0. For example, if $f(x) = x^2 - 1$ and $g(x) = 2x + 1$, we write $f(x) = x^2 + 0x - 1$ and $g(x) = 0x^2 + 2x + 1$. Then, we define the addition as

$$f(x) + g(x) = (a_n + b_n)x^n + \cdots + (a_0 + b_0)$$

and the multiplication by

$$\begin{aligned} f(x)g(x) &= (a_nx^n + \cdots + a_1x + a_0)(b_nx^n + \cdots + b_1x + b_0) \\ &= \sum_{0 \leq k \leq 2n} c_kx^k \qquad (12.1) \end{aligned}$$

where $c_k = \sum_{s+t=k} a_sb_t$. Then, $(\mathbb{R}[x], +, \cdot)$ is a ring.

[1] A cubic polynomial is of degree 3 and a quartic polynomial has degree 4. For more degrees, 5=quintic, 6=hexic, 7=heptic, 8=octic, 9=nonic, 10=docic, etc.

11. $\mathbb{R}[x]$ is called **the polynomial ring over** $\mathbb{R}$ and it is a commutative unitary ring since $1 \in \mathbb{R}[x]$ is the multiplicative identity.

12. With the same fashion, we can see that $\mathbb{Z}[x]$, $\mathbb{Q}[x]$, and $\mathbb{C}[x]$ are all polynomial rings over different rings.

13. If a different number is used instead of the indeterminate x, we have a different ring. For example, if $x = i$, then we have the Gaussian Integer Ring $\mathbb{Z}[i]$ and if $x = \sqrt{7}$, we have the ring in Example 12.1.7.

Example 12.2.1. *Let $f(x) = 3x^3 + 4x + 6$. Then, $deg\big(f(x)\big) = 3$ and $f(x)$ is consisting of three monomials: $3x^3, 4x$, and 6. If $f(x) = a_n x^n + \cdots + a_1 x + a_0$, then $n = 3$, $a_3 = 3$, $a_2 = 0$, $a_1 = 4$, $a_0 = 6$, and $a_i = 0$ for $i \geq 3$.*

Example 12.2.2. *Let $f(x) = 2x^2 + 3x - 1$ and $g(x) = x + 3$. Then, $f(x)$ is a quadratic polynomial and $g(x)$ is linear. Also, $f(x) + g(x) = 2x^2 + 4x + 2$ and $f(x)g(x) = (2x^2 + 3x - 1)(x + 3) = 2x^3 + 9x^2 + 8x - 3$.*

A polynomial ring is created from a given ring that provides coefficients. The following are other ways to make new rings from given rings.

Example 12.2.3. *Let $R_1, \cdots, R_n$ be rings. Then, $R_1 \oplus \cdots \oplus R_n = \oplus_{1 \leq i \leq n} R_i = \{(a_1, \cdots, a_n) | a_i \in R_i\}$ becomes a ring if we define the addition $+$ and the multiplication $\cdot$ as follows:*

$$\begin{aligned}(a_1, \cdots, a_n) + (b_1, \cdots, b_n) &= (a_1 + b_1, \cdots, a_n + b_n) \\ (a_1, \cdots, a_n) \cdot (b_1, \cdots, b_n) &= (a_1 b_1, \cdots, a_n b_n),\end{aligned}$$

for $(a_1, \cdots, a_n), (b_1, \cdots, b_n) \in \oplus_{1\leq i\leq n} R_i$.

Notes: $\oplus_{1\leq i\leq n} R_i$ was the external direct product of the groups and now it is the external direct product of the rings $R_1, \cdots, R_n$. Also note that $(0, \cdots, 0)$ is the additive identity and $(e_1, \cdots, e_n)$ is the unity of the ring if R_i has the unity e_i for $1 \leq i \leq n$.

Example 12.2.4. $\mathbb{Z}_2 \oplus \mathbb{Z}_3 = \{(\bar{0}, \bar{0}), (\bar{0}, \bar{1}), (\bar{0}, \bar{2}), (\bar{1}, \bar{0}), (\bar{1}, \bar{1}), (\bar{1}, \bar{2})\}$ *is the external direct product of two rings:* $\mathbb{Z}_2$ *and* $\mathbb{Z}_3$*.* $(\bar{0}, \bar{0})$ *is the additive identity and* $(\bar{1}, \bar{1})$ *is the multiplicative identity so* $\mathbb{Z}_2 \oplus \mathbb{Z}_3$ *is a commutative unitary ring.*

12.3 Properties of Ring Elements and Subrings

Theorem 12.3.1. *Let* R *be a ring and* $a, b, c \in R$*. Then,*

1. $00 = 0$.

2. $a0 = 0a = 0$.

3. $a(-b) = -(ab) = (-a)b$.

4. $a(b - c) = ab - ac$ *and* $(b - c)a = ba - ca$.

Proof.

1. $0(0 + 0) = 00 + 00$. On the other hand, since $0 + 0 = 0$, we have that $0(0 + 0) = 00$. This implies that $00 = 0$.

2. Since $a0 = a(0 + 0) = a0 + a0$, $a0 = 0$. $0a = 0$ is similar.[2]

[2] R is not necessarily commutative so we need to prove them both.

3. Since $a(-b+b) = a0 = 0$ and $a(-b+b) = a(-b) + ab$, we conclude that $a(-b) = -ab$. The other part is similar.

4. $a(b-c) = a(b+(-c)) = ab + a(-c) = ab - ac$. The other part is similar.

   □

If we apply -1 or 1 to Theorem 12.3.1, we have the following special cases.

Corollary 12.3.1. *Let R be a ring and $a \in R$. Then,*

1. $1(-1) = -1 = (-1)1$.

2. $(-1)(-1) = 1$.

3. $a(-1) = -a = (-1)a$.

4. $-(-a) = a$.

Proposition 12.3.1. *If a ring R has a unity, then it is unique.*

Proof. Let 1 and 1_1 be unities of R. Since 1 is a unity of R, we have that $1(1_1) = 1_1$. On the other hand, since 1_1 is a unity of R, we have that $1(1_1) = 1$. This shows that $1 = 1_1$. □

Note that $\mathbb{Z} \subset \mathbb{Q} \subset \mathbb{R} \subset \mathbb{C}$ are all rings under the same operations so we define a ring that is a subset of another ring.

Definition 12.3.1. *Let R be a ring. A non-empty subset S of R is called* **a subring of** *R if S is a ring with the addition and the multiplication of R.*

Example 12.3.1. *$\mathbb{Z}$ is a subring of $\mathbb{Q}$, $\mathbb{R}$ or $\mathbb{C}$. Also, $\mathbb{Q}$ is a subring of $\mathbb{R}$ or $\mathbb{C}$.*

Example 12.3.2. *$M_n(\mathbb{Z})$ is a subring of the ring $M_n(\mathbb{Q})$ and $M_n(\mathbb{R})$ is a subring of the ring $M_n(\mathbb{C})$.*

Example 12.3.3. *$\mathbb{Z}[\sqrt{7}] = \{a + b\sqrt{7} | a, b \in \mathbb{Z}\}$ is a subring of $\mathbb{R}$.*

Example 12.3.4. *$\{\overline{0}, \overline{2}, \overline{4}, \overline{6}\}$ is a subring of $\mathbb{Z}_8$.*

Example 12.3.5. *If R is a ring, $R_1 = \left\{ \begin{pmatrix} a & b \\ 0 & 0 \end{pmatrix} | a, b \in R \right\}$ is a subring of $M_2(R)$.*
Note that the unity of R_1 is $\begin{pmatrix} 1 & 0 \\ 0 & 0 \end{pmatrix}$ and the unity of $M_2(R)$ is $\begin{pmatrix} 1 & 0 \\ 0 & 1 \end{pmatrix}$.

From Example 12.3.5, we can see that a ring and its subring could have different unities.

Theorem 12.3.2. (Subring Test)
If R_1 is a non-empty subset of a ring R, then R_1 is a subring of R if and only if $r_1 - r_2 \in R_1$ and $r_1 r_2 \in R_1$ for any $r_1, r_2 \in R_1$.

Proof. If R_1 be a subring of a ring R, then $r_1 - r_2 \in R_1$ and $r_1 r_2 \in R_1$ for any $r_1, r_2 \in R_1$ obviously. Conversely, let us assume that $r_1 - r_2 \in R_1$ and $r_1 r_2 \in R_1$ for any $r_1, r_2 \in R_1$. If $r_2 \in R_1$, $r_2 - r_2 = 0 \in R_1$ and $0 - r_2 = -r_2 \in R_1$. This implies that $r_1 - (-r_2) = r_1 + r_2 \in R_1$. So R_1 is an additive group and it is Abelian since R is an additive Abelian group. Since R_1 is a subset of the ring R, the associativity and the distributivity are inherited. So R_1 is a ring and it makes it a subring of R. □

Example 12.3.6. *Show that* $n\mathbb{Z} = \{na|a \in \mathbb{Z}\} =< n >$ *is a subring of* $\mathbb{Z}$.

Solution. For $na_1, na_2 \in n\mathbb{Z}$, $na_1 - na_2 = n(a_1 - a_2) \in n\mathbb{Z}$. *We also have that* $na_1na_2 = n(na_1a_2) \in n\mathbb{Z}$. *Therefore,* $n\mathbb{Z}$ *is a subring of* $\mathbb{Z}$ *by the Subring Test.*

□

Example 12.3.7. *Since all the subrings of* $\mathbb{Z}_{10}$ *are* $< 2 >= \{\overline{0}, \overline{2}, \overline{4}, \overline{6}, \overline{8}\}$, $< 5 >= \{\overline{0}, \overline{5}\}$, *and* $\{0\}$, *we can construct the following subring lattice.*

Figure 12.3.1.

$$\mathbb{Z}_{10}$$

$$\nearrow \qquad\qquad \nwarrow$$

$$< \overline{2} > \qquad\qquad < \overline{5} >$$

$$\nwarrow \qquad\qquad \nearrow$$

$$\{\overline{0}\}$$

Chapter 12 Exercises.

1. Show that $\mathbb{Z}^-$, the set of all negative integers is not a ring under $+$ and $\cdot$. Also show why $\mathbb{Q}^-$ is not a ring.

2. Prove that $\mathbb{Z}[\sqrt{5}]$ is a ring and identify all the elements of $\mathbb{Z}[\sqrt{5}]$

3. Let $\mathbb{R}_5[x]$ be the set of all the polynomials (over $\mathbb{R}$) whose degrees are 5 or less. Prove that $\mathbb{R}_5[x]$ is a ring.

4. Prove that $\mathbb{Z}_6$ is a ring.

5. Prove that $\mathbb{Z}[i]$ is a subring of $\mathbb{Q}[i]$.

6. Prove that $\mathbb{Z}[\sqrt{11}]$ is a subring of $\mathbb{R}$. Is $\mathbb{Z}[\sqrt{11}]$ a subring of $\mathbb{Q}$?

7. Is $\mathbb{Q}[\sqrt{2}, \sqrt{5}] = \{a + b\sqrt{2} + c\sqrt{5} | a, b, c \in \mathbb{Q}\}$ a ring? Why or why not?

8. (a) Prove that $\mathbb{Z}[\sqrt{2}, \sqrt{5}, \sqrt{10}] = \{a + b\sqrt{2} + c\sqrt{5} + d\sqrt{10} | a, b, c, d \in \mathbb{Z}\}$ is a ring.

   (b) Does the equality $\mathbb{Z}[\sqrt{2}, \sqrt{5}, \sqrt{10}] = \mathbb{Z}[\sqrt{2}, \sqrt{5}]$ hold?

9. (a) Prove that $\mathbb{Q}[\sqrt{2}, \sqrt{5}, \sqrt{10}, \sqrt{30}, \sqrt{150}]$ is a ring.

   (b) Prove that $\mathbb{Q}[\sqrt{2}, \sqrt{3}, \sqrt{5}, \sqrt{10}]$ is not a ring.

10. Prove that $\sqrt{3}$ is not an element of $\mathbb{Q}[\sqrt{2}]$.

11. Suppose R be a ring. Prove that the following subsets are subrings of $M_2(R)$.

    (a) $S_1 = \left\{\begin{pmatrix} a & 0 \\ b & 0 \end{pmatrix} | a, b \in R\right\}$. (b) $S_2 = \left\{\begin{pmatrix} 0 & a \\ 0 & b \end{pmatrix} | a, b \in R\right\}$.

    (c) $S_3 = \left\{\begin{pmatrix} 0 & 0 \\ a & b \end{pmatrix} | a, b \in R\right\}$. (d) $S_4 = \left\{\begin{pmatrix} a & b \\ 0 & c \end{pmatrix} | a, b \in R\right\}$.

    (e) Find a non-trivial subring of S_4.

12. (a) Prove that $M_n(\mathbb{Z}_5)$ is non-commutative.

    (b) Prove that $M_n(\mathbb{Q}[\sqrt{2}])$ is non-commutative.

13. For a given positive integer $n > 1$, find a subring S of $\mathbb{Z}_n$ such that $\overline{1}$ is not a unity of S.

14. If e is a unity, prove $(mn)e = (me)(ne)$ for any $m, n \in \mathbb{Z}$.

15. Draw subring lattices for the following rings.

    (a) $\mathbb{Z}_8$ (b) $\mathbb{Z}_{12}$ (c) $\mathbb{Z}_{30}$ (d) $\mathbb{Z}_{36}$ (e) $\mathbb{Z}_{60}$

16. Find $A, B, C, D \in M_3(\mathbb{Q})$ such that

    (a) $AB = 0$ but $A \neq 0$ and $B \neq 0$.

    (b) $CD \neq DC$.

    (c) Do the same things for $M_3(\mathbb{Z}_6)$.

    (d) Do the same things for $M_3(\mathbb{Z}_5)$.

17. (a) Prove that $\mathbb{Z}_2 \oplus \mathbb{Z}_3 \oplus \mathbb{Z}_5$ is a ring.

    (b) Find the additive identity and the unity of $\mathbb{Z}_2 \oplus \mathbb{Z}_3 \oplus \mathbb{Z}_5$.

    (c) Find the order of $\mathbb{Z}_2 \oplus \mathbb{Z}_3 \oplus \mathbb{Z}_5$.

    (d) Find all the subrings of $\mathbb{Z}_2 \oplus \mathbb{Z}_3 \oplus \mathbb{Z}_5$.

    (e) Do the same things for $\mathbb{Z}_2 \oplus \mathbb{Z}_3 \oplus \mathbb{Z}_6$

18. Let $\overline{M_\infty(\mathbb{R})}$ is the set of all $\infty \times \infty$ matrices over $\mathbb{R}$ such that for any A of $\overline{M_\infty(\mathbb{R})}$, A has only finitely many non-zero entries.

    (a) Is $\overline{M_\infty(\mathbb{R})}$ a ring under the usual addition and the multiplication?

    (b) Does $\overline{M_\infty(\mathbb{R})}$ hold the associative law under the usual addition?

    (c) Does $\overline{M_\infty(\mathbb{R})}$ hold the associative law under the usual multiplication?

19. Let $M_\infty(\mathbb{R}) = \Big\{(a_{ij}) | (a_{ij}) \text{ is an } \infty \times \infty \text{ matrix with } |a_{ij}| < \infty\Big\}$.

    (a) Is $M_\infty(\mathbb{R})$ a ring like $M_n(\mathbb{R})$?

    (b) Does $M_\infty(\mathbb{R})$ hold the associative law under the usual addition?

    (c) Find $A, B, C \in M_\infty(\mathbb{R})$ such that $A(BC) \neq (AB)C$. (This means that this ring is not associative under the usual multiplication.)

Hint: Use $\begin{pmatrix} 1 & -1 & 1 & -1 & \cdots \\ 0 & 0 & 0 & 0 & \cdots \\ & & \cdots & & \end{pmatrix} \in M_\infty(\mathbb{R})$.

20. The center of a ring R is the set $C = \{r \in R | rr_1 = r_1 r \text{ for } r_1 \in R\}$.

    (a) Prove that C is not the empty set.

    (b) Prove that C is a subring of R.

Chapter 13

More Rings

If two nonzero integers are multiplied, the result is never 0 but this is not always true for a different ring. For example, $2 \cdot 3 = 0$ in $\mathbb{Z}_6$ but $2 \neq 0 \neq 3$. A "usual" ring like $\mathbb{Z}$ is called an **integral domain**. We will see a formal definition of this type in this chapter and see how it is related to units, i.e. invertible elements of a ring. And then a division will be induced to define a **field** from there.

13.1 Integral Domains

Definition 13.1.1. *Let R be a ring and $a(\neq 0) \in R$. Then this a is called* a **zero divisor** *if there is a non-zero element $b \in R$ such that $ab = 0$ or $ba = 0$.*

Note that our ring might not be commutative and remember that 0 is not a zero divisor.

Example 13.1.1. *Find the zero divisors of $\mathbb{Z}_{12}$.*

Solution. $2, 3, 4, 6, 8, 9$ and 10 are zero divisors since $2 \cdot 6 = 0 = 3 \cdot 4 = 3 \cdot 8 = 6 \cdot 10 = 4 \cdot 9$, but $1, 5, 7$, and 11 are not.

Example 13.1.2. *Find a condition of a zero divisors of* $\mathbb{Z}_n$.

Solution. $m \in \mathbb{Z}_n$ is not a zero divisor if and only if $(m, n) = 1$.

Example 13.1.3. *Find a zero divisors of* $\mathbb{Z}_p$ *where* p *is a prime.*

Solution. There is no zero divisor in $\mathbb{Z}_p$ where p is a prime.

Definition 13.1.2. *A commutative ring with unity is called* **an integral domain** *if there is no zero divisor in the ring.*

Obvious examples of integral domains are $\mathbb{Z}, \mathbb{Q}, \mathbb{R}$, and $\mathbb{C}$ because if two nonzero complex numbers are multiplied, the result is never 0. Here is another way to look at these number rings. First we define a particular ring element.

Definition 13.1.3. *For a given element* r *of a ring* R *is called* **idempotent** *if* $r^2 = r$.

Example 13.1.4. *Find the idempotent elements of* $\mathbb{Z}_6$.

Solution. $0, 1, 3$, and 4 are idempotent in $\mathbb{Z}_6$ since $0^2 = 0, 1^2 = 1, 2^2 = 4, 3^2 = 3, 4^2 = 4$, and $5^2 = 1$.

Now we connect idempotent elements and zero divisors in number rings.

Example 13.1.5. *Find all idempotents of* $\mathbb{C}$.

Solution. Let x be a non-zero idempotent element of $\mathbb{C}$. Recall that $x^2 = x$ when x is idempotent. This implies that $x^2 - x = 0$ or $x(x-1) = 0$. But since there is no zero divisor in $\mathbb{C}$, we conclude that $x - 1 = 0$. So all the idempotents of $\mathbb{C}$ are 0 and 1. □

Note that all idempotents of $\mathbb{Z}, \mathbb{Q}$ or $\mathbb{R}$ are 0 and 1 as well. Now we look at other integral domains.

Example 13.1.6. $\mathbb{Q}[\sqrt{2}]$ *is an integral domain. (Why?)*

Example 13.1.7. *Suppose R is a unitary ring. Then $R \oplus R$ is not an integral domain.*

Solution. R is a ring with unity 1 and $(1,0)(0,1) = (0,0)$ but $(1,0) \neq (0,0) \neq (0,1)$. □

Example 13.1.8. $M_n(\mathbb{R})$ *is not an integral domain since*

$$\begin{pmatrix} 0 & 0 & \cdots & 0 & 1 \\ 0 & 0 & \cdots & 0 & 0 \\ & & \cdots & & \\ 0 & 0 & \cdots & 0 & 0 \end{pmatrix} \begin{pmatrix} 0 & 0 & \cdots & 0 & 1 \\ 0 & 0 & \cdots & 0 & 0 \\ & & \cdots & & \\ 0 & 0 & \cdots & 0 & 0 \end{pmatrix} = \begin{pmatrix} 0 & 0 & \cdots & 0 & 0 \\ 0 & 0 & \cdots & 0 & 0 \\ & & \cdots & & \\ 0 & 0 & \cdots & 0 & 0 \end{pmatrix}$$

Example 13.1.9. *Show that* $\mathbb{Z}_{13}[i] = \{\overline{a} + \overline{b}i | a, b \in \mathbb{Z}_{13}, i^2 = \overline{-1}\}$ *is not an integral domain.*

Solution. $(\overline{2}+\overline{3}i)(\overline{2}-\overline{3}i) = \overline{4}+\overline{9} = \overline{13} = \overline{0}$ shows that there is a zero divisor of $\mathbb{Z}_{13}[i]$. Thus it is not an integral domain. □

13.2 Fields

The main property of integral domains is the existence of zero divisors. Now we turn our attention to divisors for unity 1.

Definition 13.2.1. *Suppose R is a ring with unity 1 and $r \in R$. Then r is called* **a unit** *if there is $r' \in R$ such that $rr' = 1$ or $r'r = 1$.*[1]

Note that 0 is not a unit and 1 is a unit. Easiest way to remember a unit of a ring is that it is *invertible* under multiplication.

Example 13.2.1. *Since $3 \cdot \frac{1}{3} = 1$, 3 is unit of $\mathbb{Q}$. In fact, every non-zero elements of $\mathbb{Q}$ are units.*

We will denote the set of all units of a ring R with $U(R)$.[2] For example $U(\mathbb{Q}) = \mathbb{Q}^*$.

Example 13.2.2. $U(\mathbb{Z}) = \{1, -1\}$.

Example 13.2.3. $U(\mathbb{Z}_{12}) = \{1, 5, 7, 11\}$.

Note that units of $U(\mathbb{Z}_{12})$ are all non-zero divisors.

Example 13.2.4. *Neither* $\begin{pmatrix} 1 & 1 \\ 1 & 1 \end{pmatrix}$ *nor* $\begin{pmatrix} 1 & 0 \\ -1 & 0 \end{pmatrix}$ *is unit since* $\begin{pmatrix} 1 & 1 \\ 1 & 1 \end{pmatrix} \begin{pmatrix} 1 & 0 \\ -1 & 0 \end{pmatrix} = \begin{pmatrix} 0 & 0 \\ 0 & 0 \end{pmatrix}$.

Lemma 13.2.1. *Every unit is not a zero divisor.*

Proof. Suppose r is a unit. Then there exist r' such that $rr' = 1$ or $r'r = 1$. If $ra = 0$ for some a, we conclude that $a = 0$ by multiplying r' on the left. This

[1]It is possible to have $rr' \neq r'r$ since the ring might not be commutative.
[2]It is not ambiguous with $U(n)$, the set of all natural number less than n and relative prime to n

implies that r is not a zero divisor. □

Using this divisibility of unity, we define a field.

Definition 13.2.2. *A commutative integral domain $\mathbb{F}$ is called a* **field** *if every non-zero element of $\mathbb{F}$ is unit.*

Example 13.2.5. *$\mathbb{Z}$ is not a field.*

Example 13.2.6. *$\mathbb{Q}$, $\mathbb{R}$ and $\mathbb{Z}$ are fields.*

Suppose $\mathbb{F}$ is a field. Then $\mathbb{F}(a)$ denotes the smallest field that includes $\mathbb{F}$ and a.

Example 13.2.7. *$\mathbb{Q}(\sqrt{2})$ is the smallest field that includes $\mathbb{Q}$ and $\sqrt{2}$.*

Obviously $\mathbb{Q}(\sqrt{a}) = \left\{\frac{a+b\sqrt{2}}{c+d\sqrt{2}}|a,b,c,d \in \mathbb{Q}, c+d\sqrt{2} \neq 0\right\}$. It is also obvious that $\mathbb{Q}(\sqrt{2}) = \mathbb{Q}[\sqrt{2}] = \{a+b\sqrt{2}|a,b \in \mathbb{Q}\}$. Let's take a look at a similar example.

Example 13.2.8. *$\mathbb{Q}[i]$ is an integral domain. Furthermore it is a field. Since it is a field and it is denoted as $\mathbb{Q}(i)$, i.e., $\mathbb{Q}[i] = \mathbb{Q}(i)$.*

Example 13.2.9. *Since x is not invertible in the ring $\mathbb{Q}[x]$, $\mathbb{Q}[x]$ is not a field.*[3]

[3]This is why we call $F[x]$ a polynomial ring, not a polynomial field even though F is a field

Example 13.2.10. *Suppose R is a unitary ring. Since* $(1,0)(0,1) = (0,0)$, $(1,0)$ *is a zero divisor of* $R \oplus R$. *So the ring* $R \oplus R$ *is neither an integral domain nor a field.*

Example 13.2.11. $M_n(\mathbb{R})$ *is nether an integral domain nor a field.*

We start to see that if a ring is not a field, it is not an integral domain. To confirm this argument, we consider the following cancelation law of an integral domain.

Lemma 13.2.2. (Cancelation Law) *Let R be an integral domain and $r_1, r_2, r_3 \in R$. Then if $r_1r_2 = r_1r_3$, then $r_1 = 0$ or $r_2 = r_3$.*

Proof. If $r_1 = 0$, then there is nothing to prove. Assume that $r_1 \neq 0$. By $r_1r_2 = r_1r_3$, we have that $r_1r_2 - r_1r_3 = 0$. This implies that $r_1(r_2 - r_3) = 0$. Since R is an integral domain and $r_1 \neq 0$, we have that $r_2 - r_3 = 0$. This implies that $r_2 = r_3$. □

Theorem 13.2.1. *A finite integral domain R is a field.*

Proof. Since R is an integral domain, it is a commutative ring with unity. So it is enough to show that every non-zero element u of R is unit. Let S be the set $\{u^i | i \in \mathbb{N}\}$. Since S is a subset of R, it is finite. This means that $u^{j_1} = u^{j_2}$ for $j_1 > j_2$. By $u^{j_1} - u^{j_2} = 0$, we have that $u^{j_2}(u^{j_1-j_2} - e) = 0$. Since u is a non-zero element, we have that $u^{j_1-j_2} = e$. This means that $u^{j_1-j_2-1}u = e$, i.e., $u^{j_1-j_2-1}$ is the inverse of u. This implies that R is a field. □

The following is a typical example of a finite integral domain.

Example 13.2.12. *Prove that if p is prime, then $\mathbb{Z}_p$ is a field.*

Proof. Since $\mathbb{Z}_p$ is a commutative ring with unity, it is enough to prove that $\mathbb{Z}_p$ does not have a zero divisor. Let us assume that $\overline{a}\overline{b} = \overline{0}$. This implies that $ab = rp$ for $r \in \mathbb{N}$. This implies that p divides ab. Since p is prime, p divides a or p divides b. We have that $\overline{a} = \overline{0}$ or $\overline{b} = \overline{0}$. This implies that $\mathbb{Z}_p$ does not have a zero divisor. So $\mathbb{Z}_p$ is a finite integral domain and by Theorem 13.2.1, it is a field.[4] □

The following is a generalization of Example 13.2.12.

Proposition 13.2.1. *A field R is an integral domain.*

Proof. Since R is a field, it is a commutative ring with the unity. For elements a, b of R, let us assume that $ab = 0$. If a is zero, then there is nothing to prove. Let us assume that $a \neq 0$. This implies that $a^{-1}(ab) = (a^{-1}a)b = b = 0$. So R does not have a zero divisor and this makes R an integral domain. □

Example 13.2.13. *$\mathbb{Z}_p$ is an integral domain because it is a field by Example 13.2.12.*

13.3 Characteristics of Integral Domains

We want to look at one way to classify integral domains and this tool is a key to distinguish infinite fields and finite fields. First we define the characteristic of an element of a ring.

[4]This $\mathbb{Z}_p$ is called a prime field.

Definition 13.3.1. *Suppose R be a ring and $0 \neq r \in R$. A positive integer n is called* **the characteristic** *of r if n is the smallest positive integer such that $nr = 0$. If there is not such a number, we say that the characteristic is zero.*

Example 13.3.1. *Find the characteristics of elements of $\mathbb{Z}_{12}$.*

Solution. The characteristic of $\overline{1}, \dots, \overline{11}$ are $12, 6, 4, 3, 12, 2, 12, 3, 4, 6$, and 12 respectively. For example, the characteristic of $\overline{3}$ is 4 since $4(\overline{3}) = \overline{3}+\overline{3}+\overline{3}+\overline{3} = \overline{12} = \overline{0}$. □

Example 13.3.2. *If p is prime, then for $\overline{0} \neq \overline{a} \in \mathbb{Z}_p$, $p(\overline{a}) = \overline{pa} = \overline{0}$. Thus the characteristic of $\overline{a}$ is p.* □

Using the characteristics of elements on a ring, we now define the characteristic of the ring.

Definition 13.3.2. *The least positive integer n such that $nr = 0$ for any non-zero element r in a ring R is called* **the characteristic of** *R denoted by $char(R) = n$. If there is not such a positive number, we say that $char(R) = 0$.*

Even though the characteristic of a ring depends on every elements of the ring, we only need to look at unity if the ring is unitary. We will use e for unity on the next theorem.

Theorem 13.3.1. *A ring with unity e has characteristic $n < \infty$ if and only if $ne = 0$.*

Proof. If $char(R) = n$, then $ne = 0$ holds. Conversely, let us assume that $ne = 0$ holds. For any $r \in R$, $r = er$. This implies that $nr = n(er) = (ne)r = 0r = 0$ and this completes the proof. □

Example 13.3.3. *Since* $(12)\overline{1} = \overline{12} = \overline{0}$, *the characteristic of* $\mathbb{Z}_{12}$ *is* 12. □

The characteristic of a ring could be any whole number but there are only two choices for integral domains.

Theorem 13.3.2. *The characteristic of an integral domain* R *is prime or zero.*

Proof. If $char(R) = 0$, then there is nothing to prove. Suppose that $char(R) = n < \infty$. This means that $nr = 0$ for any $r \in R$. In particular, $ne = 0$ where e is the unity of R. Now assume that n is not prime, i.e., $n = n_1n_2$ where $1 < n_1, n_2 < n$. This implies that $ne = (n_1n_2)e = (n_1e)(n_2e) = 0$. By Lemma 13.2.2, $n_1e = 0$ or $n_2e = 0$. So either n_1 or n_2 is the characteristic of R, a contradiction by Theorem 13.3.1 since n is the characteristic of R, the smallest number. This implies that n is prime and it proves the theorem. □

Corollary 13.3.1. *The characteristic of a field is prime or zero.*

Example 13.3.4. $char(\mathbb{R})$ *is zero.*

Example 13.3.5. $char(\mathbb{Z}_p)$ *is* p *where* p *is prime.*

13.4 Localizations of Integral Domains

A field like $\mathbb{Q}$ has something that an integral domain like $\mathbb{Z}$ doesn't have: reciprocals. Now we develop a construction to build a field from an integral domain and it is called **a localization of an integral domain**. We start with an integral domain R to define the quotient field R^{-1} of R.

Let $\overline{R} = \{(a,b)|a,b \in R, b \neq 0\}$ and let us define a relation $\sim$ on $\overline{R}$ as follows: $(a,b) \sim (c,d)$ if and only if $ad = bc$ for $(a,b),(c,d) \in \overline{R}$. It is easy to check that $\sim$ is an equivalence relation and the equivalence class of (a,b) is $[(a,b)] = \{(c,d) \in \overline{R}|(a,b) \sim (c,d)\}$. Let's denote $[(a,b)]$ by $\frac{a}{b}$ and let $R^{-1} = \{\frac{a}{b}|a,b \in R, b \neq 0\}$. We will use $0 = \frac{0}{a}$ and $1 = \frac{b}{b}$ for non-zero elements a,b of R. Now we define an the addition $+$ and a multiplication $\cdot$ on R^{-1} as follows:

$$\frac{a}{b} + \frac{c}{d} = \frac{ad+bc}{bd} \quad \text{and} \quad \frac{a}{b} \cdot \frac{c}{d} = \frac{ac}{bd} \tag{13.1}$$

for $\frac{a}{b}, \frac{c}{d} \in R^{-1}$.

Lemma 13.4.1. *The addition $+$ and the multiplication $\cdot$ of (13.1) are well defined on R^{-1}.*

Proof. Suppose $\frac{a}{b} = \frac{a_1}{b_1}, \frac{c}{d} = \frac{c_1}{d_1} \in R^{-1}$. We need to show that

$$\frac{a}{b} + \frac{c}{d} = \frac{a_1}{b_1} + \frac{c_1}{d_1} \tag{13.2}$$

and

$$\frac{a}{b} \cdot \frac{c}{d} = \frac{a_1}{b_1} \cdot \frac{c_1}{d_1}. \tag{13.3}$$

We have that

$$\frac{a}{b}+\frac{c}{d}=\frac{ad+bc}{bd}. \tag{13.4}$$

We also have that

$$\frac{a_1}{b_1}+\frac{c_1}{d_1}=\frac{a_1d_1+b_1c_1}{b_1d_1}. \tag{13.5}$$

$\frac{a}{b}=\frac{a_1}{b_1}$ and $\frac{c}{d}=\frac{c_1}{d_1}$ imply that

$$ab_1=a_1b \quad \text{and} \quad cd_1=c_1d \tag{13.6}$$

and from (13.4) and (13.5), we derive that $bd(a_1d_1+b_1c_1)=a_1bdd_1+bb_1c_1d=ab_1dd_1+cd_1bb_1$. This implies that $\frac{a}{b}+\frac{c}{d}=\frac{a_1}{b_1}+\frac{c_1}{d_1}$. So the addition is well defined. For the multiplication,

$$\frac{a}{b}\cdot\frac{c}{d}=\frac{ac}{bd} \quad \text{and} \quad \frac{a_1}{b_1}\cdot\frac{c_1}{d_1}=\frac{a_1c_1}{b_1d_1}$$

give us $acb_1d_1=a_1bcd_1=c_1da_1b$. This implies that $\frac{a}{b}\cdot\frac{c}{d}=\frac{a_1}{b_1}\cdot\frac{c_1}{d_1}$. Thus the multiplication is well defined. □

Obviously 0 is the additive identity of R^{-1}. Since $\frac{a}{b}+\frac{c}{d}=\frac{ad+bc}{bd}=\frac{cb+da}{bd}=\frac{c}{d}+\frac{a}{b}$ and $\frac{a}{b}\cdot\frac{c}{d}=\frac{ac}{bd}=\frac{ca}{db}=\frac{c}{d}\cdot\frac{a}{b}$, R^{-1} is a commutative ring. By $1\cdot\frac{a}{b}=\frac{a}{b}$, we see that 1 is the unity of R^{-1} and it makes R^{-1} a commutative ring with the unity. At last, any non-zero element $\frac{a}{b}$ of R^{-1} is a unit since $\frac{a}{b}\cdot\frac{b}{a}=1$. Therefore, we have the following conclusion.

Theorem 13.4.1. *If R is an integral domain, then R^{-1} is a field.*

The following is an interesting example that shows that the commutativity is required for a field.

Definition 13.4.1. **A skew field** *(or* **a division ring***) holds all the axioms of a field except the multiplicative commutativity.*

In other words, a field is a commutative division ring. Then a natural question is "What is an example of a non-commutative division ring?" At this point, we know only one example made by an Irish mathematician **William Rowan Hamilton**.

Example 13.4.1. $\mathbb{H} = \{a + b\mathbf{i} + c\mathbf{j} + d\mathbf{k} | a, b, c, d \in \mathbb{R}\}$ *is a skew field under proper operations.*

Solution. Suppose $\mathbf{a} = a_1 + a_2\mathbf{i} + a_3\mathbf{j} + a_4\mathbf{k}$ and $\mathbf{b} = b_1 + b_2\mathbf{i} + b_3\mathbf{j} + b_4\mathbf{k} \in \mathbb{H}$. We define an addition on $\mathbb{H}$ as

$$\mathbf{a} + \mathbf{b} = a_1 + b_1 + (a_2 + b_2)\mathbf{i} + (a_3 + b_3)\mathbf{j} + (a_4 + b_4)\mathbf{k}$$

For the multiplication, we set $\mathbf{i}^2 = \mathbf{j}^2 = \mathbf{k}^2 = -1$, $\mathbf{ij} = \mathbf{k}$, $\mathbf{jk} = \mathbf{i}$, $\mathbf{ki} = \mathbf{j}$, $\mathbf{ji} = -\mathbf{k}$, $\mathbf{kj} = -\mathbf{i}$, and $\mathbf{ik} = -\mathbf{j}$. Then we are able to see that $\mathbb{H}$ is a non-commutative division ring. □

This $\mathbb{H}$ is called **the Quaternion** or **the Hamiltonian** and it is a useful tool in quantum mechanics. The following will confirm that every non-zero element of $\mathbb{H}$ is invertible.

Example 13.4.2. *Find the inverse of*

1. $a + b\mathbf{i} + c\mathbf{j} + d\mathbf{k}$ *of* $\mathbb{H}$ *where at least one of* $a, b, c,$ *and* d *is not zero*

2. $2 - 3\mathbf{i} + 4\mathbf{j}$ *of* $\mathbb{H}$

Solution. 1. $\dfrac{a - b\mathbf{i} - c\mathbf{j} - d\mathbf{k}}{a^2 + b^2 + c^2 + d^2}$ is the inverse of $a + b\mathbf{i} + c\mathbf{j} + d\mathbf{k}$ from a straight calculation.

2. $(2 - 3\mathbf{i} + 4\mathbf{j})(2 + 3\mathbf{i} - 4\mathbf{j}) = 29$. So $\frac{2+3\mathbf{i}-4\mathbf{j}}{29}$ is the inverse of $2 - 3\mathbf{i} + 4\mathbf{j}$.

Chapter 13 Exercises.

1. (a) Prove that $\mathbb{Q}[\sqrt{7}]$ is an integral domain.

   (b) Prove that $\mathbb{Q}[\sqrt{7}]$ is a field, i.e.
   $\mathbb{Q}[\sqrt{7}] = \mathbb{Q}(\sqrt{7}) = \left\{ \dfrac{a + b\sqrt{7}}{c + d\sqrt{7}} | a, b, c, d \in \mathbb{Q}, c + d\sqrt{7} \neq 0 \right\}.$

2. (a) Prove that $\mathbb{Q}[i]$ is an integral domain.

   (b) Prove $\mathbb{Q}[i] = \mathbb{Q}(i)$ directly.

3. Prove that the polynomial ring $\mathbb{Q}[x]$ is an integral domain.

4. Prove that $R_1 \oplus R_2$ is not an integral domain even though R_1 and R_2 are rings.

5. (a) Prove that $\mathbb{Z}$ is not a field.

   (b) Find all the units of $\mathbb{Z}_8$.

   (c) Find all the units of $\mathbb{Z}_{30}$.

   (d) Find all the zero divisors of $\mathbb{Z}_8$.

   (e) Find all the zero divisors of $\mathbb{Z}_{30}$.

6. (a) Find all the units of $\mathbb{Z}[\sqrt{2}]$.

(b) Find all the units of $\mathbb{Z}[\sqrt{3}]$.

7. Show that $\mathbb{Z}_5[i]$ is not a field.

8. Find the characteristics of the following rings.

   (a) $\mathbb{Z}_6$

   (b) $\mathbb{Z}_{11}$

   (c) $\mathbb{Z}_4 \oplus \mathbb{Z}_2$

   (d) $\mathbb{Z}_p \oplus \mathbb{Z}_2$ where p is an odd prime.

   (e) $\mathbb{Z}_p \oplus \mathbb{Z}_2$ where p, q are primes.

9. Prove that $\mathbb{Q}[\sqrt{13}]$ is a field.

10. (a) Show that $\mathbb{Z}[\sqrt{13}]^{-1} = \mathbb{Q}[\sqrt{13}]$.

    (b) Show that $\mathbb{Q}[\sqrt{7}]^{-1} = \mathbb{Q}(\sqrt{7})$.

    (c) Show that $\mathbb{Z}^{-1} = \mathbb{Q}$.)

11. Prove that $\{\frac{f(x)}{g(x)} | f(x), g(x) \in \mathbb{C}[x], g(x) \neq 0\}$ with the usual addition and the multiplication is a field. It is denoted as $\mathbb{C}(x)$.

12. Prove that $\mathbb{C}$ is a subring of the Hamiltonian $\mathbb{H}$.

13. Prove that $\mathbb{C}[x]$ is not a field.

14. Find the inverses of the following in $\mathbb{H}$.

    (a) $3 + 4\mathbf{i}$

    (b) $2 + 3\mathbf{i} + 4\mathbf{j}$

    (c) $2 + 3\mathbf{i} + 4\mathbf{j} - 2\mathbf{k}$

    (d) $2\mathbf{i} + 3\mathbf{j} - 2\mathbf{k}$

15. Prove that every subring of an integral domain is an integral domain as well.

16. Find an example of an integral domain which is not a field.

17. If R is an integral domain such that $char(R) = 0$, prove that $char(R^{-1}) = 0$ as well.

18. (a) Prove that $\mathbb{Q}^{-1} = \mathbb{Q}$.

    (b) Prove that $\mathbb{R}^{-1} = \mathbb{R}$.

    (c) Prove that $\mathbb{C}^{-1} = \mathbb{C}$.

19. Let $t = x^3 + 3x + 6 \in \mathbb{Q}[x]$.

    (a) Prove that $\mathbb{Q}[t]$ is a subring of $\mathbb{Q}[x]$.

    (b) Does the equality $\mathbb{Q}[t] = \mathbb{Q}[x]$ hold? Why or why not?

    (c) Is $\mathbb{Q}[t]$ an integral domain? Why or why not?

    (d) Is $\mathbb{Q}[t]$ a field? Why or why not?

20. (a) Prove that $\mathbb{Z}[\sqrt{2}, \sqrt{3}] = \{a + b\sqrt{2} + c\sqrt{3} + d\sqrt{6} | a, b, c, d \in \mathbb{Z}\}$ is a subring of $\mathbb{R}$.

    (b) Find all the units of $\mathbb{Z}[\sqrt{2}, \sqrt{3}]$.

21. Prove that a unit of a ring can not be a zero divisor.

22. Show that the Hamiltonian $\mathbb{H} = \{a+b\mathbf{i}+c\mathbf{j}+d\mathbf{k} | a, b, c, d \in \mathbb{R}, \mathbf{i}^2 = \mathbf{j}^2 = \mathbf{k}^2 = -1, \mathbf{ij} = \mathbf{k}, \mathbf{jk} = \mathbf{i}, \mathbf{ki} = \mathbf{j}, \mathbf{ji} = -\mathbf{k}, \mathbf{kj} = -\mathbf{i}, \mathbf{ik} = -\mathbf{j}\}$ is a non-commutative division ring.

Chapter 14

Polynomial Rings

Recall that $\mathbb{Q}$, $\mathbb{R}$, and $\mathbb{C}$ are fields of characteristic zero. Throughout this chapter, $\mathbb{F}$ will denote a field of characteristic zero. A polynomial is applicable to so many areas of mathematics and other areas of sciences. It is continuous everywhere, and infinitely differentiable and integrable so a polynomial is a useful approximation of a complex function. In this chapter, we will study the polynomial ring $\mathbb{F}[x_1, \cdots, x_n]$ using the basic concepts that were studied previous chapters.

14.1 Addition and Multiplication

Example 14.1.1. *For a trinomial* $f(x) = 2x^2 + 3x - 1$ *and a binomial* $g(x) = x + 3$, *find their sum and product.*

Solution. $f(x) + g(x) = 2x^2 + 4x + 2$ and $f(x)g(x) = (2x^2 + 3x - 1)(x + 3) = 2x^3 + 9x^2 + 8x - 3$. $\square$

Note that the sum and product of polynomials are also polynomials. The following are basic properties of them.

Theorem 14.1.1. *Suppose $f(x)$ and $g(x) \in \mathbb{F}[x]$. Then*

1. $deg(f(x) + g(x)) \leq max\{deg(f(x)), deg(g(x))\}$

2. $deg(f(x)g(x)) = deg(f(x)) + deg(g(x))$

Proof. If $deg(f(x)) = r$ and $deg(g(x)) = s$, then $f(x) = a_r x^r + \cdots + a_0$ and $g(x) = b_s x^s + \cdots + b_0$ for some $a_r, b_s \neq 0$ and $a_r, \cdots, b_0 \in \mathbb{F}$. Let $t = max\{r, s\}$. Then their sum is $f(x) + g(x) = (a_t + b_t)x^t + \cdots + (a_0 + b_0)$ and this implies that $deg(f(x) + g(x)) \leq max\{deg(f(x)), deg(g(x))\}$. Similarly, their product is $f(x)g(x) = a_r b_s x^{r+s} + \cdots + a_0 b_0$ and since $a_r b_s$ is not zero, we have that $deg(f(x)g(x)) = r + s = deg(f(x)) + deg(g(x))$. □

Example 14.1.2. *Check Theorem 14.1.1 for $f(x) = 2x^2 + 3x - 1$ and $g(x) = -2x^2 + 4x - 7 \in \mathbb{F}[x]$.*

Solution. $f(x) + g(x) = 7x - 8$ and $f(x)g(x) = (2x^2 + 3x - 1)(-2x^2 + 4x - 7) = -4x^4 + 2x^3 - 25x + 7$. This implies that $deg(f(x) + g(x)) = 1 \leq 2$ and $deg(f(x)g(x)) = 4 = deg(f(x)) + deg(g(x))$ hold. □

Theorem 14.1.1 holds because $\mathbb{F}$ is a field. Check the next example for a different case.

Example 14.1.3. *Check Theorem 14.1.1 for $f(x) = \overline{3}x^2 + x + \overline{2}$ and $g(x) = \overline{4}x + \overline{3} \in \mathbb{Z}_{12}[x]$*

Solution. $f(x)g(x) = (\overline{3}x^2 + x + \overline{2})(\overline{4}x + \overline{3}) = x^2 + \overline{5}x + \overline{5}$. This implies that $deg(f(x)g(x)) = 2 \neq 3 = deg(f(x)) + deg(g(x))$. Thus the equality $deg(f(x)g(x)) = deg(f(x)) + deg(g(x))$ does not hold. □

Theorem 14.1.1 implies that if $f(x)g(x) = 0$, then $f(x) = 0$ or $g(x) = 0$. Note that if $deg(f(x)) \neq 0$, then there is no polynomial $g(x)$ of $\mathbb{F}[x]$ such that $f(x)g(x) = 1$. Thus $\mathbb{F}[x]$ is not a field.

14.2 Division and Factors

Now, let's talk about divisions of polynomials.

Theorem 14.2.1. (Division Algorithm) *Let $f(x), g(x) \in \mathbb{F}[x]$. If $g(x) \neq 0$, then there are unique $q(x)$ and $r(x)$ of $\mathbb{F}[x]$ such that*

$$f(x) = q(x)g(x) + r(x) \tag{14.1}$$

where $r(x) = 0$ or $deg(r(x)) < deg(g(x))$.

Proof. Let's assume that the degree of $f(x)$ and $g(x)$ are n and t respectively. This means that $f(x) = a_n x^n + \cdots + a_1 x + a_0$ and $g(x) = b_t x^t + \cdots + b_1 x + b_0$ with appropriate coefficients. If $n < t$, then by takeing $q(x) = 0$ and $r(x) = f(x)$ we prove the theorem since $f(x) = 0g(x) + r(x)$. Now, let's assume that $n > t$ and we prove the theorem by induction on the degree n of $f(x) = a_n x^n + \cdots + a_1 x + a_0$. Assume that the theorem holds for polynomials with degrees less than or equal to $n - 1$. Since $deg(b_t f(x) - a_n g(x) x^{n-t}) < n$, by induction, there are $q_1(x)$ and $r_1(x)$ of $\mathbb{F}[x]$ such that $b_t f(x) - a_n g(x) x^{n-t} = g(x) q_1(x) + r_1(x)$ where $r_1(x) = 0$ or $deg(r_1(x)) < deg(g(x))$. This implies that $f(x) = g(x)(\frac{a_n x^{n-t} + q_1(x)}{b_t}) + \frac{r_1(x)}{b_t}$ and it is exactly the equation 14.1) since $q(x) = \frac{a_n x^{n-t} + q_1(x)}{b_t}$ and $r(x) = \frac{r_1(x)}{b_t}$. Now, let's prove the uniqueness of the quotient and the remainder. Suppose that there are $q_2(x)$ and $r_2(x)$ such that

$$f(x) = g(x)q_2(x) + r_2(x) \tag{14.2}$$

where $r_2(x) = 0$ or $deg(r_2(x)) < deg(g(x))$. By (14.1) and (14.2), we have that

$$0 = (q(x) - q_2(x))g(x) + r(x) - r_2(x). \tag{14.3}$$

This implies that $(q(x) - q_2(x))g(x) = r_2(x) - r(x)$. If $q(x) - q_2(x) \neq 0$, we have a contradiction by Theorem 14.1.1 since $deg(q(x)) > 0$. So $q(x) - q_2(x) = 0$, i.e., $q(x) = q_2(x)$ and this makes $r_2(x) = r(x)$ by (14.3). Thus we have proven the theorem. □

In Theorem 14.2.1, $f(x), g(x), q(x)$, and $r(x)$ are called **the dividend, the divisor, the quotient,** and **the remainder** respectively.

Example 14.2.1. *Find the quotient and the remainder of the division:* $(2x^3 + 6x^2 + x - 2) \div (x^2 - 3x + 1)$.

Solution. Using the long division, we can derive that $(2x^3 + 6x^2 + x - 2) \div (x^2 - 3x + 1) = (2x + 12)(x^2 - 3x + 1) + 35x - 14$. Therefore, the quotient is $(2x + 12)$ and the remainder is $35x - 14$ of the division. □

When the remainder is 0, a polynomial can be factored by the divisor and the quotient, i.e. $f(x) = q(x)g(x)$. In this situation, $q(x)$ and $g(x)$ are factors of $f(x)$, and here comes a formal definition.

Definition 14.2.1. *1. Suppose $f(x) = g(x)h(x)$. We call $g(x)$ and $h(x)$* **divisors** *or* **factors** *of $f(x)$, and $f(x)$* **a multiple** *of $g(x)$ and/or $h(x)$. In this case, we say that $g(x)$* **divides** *$f(x)$ and it is denoted by $g(x)|f(x)$.*

2. *Suppose $h(x)|f(x)$ and $h(x)|g(x)$. Then $h(x)$ is called* **a common divisor of $f(x)$ and $g(x)$**.

3. **(GCD)**$h(x)$ *is called* **a[1] greatest common divisor of $f(x)$ and $g(x)$** *if the degree of $h(x)$ is maximal among all the common divisors of $f(x)$ and $g(x)$ and it is denoted by $gcd(f(x), g(x))$ or simply $(f(x), g(x))$. Similarly, $(f_1(x), \cdots, f_n(x))$ denotes a greatest common divisor of $f_1(x), \cdots, f_n(x)$.*

4. *If $(f_1(x), \cdots, f_n(x)) = 1$, then we say that $f_1(x), \cdots, f_n(x)$ are relatively prime.*

Example 14.2.2. *Find a GCD of $f(x) = x^3 + x^2 - 4x - 4$ and $g(x) = x^4 + x^3 - 3x^2 - 4x - 4$*

Solution. Note that $f(x) = x^3 + x^2 - 4x - 4 = (x - 2)(x + 2)(x + 1)$ and $g(x) = x^4+x^3-3x^2-4x-4 = (x-2)(x+2)(x^2+x+1)$ so $gcd(f(x), g(x)) = x^2-4$. □

In Example 14.2.2, $x - 2$ and $x + 2$ are common divisors, not a GCD. The following is a useful tool that related to a GCD.

Theorem 14.2.2. *For given polynomials $f_1(x), \cdots, f_n(x) \in \mathbb{F}[x]$, there are $s_1(x), \cdots, s_n(x) \in \mathbb{F}[x]$ such that $f_1(x)s_1(x)+\cdots+f_n(x)s_n(x) = gcd(f_1(x), \cdots, f_n(x))$.*

Proof. Let D be the set $\{f_1(x)h_1(x) + \cdots + f_n(x)h_n(x)|h_1(x), \cdots, h_n(x) \in \mathbb{F}[x]\}$. Note that $f_1(x), \cdots, f_n(x) \in D$. Since D is the non-empty set of polynomials, there is a polynomial $d(x) \in \mathbb{F}[x]$ such that $d(x)$ has the minimal degree in D. Since $d(x) \in D$, we have that

[1] It is not "the" GCD but "a" GCD

$$d(x) = f_1(x)h_1'(x) + \cdots + f_n(x)h_n'(x) \tag{14.4}$$

for some $h_1'(x), \cdots, h_n'(x) \in \mathbb{F}[x]$. Now we will show that $d(x)$ divides $f_i(x)$, $1 \le i \le n$. Let's say that $f_i(x) = d(x)q(x) + r(x)$ with $r(x) = 0$ or $deg(r(x)) < deg(d(x))$ using the Division Algorithm. Then $f_i(x) = f_1(x)h_1'(x)q(x) + \cdots + f_n(x)h_n'(x)q(x) + r(x)$. This makes that $-f_1(x)h_1'(x)q(x) + \cdots + f_i(x)(1 - h_i'(x)q(x)) - \cdots - f_n(x)h_n'(x)q(x) = r(x) \in D$ and it means that $r(x) = 0$ since $d(x)$ has the minimal degree so $d(x)$ divides $f_i(x)$. Let $p(x)$ be a divisor of $f_i(x)$, $1 \le i \le n$. This implies that $f_i(x) = p(x)q_i(x)$, $1 \le i \le n$, for some $q_i(x) \in \mathbb{F}[x]$. By (14.4), we have that $d(x) = p(x)q_1(x)h_1'(x) + \cdots + p(x)q_n(x)h_n'(x) = p(x)(q_1(x)h_1'(x) + \cdots + q_n(x)h_n'(x))$. This implies that $p(x)$ divides $d(x)$ and it makes $d(x)$ a GCD of $f_1(x), \cdots, f_n(x)$. Thus we have proven the theorem. □

Theorem 14.4 says that GCD of $f_1(x), \cdots, f_n(x)$ is a polynomial combination of $f_1(x), \cdots, f_n(x)$ and the following are simple examples of the Theorem.

Corollary 14.2.1. *For $f(x), g(x) \in \mathbb{F}[x]$, there are $s(x)$ and $t(x) \in \mathbb{F}[x]$ such that $f(x)s(x) + g(x)t(x) = (f(x), g(x))$.*

Corollary 14.2.2. *For $f_1(x), \cdots, f_n(x) \in \mathbb{F}[x]$, if $f_1(x), \cdots, f_n(x)$ are relatively prime, then there are polynomials $s_1(x), \cdots, s_n(x) \in \mathbb{F}[x]$ such that $f_1(x)s_1(x) + \cdots + f_n(x)s_n(x) = 1$.*

Example 14.2.3. *Find a combination for a GCD of $f(x) = x^3 - 1$ and $g(x) = x^3 + 4x^2 + 4x + 3$ as in Corollary 14.2.1.*

Solution. From $f(x) = (x-1)(x^2+x+1)$ and $g(x) = (x+3)(x^2+x+1)$, we can see that $gcd(f(x), g(x)) = x^2+x+1$. Then we can make $x^2+x+1 = \dfrac{-1}{4}(x^3-1) + \dfrac{1}{4}(x^3+4x^2+4x+3)$ using $s(x) = \frac{-1}{4}$ and $t(x) = \frac{1}{4}$. $\square$

Theorem 14.2.3. (Remainder Theorem) *If we divide $f(x)$ by $x-c$, then $f(c)$ is the remainder of the division.*

Proof. By the Division Algorithm, we have that $f(x) = (x-c)q(x) + r(x)$ where $q(x), r(x) \in \mathbb{F}[x]$ where $r(x) = 0$ or $deg(r(x)) < 1$. Since the degree of $x-c$ is one, $r(x)$ is a constant. This implies that $f(c) = (c-c)q(c) + r$ and we can conclude that $f(c) = r$. $\square$

14.3 Zeros Of A Polynomial

We learned the following definitions at early ages but it is a good time to see connections among polynomials, polynomial equations, and polynomial functions clearly.

Definition 14.3.1. *Let $f(x) \in \mathbb{F}[x]$ and suppose $f(a) = 0$.*

1. *a is* **a zero** *or* **a root** *of the polynomial $f(x)$.*
2. *a is* **a solution** *of the polynomial equation $f(x) = 0$.*
3. *a is* **an x-intercept** *of the polynomial function $y = f(x)$.*

Example 14.3.1. *Explain significance of 3 and 1 for $f(x) = x^3 - 3x^2 + 3x - 9 \in \mathbb{R}[x]$.*

Solution. Since $f(3) = 0$, we say that 3 is a zero of $x^3 - 3x^2 + 3x - 9$, a solution of $x^3 - 3x^2 + 3x - 9 = 0$, and an x-intercept of $y = x^3 - 3x^2 + 3x - 9$. Since $f(1) = -8$, 1 is not a zero of $f(x)$ but $f(1) = -8$ shows that the remainder is -8 if $x^3 - 3x^2 + 3x - 9$ is divided by $x - 1$. □

Corollary 14.3.1. *$c \in \mathbb{F}$ is a zero of $f(x) \in \mathbb{F}[x]$ if and only if $f(x) = (x - c)q(x)$ for some $q(x) \in \mathbb{F}[x]$.*

Proof. The proof of the corollary is straightforward by the Remainder Theorem 14.2.3. □

Example 14.3.2. *If we divide $f(x) = 3x^3 - 4x^2 + 6x - 5$ by $g(x) = x - 1$, the remainder is $f(1) = 3(1)^3 - 4(1)^2 + 6(1) - 5 = 0$. This shows that $g(x)$ divides $f(x)$.*

Theorem 14.3.1. *$f(x)$ has at most n zeros if $deg(f(x)) = n \geq 1$ for any polynomial $f(x) \in \mathbb{R}[x]$.*

Proof. We'll prove it by induction on the degree of $f(x)$. If $deg(f(x)) = 1$, then $f(x) = ax + b \in \mathbb{R}[x]$ for $a, b \in \mathbb{R}$ and $a \neq 0$. This implies that $\frac{-b}{a}$ is a zero of $f(x)$ so $f(x)$ has only one zero in $\mathbb{R}$. Now let's assume that the theorem is true for $f(x)$ with degree less than n. If $f(x)$ does not have a zero in $\mathbb{R}$, then there is nothing to prove. Suppose a is a zero of $f(x)$. Then $f(x) = (x - a)g(x)$ for some $g(x) \in \mathbb{R}[x]$. Since $deg(g(x)) < deg(f(x)) = n$, $g(x)$ has at most $n - 1$ zeros by the induction assumption. Therefore, $f(x)$ has at most n zeros and this proves the theorem by induction. □

Example 14.3.3. *Find the zeros of $f(x) = x^3 - 6x^2 + 11x - 6$.*

Solution. Since the sum of the coefficients of the polynomial is zero, 1 is a zero of the polynomial, i.e. $f(x) = 1^3 - 6(1^2) + 11(1) - 6 = 0$. By the long division or the synthetic division, we have that $x^3 - 6x^2 + 11x - 6 = (x-1)(x^2 - 5x + 6)$ and $x^2 - 5x + 6$ is easy to factor: $(x-2)(x-3)$. So $f(x) = (x-1)(x-2)(x-3)$ and this shows that 1, 2, and 3 are the three zeros of the polynomial. Since $deg(f(x)) = 3$, we can conclude that these are only zeros by Theorem 14.3.1. □

14.4 Formal Power Series Rings

Even though a polynomial ring has infinite elements, a polynomial has finite terms. We can still define a ring with elements of infinite terms. In this section, we'd like to present a definition of such a ring.

Let $\mathbb{F}[[x]]$ denote the set of all the expressions in the following form:

$$f(x) = \sum_{0 \le k \le \infty} a_k x^k$$

where $a_k \in \mathbb{F}$. For $f(x) = \sum_{0 \le k \le \infty} a_k x^k, g(x) = \sum_{0 \le k \le \infty} b_k x^k \in \mathbb{F}[[x]]$, we define the addition and the multiplication as follows:

$$f(x) + g(x) = \sum_{0 \le k \le \infty} a_k x^k + \sum_{0 \le k \le \infty} b_k x^k = \sum_{0 \le k \le \infty} (a_k + b_k) x^k$$

and

$$f(x)g(x) = \sum_{0 \le k \le \infty} a_k x^k \sum_{0 \le k \le \infty} b_k x^k = \sum_{0 \le k \le \infty} (c_k) x^k$$

where $c_k = \sum_{i+j=k} a_i b_j$, $0 \le k \le \infty$. Then $\mathbb{F}[[x]]$ is a ring and it is called the formal power series ring.

Chapter 14 Exercises.

1. Find the degrees of

   (a) $f(x) = 3 - x$

   (b) $g(x) = x^2 - \pi x + e$

2. Calculate $f(x) - g(x)$ and $f(x)g(x)$.

   (a) $f(x) = x^3 + 4x^2 - x - 3$ and $g(x) = x^3 + 3$

   (b) $f(x) = -3x^3 + 4x^2 - x - 3$ and $g(x) = x^2 - x + 3$

3. Find the quotient and the remainder.

   (a) $(-x - 3) \div (x^2 - x + 3)$

   (b) $(x^5 + 4x^2 - x - 3) \div (x + 2)$

   (c) $(2x^4 - 2x^3 + 4x^2 - x - 3) \div (-x^2 + 3)$

   (d) $(x^4 + 3x^3 + 4x^2 - x - 3) \div (x^3 - x + 3)$

4. Find greatest common divisors of the following in $\mathbb{R}$

   (a) $x^2 - 1$ and $x^3 - 1$

   (b) $2x^2 + 3x + 1$ and $(2x + 1)^3$

   (c) $3x^2 + 6x + 1$ and $x^2 + 9$

   (d) $x^3 + 6x^2 + 9x - 3$ and $x^3 + 5x^2 + 8x + 4$

(e) $x^4 + 3x^3 - 11x^2 - 3x + 10, x^3 + x^2 - 10x + 8$, and $x^3 - 6x^2 + 11x - 6$

(f) $x^3 + x^2 - 9x - 9, x^3 + 2x^2 - 5x - 6$, and $x^3 + 6x^2 + 11x + 6$

5. Without using long divisions, check if $g(x)$ is a divisor of $f(x)$ or not.

   (a) $f(x) = x^3 - x^2 + x - 1$ and $g(x) = x - 1$

   (b) $f(x) = x^4 - 4x^2 - 7x + 14$ and $g(x) = x - 2$

   (c) $f(x) = 2x^3 - 2x^2 + 2x + 1$ and $g(x) = 2x - 1$

   (d) $f(x) = 3x^4 + 3x^2 - 5x - 2$ and $g(x) = 2x + 3$

6. Find all zeros of each polynomial. Are your answers in $\mathbb{R}$ or $\mathbb{C}$?

   (a) $x^2 - 3x - 4$

   (b) $x^3 - 7x + 6$

   (c) $x^4 - 4x^3 - 7x^2 + 34x - 24$

   (d) $x^5 - 3x^4 - 11x^3 + 27x^2 + 10x - 24$

   (e) $x^4 - 1$

   (f) $x^4 - 2$

   (g) $x^6 - 1$

   (h) $x^6 - 7$

7. (a) Prove that $\mathbb{F}[[x]]$ is a ring.

   (b) Prove that $\mathbb{F}[x]$ is a subring of $\mathbb{F}[[x]]$.

   (c) Prove that there is a bijection from $\mathbb{F}[[x]]$ to itself other than the identity function.

8. Prove Corollary 14.2.1 without referring to Theorem 14.2.2.

9. Prove Corollary 14.2.2 without referring to Theorem 14.2.2.

Chapter 15

Ideals and Factor Rings

A normal subgroup of a group is important to study the structure of the group. For a ring, there is a subring that has similar role to a normal subgroup of a group and it is called **an ideal**. Therefore, it is crucial to study an ideal of a ring to investigate the structure of the ring and this will make a quotient ring possible just like a normal subgroup. We start this chapter with a formal definition of an ideal of a ring.

15.1 Ideals

Definition 15.1.1. *Let R be a ring. A subring I of R is called* **a right (resp. left) ideal of** R *if I is a subring of R and $ri \in I$ (resp. $ir \in I$) for any $r \in R$ and $i \in I$. An ideal I of a ring R is called* **a two-sided ideal** *if I is a right and a left ideal of the ring.*

If R is a commutative ring, then a right or left ideal I of R is a two-sided ideal. From now on, an ideal means a two sided ideal and a two sided ideal I

of a ring R is denoted by $I \lhd R$. Note that generally a subring of a ring is not an ideal of the ring, but an ideal of a ring is always a subring of the ring.

Example 15.1.1. *R itself and $\{0\}$ are ideals of any ring R, and they are called* **trivial ideals** *of R.*

Example 15.1.2. *For a commutative ring R with unity and $a \in R$, $I = \{ar | r \in R\}$ is an ideal of R.*

Solution. It is easy to check that I is a subring of R. Suppose r and $ar_1 \in I$. Then $(ar_1)r = a(r_1 r) \in I$ since R is commutative. This implies that $I \lhd R$ □

The ideal in Example 15.1.2 is called **the principal ideal generated by** a and denoted by $< a >$. a is called **a generator**. Note that $< a >= aR$.

Example 15.1.3. *Prove that $n\mathbb{Z} =< n >$ is an ideal of $\mathbb{Z}$.*

Solution. By Example 12.3.6, we know that $n\mathbb{Z}$ is a subring of $\mathbb{Z}$. For $m \in \mathbb{Z}$ and $na \in n\mathbb{Z}$, we have that $mna = n(ma) \in n\mathbb{Z}$. This implies that $n\mathbb{Z}$ is an ideal of $\mathbb{Z}$. Thus we have solved the example. □

This concept of a principal ideal $< a >$ can be generalized to ideals with several generators. For a commutative ring R with unity and $a_1, \cdots, a_n \in R$, we can define the ideal $\{a_1 r_1 + \cdots + a_n r_n | r_i \in R, 1 \leq i \leq n\}$ generated by $a_1, \cdots, a_n$. It is denoted by $< a_1, \cdots, a_n >$ and it is actually an ideal by the following example. Since $< a >= aR$, $< a_1, \cdots, a_n >$ can be written ab $a_1 R + \cdots + a_n R$. Note that $< a_1, \cdots, a_n >$ is usually not a principal ideal.

Example 15.1.4. *Let $R_1, \cdots, R_n$ be rings and $I_1, \cdots, I_n$ are ideals of $R_1, \cdots, R_n$ respectively. Prove that $\oplus_{1\leq j\leq n} I_j$ is an ideal of the ring $\sum_{1\leq j\leq n} R_j$.*

Solution. It is obvious that $\oplus_{1\leq j\leq n} I_j$ is a subring of $\oplus_{1\leq j\leq n} R_j$. For any $\oplus_{1\leq j\leq n} a_j \in \oplus_{1\leq j\leq n} I_j$ and $\oplus_{1\leq j\leq n} r_j \in \oplus_{1\leq j\leq n} R_j$, we have that $(\oplus_{1\leq j\leq n} r_j)(\oplus_{1\leq j\leq n} a_j) = \oplus_{1\leq j\leq n} r_j a_j \in \oplus_{1\leq j\leq n} I_j$. This means that $\oplus_{1\leq j\leq n} I_j$ is an ideal of $\oplus_{1\leq j\leq n} R_j$. □

Even though we will consider two sided ideals most of times, it is worth mentioning one sided ideals at this point. We can see differences through the following example.

Example 15.1.5. *Prove that $I = \left\{ \begin{pmatrix} a & 0 \\ b & 0 \end{pmatrix} | a, b \in R \right\}$ is a right ideal of $M_2(R)$, not a left ideal.*

Solution. Obviously I is an additive Abelian subgroup of $M_2(R)$. For any $\begin{pmatrix} c & d \\ e & f \end{pmatrix} \in M_2(R)$ and for any $\begin{pmatrix} a & 0 \\ b & 0 \end{pmatrix} \in I$, we have that $\begin{pmatrix} c & d \\ e & f \end{pmatrix} \begin{pmatrix} a & 0 \\ b & 0 \end{pmatrix} = \begin{pmatrix} ac+bd & 0 \\ ae+bf & 0 \end{pmatrix} \in I$. This shows that I is a right ideal of $M_2(R)$. Since $\begin{pmatrix} a & 0 \\ b & 0 \end{pmatrix} \begin{pmatrix} c & d \\ e & f \end{pmatrix} = \begin{pmatrix} ac & ad \\ be & bf \end{pmatrix} \notin I$, I is not a left ideal of $M_2(R)$. □

Example 15.1.6. *Identify all elements of the ideal $< x^2+1 >$ of $\mathbb{R}[x]$.*

Solution. By the definition, $< x^2+1 >= (x^2+1)\mathbb{R}[x] = \{(x^2+1)f(x) | f(x) \in \mathbb{R}[x]\}$. Thus, any element of this ideal is a multiple of x^2+1. □

15.2 Quotient Rings

Before we construct quotient (or factor) rings, we need to understand operations among ideals. We only consider commutative rings with unity from now on.

For ideals I_1 and I_2 of a ring R, we define $I_1 + I_2 = \{i_1 + i_2 | i_1 \in I_1, i_2 \in I_2\}$ and $I_1 I_2 = \{\sum_k i_k i'_k | i_k \in I_1, i'_k \in I_2\}$, and these are ideals as well.

Proposition 15.2.1. *Let I_1 and I_2 be ideals of a ring R. Then the following holds.*

1. *$I_1 \cap I_2$ is an ideal of R.*
2. *$I_1 + I_2$ is an ideal of R.*
3. *$I_1 I_2$ is an ideal of R.*

Proof.

1. For any $i_1, i_2 \in I_1 \cap I_2$, we have that $i_1, i_2 \in I_1$ and $i_1, i_2 \in I_2$. So we have that $i_1 - i_2, i_1 i_2 \in I_1$ and $i_1 - i_2, i_1 i_2 \in I_2$. This implies that $i_1 - i_2 \in I_1 \cap I_2$ and $i_1 i_2 \in I_1 \cap I_2$. This makes $I_1 \cap I_2$ a subring of R. For any $r \in R$ and $i \in I_1 \cap I_2$, we have that $ir \in I_1$ and $ir \in I_2$. This implies that $ir \in I_1 \cap I_2$. So $I_1 \cap I_2$ is an ideal of R.
2. For any $i_1 + i_2, i_3 + i_4 \in I_1 + I_2$, $i_1 + i_2 - (i_3 + i_4) = i_1 - i_3 + i_2 - i_4 \in I_1 + I_2$. We also have that $(i_1 + i_2)(i_3 + i_4) = i_1 i_3 + i_1 i_4 + i_2 i_3 + i_2 i_4 = i_1(i_3 + i_4) + i_2(i_3 + i_4) \in I_1 + I_2$ and this makes $I_1 + I_2$ a subring. For any $r \in R$, we have that $(i_1 + i_2)r = i_1 r + i_2 r \in I_1 + I_2$. So $I_1 + I_2 \lhd R$.
3. The proof of this part is similar to 2, so it is omitted. □

Let I be an ideal of a ring R. Since I is a normal subgroup of this group R, the set of all cosets, $R/I = \{r + I | r \in R\}$ is a group. The following lemma explains a relation between a coset representative and an ideal, and we will define operations on R/I using it.

Lemma 15.2.1. *Suppose $I \lhd R$. Then $i + I = I$ if and only if $i \in I$.*

Proof. Suppose that $i+I = I$. Then there are $i_1, i_2 \in I$ such that $i+i_1 = i_2$. This implies that $i = i_2 - i_1 \in I$. Conversely, suppose that $i \in I$. Note that $i + i_3 \in I$ for any $i_3 \in I$. This implies that $i + I \subset I$. Also, for any $i_4 \in I$, $i_4 = i + (i_4 - i) \in i + I$ and this implies that $I \subset i + I$. So $i + I = I$. Thus we have proven the lemma. □

Example 15.2.1. *If we consider $7\mathbb{Z} \lhd \mathbb{Z}$, $7 + 7\mathbb{Z} = 7\mathbb{Z}$ since $7 \in 7\mathbb{Z}$. In fact,* $7+7\mathbb{Z} = \{\cdots, 7-7, 7+0, 7+7, 7+14, \cdots\} = \{\cdots, -7, 0, 7, 14, \cdots\} = 7\mathbb{Z}$.

Example 15.2.2. *From $7\mathbb{Z} \lhd \mathbb{Z}$, we have that* $\mathbb{Z}/7\mathbb{Z} = \{0+7\mathbb{Z} = 7\mathbb{Z}, 1+7\mathbb{Z}, 2+7\mathbb{Z}, 3+7\mathbb{Z}, 4+7\mathbb{Z}, 5+7\mathbb{Z}, 6+7\mathbb{Z}\}$.

Now define the addition $+$ and $\cdot$ on R/I as follows:

$$
\begin{aligned}
(r_1 + I) + (r_2 + I) &= r_1 + r_2 + I \\
(r_1 + I) \cdot (r_2 + I) &= r_1 r_2 + I \qquad (15.1)
\end{aligned}
$$

for $r_1 + I, r_2 + I \in R/I$.

Lemma 15.2.2. *The addition $+$ and the multiplication $\cdot$ of (15.1) are well defined. Recall that the addition is same as the one in additive groups.*

Proof. For $r_1+I=r_1'+I, r_2+I=r_2'+I \in R/I$, we need to prove that $(r_1+I)+(r_2+I)=(r_1'+I)+(r_2'+I)$ and $(r_1+I)(r_2+I)=(r_1'+I)(r_2'+I)$. Since $r_1+I=r_1'+I$, we have that $r_1=r_1'+i_1$ for $i_1 \in I$ and $r_2+I=r_2'+I$, we have that $r_2=r_2'+i_2$ for $i_2 \in I$. $(r_1+I)+(r_2+I)=r_1'+i_1+I+r_2'+i_2+I=r_1'+r_2'+I$, i.e., the addition is well defined. $(r_1+I)(r_2+I)=(r_1'+i_1+I)(r_2'+i_2+I)=r_1'r_2'+r_1'i_1+i_1r_2'+i_1i_2+I=r_1'r_2'+I$, i.e., the multiplication is well defined. □

Theorem 15.2.1. *Let $I \lhd R$. Then R/I is a ring under the addition $+$ and the multiplication $\cdot$ of (15.1).*

Proof. The proof of the theorem is straightforward by Lemma 15.2.2. □

In Theorem 15.2.1, R/I is called **a factor ring** or **a quotient ring**. If a ring R has the unity e (or 1), then $e+I$ (or $1+I$) is the unity of R/I. Recall that I is the identity for addition in R/I.

Example 15.2.3. *Find all the elements of $\mathbb{Z}/(5\mathbb{Z})$.*

Solution. $\mathbb{Z}/(5\mathbb{Z})=\{5\mathbb{Z}, 1+5\mathbb{Z}, 2+5\mathbb{Z}, 3+5\mathbb{Z}, 4+5\mathbb{Z}\}$. □

Example 15.2.4. *Find all the elements of $\mathbb{R}[x]/<x^3+1>$.*

Solution. Note that $<x^3+1>=\{f(x)<x^3+1>|f(x)\in\mathbb{R}[x]\}$ and $\mathbb{R}[x]/<x^3+1> = \{g(x)+<x^3+1>|g(x)\in\mathbb{R}[x]\}$. By the Division Algorithm, we have that $g(x)=(x^3+1)q(x)+r(x)$ where $q(x), r(x)\in\mathbb{R}[x]$ with $r(x)=0$ or $deg(r(x))<3$. This implies that

$$\mathbb{R}[x]/<x^3+1>=\{ax^2+bx+c+<x^3+1>|a,b,c\in\mathbb{R}\}. \quad \square$$

15.3 Principal Ideal Domain (PID)

We'd like to find another excellent property of a polynomial ring and this property is related to special ideals: principal ideals. So first, we restate the definition.

Definition 15.3.1. *An ideal I of a ring R is called* **principal** *if there is $r \in R$ such that $< r >= I$. This r is called* **a generator** *of I or we say that r generates the ideal I.*

Example 15.3.1. *Show that the ideal $5\mathbb{Z}$ of $\mathbb{Z}$ is a principal ideal of $\mathbb{Z}$.*

Solution. Since $5 \in 5\mathbb{Z}$, we have that $< 5 >\subset 5\mathbb{Z}$. Conversely for any $a \in 5\mathbb{Z}$, since $a = 5b$ for some $b \in \mathbb{Z}$, $a \in< 5 >$. This means that $5\mathbb{Z} \subset< 5 >$ so $< 5 >= 5\mathbb{Z}$. Thus we have shown that 5 generates $5\mathbb{Z}$. □

Note that there is another generator for $5\mathbb{Z}$: -1. Also, it is obvious that every ideal of $\mathbb{Z}$ has a generator.

Lemma 15.3.1. *Every ideal of the ring $\mathbb{Z}$ is principal.*

$\mathbb{Z}$ is not the only one with the property so we start with a definition for such rings.

Definition 15.3.2. *A ring is called* **a principal ideal ring (PIR)** *if every ideal of the ring is principal. If the ring is an integral domain with the property, then it is called* **a principal ideal domain (PID)**.

Now we are ready to state the main theorem of the section.

Theorem 15.3.1. *The polynomial ring $\mathbb{F}[x]$ is a principal ideal domain.*

Proof. Let I be a non-zero ideal of $\mathbb{F}[x]$. We want to show that I is generated by one polynomial of $\mathbb{F}[x]$. By the Well Ordering Principle, there exists a polynomial $m(x) \in I$ whose degree is minimal in I. Suppose $f(x) \in I$. Then there are $q(x), r(x) \in \mathbb{F}[x]$ such that $f(x) = m(x)q(x) + r(x)$ with $r(x) = 0$ or $0 \leq deg(r(x)) < deg(m(x))$ by the Division Algorithm. If $r(x) \neq 0$, then we have that $f(x) - m(x)q(x) = r(x)$. Since $f(x), m(x) \in I$, $r(x)$ is an element of I. This contradicts the fact that $m(x)$ has the minimal degree of I. So $r(x) = 0$. This implies that $I =< m(x) >$. Since $\mathbb{F}[x]$ is an integral domain, it is a principal ideal domain. $\square$

From the proof of Theorem 15.3.1, any ideal of $\mathbb{F}[x]$ is generated by a polynomial which has the smallest degree of all polynomials of the ideal.

Example 15.3.2. *Let $I =< x^4 - 1, x^3 - 1 > \lhd \mathbb{R}[x]$. Then show that $I =< x - 1 >$.*

Solution. Note that $x^4 - 1 = (x-1)(x+1)(x^2+1)$ and $x^3 - 1 = (x-1)(x^2+x+1)$. Thus since I is generated by $x^4 - 1$ and $x^3 - 1$, we have that is I is a subset of $< x - 1 >$. Conversely, since $x - 1$ is the greatest common divisor of $x^4 - 1$ and $x^3 - 1$, there are $s(x), t(x) \in \mathbb{F}[x]$ such that $x - 1 = s(x)(x^4 - 1) + t(x)(x^3 - 1) \in I$ by Theorem 14.2.2. This shows that $< x - 1 > \subset I$. So we have that $< x - 1 >= I$ and this solves the example. $\square$

The following is another example related to a degree of a polynomial generator.

Example 15.3.3. *Show that* $\mathbb{Q}[x]/ < x^2+5x+1 >= \{ax+b+ < x^2+5x+1 > |a, b \in \mathbb{Q}\}$

Solution. Let $I =< x^2 + 5x + 1 > \lhd \mathbb{Q}[x]$ and consider a quotient ring $\mathbb{Q}[x]/I = \{f(x) + I|f(x) \in \mathbb{Q}[x]\}$. Note that $x^2 + 5x + 1 + I = I$. Then $x^2 + I = -5x - 1 + I$ and this means that $\mathbb{Q}[x]/ < x^2 + 5x + 1 >= \{ax + b+ < x^2 + 5x + 1 > |a, b \in \mathbb{Q}\}$. For example, $x^3 + I = 24x + 5 + I$ since $x^3 = (x^2 + 5x + 1)(x - 5) + 24x + 5$ by a long division. With a same fashion, $x^5 + 2x^4 + I = (x^2 + 5 + 1)(x^3 - 3x + 14) - 67x - 14 + I = -67x - 14 + I$. □

The solution of this way is not very rigorous but still understandable. In general,
$\mathbb{F}[x]/ < f(x) >= \{r(x)+ < f(x) > |r(x) \in \mathbb{F}[x] \text{ and } deg(r(x)) < deg(f(x))\}$.
Let's take a look at another example.

Example 15.3.4. *Let* $f(x) = 3x^3 + 2x^2 - 4x + 6$. *Then find the multiplicative inverse of* $x + 1+ < f(x) >$ *in the field* $\mathbb{Q}[x]/ < f(x) >$[1]

Solution. Using a long division (or a synthetic division), we can compute that $f(x) = (x + 1)(3x^2 - 3x - 1) + 9$. This implies that

$$(x + 1)(3x^2 - 3x - 1) + 9+ < f(x) >=< f(x) >$$

and this makes

$$(x + 1)(3x^2 - 3x - 1)+ < f(x) >= -9+ < f(x) >$$

[1] The fact that $\mathbb{Q}[x]/ < f(x) >$ is a field will be shown in Chapter 22.

So we have

$$(x+1)\left(\frac{3x^2-3x-1}{-9}\right)+<f(x)>=1+<f(x)>$$

Therefore, $-\frac{1}{3}x^2+\frac{1}{3}x+\frac{1}{9}+<f(x)>$ is the multiplicative inverse of $x+1+<f(x)>$. □

Now, let's go back to $\mathbb{Z}$, the integer ring.

Theorem 15.3.2. *Prove that $\mathbb{Z}$ is a PID.*

Solution. Let I be an ideal of $\mathbb{Z}$. If I is the zero ideal, then there is nothing to prove. Let us assume that I is not the zero ideal and suppose a is a non-zero element of I. Since $-a \in I$ as well, I contains a positive integer. Then by the Well Ordering Principle, there is the minimal positive integer d in I. We will show that $<d>=I$. Since $a \in I$, we have that $<a>\subseteq I$. For any non-zero element b of I, we have that $b=dq+r$ where q is the quotient and $0 \leq r < d$. Then $b-dq=r$ and since $b \in I$ and $d \in I$, we can say that $b-dq$ is an element of I, i.e., $r \in I$. If $r \neq 0$, we have a contradiction since d is the minimal positive integer in I, not r. So, $r=0$ and this means that $b=dr \in <d>$ and it implies that $I \subseteq <d>$. So $I=<d>$ and since $\mathbb{Z}$ is an integral domain, $\mathbb{Z}$ is a PID. □

15.4 Prime and Maximal Ideals

We want to look at two special ideals of rings and these ideals will produce integral domains and fields.

Definition 15.4.1. *An ideal P of a ring R is called* **prime** *if $p_1p_2 \in P$ implies that $p_1 \in P$ or $p_2 \in P$. An ideal M of R is called* **maximal**, *if $M_1 \lhd R$ such that $M \subseteq M_1 \subseteq R$, then $M_1 = M$ or $M_1 = R$.*

Briefly, a prime ideal provides a kind of an unique factorization and there is no intermediate ideal between a maximal ideal and the ring. The following are good examples to understand the concept. We start with primes.

Theorem 15.4.1. *$n\mathbb{Z}$ is prime if and only if n is prime.*

Solution. Suppose that n is prime and let ab be an element of $n\mathbb{Z}$. This implies that $ab = nz$ for $z \in \mathbb{Z}$. Since n is prime and n divides ab, we have that $n|a$ or $n|b$. If $n|a$, then $a \in n\mathbb{Z}$, and if $n|b$, then $b \in n\mathbb{Z}$. So $n\mathbb{Z}$ is prime. Conversely, suppose that $n\mathbb{Z}$ is prime. If n is not prime, $n = n_1n_2$ such that $1 < n_1, n_2 < n$. Since $n \in n\mathbb{Z}$, we have that $n_1n_2 \in n\mathbb{Z}$. Since $1 < n_1, n_2 < n$, we have that $n_1, n_2 \notin n\mathbb{Z}$. This implies that $n\mathbb{Z}$ is not prime, a contradiction. Therefore, n is prime and it completes the proof. □

Example 15.4.1. *$< 8 > \lhd \mathbb{Z}$ is not a prime ideal by Theorem 15.4.1 since $< 8 >= 8\mathbb{Z}$ and 8 is not a prime number.*

Example 15.4.2. *$< 7 >= 7\mathbb{Z}$ is a prime ideal of $\mathbb{Z}$ by Theorem 15.4.1.*

Now, the following examples are about maximal ideals.

Example 15.4.3. *Show that the ideal $< \overline{2} >$ of $\mathbb{Z}_{12}$ is maximal.*

Solution. Suppose I is an ideal of $\mathbb{Z}_{12}$ such that $<\overline{2}>\subset I$ but $<\overline{2}>\neq I$. This means that there is an element $z\in I$ such that $z\neq<\overline{2}>$ and $(2,z)=1$. Then there are $s,t\in\mathbb{Z}$ such that $2s+tz=1$. This implies that $\overline{1}=\overline{2s+tz}\in I$ so for any $z\in\mathbb{Z}_{12}$, $\overline{1}\overline{z}\in I$. This makes $\mathbb{Z}_{12}\subset I$. Therefore, $I=\mathbb{Z}_{12}$ so $<\overline{2}>$ is maximal. □

Example 15.4.4. *Since $<12>$ is a proper subset of $<2>$, the ideal $<12>$ is not a maximal ideal of $\mathbb{Z}$. But $<2>$ is a maximal ideal of $\mathbb{Z}$.*

Note that for a given ring R with unity and an ideal I of R, $1\in I$ if and only if $I=R$. Next two theorems are important and convenient in many places.

Theorem 15.4.2. *Let P be an ideal of a commutative ring R with unity 1. Then, I is prime if and only if R/P is an integral domain.*

Proof. First assume that R/P is an integral domain. For $ab\in P$, we have that $ab+P=P$. This implies that $(a+P)(b+P)=P$. Since P is an integral domain, $a+P=P$ or $b+P=P$ since P is the additive identity 0 in R/P. This implies that $a\in P$ or $b\in P$. So P is a prime ideal. Next, assume that P is prime. Since R is a commutative ring with unity 1, R/P is a commutative ring with unity $1+P$. It is enough to prove that R/P does not have a zero divisor. Let us assume that $(a+P)(b+P)=P$. This implies that $ab+P=P$. This also implies that $ab\in P$. Since P is prime $a\in P$ or $b\in P$. This implies that $a+P=P$ or $b+P=P$, i.e., there is no zero divisor and this completes the proof. □

Example 15.4.5. *Since $7\mathbb{Z}$ is a prime ideal of $\mathbb{Z}$, by Theorem 15.4.2, $\mathbb{Z}/(7\mathbb{Z})$ is an integral domain. Since $\mathbb{Z}/(7\mathbb{Z})$ is a finite integral domain, it is also a field.*

Next theorem will be useful when a field theory is studied deeply.

Theorem 15.4.3. *Let M be an ideal of a commutative ring R with unity* 1. *Then, I is maximal if and only if R/M is a field.*

Proof. First assume that R/M is a field. If I is an ideal of R such that M is a proper subset of I, i.e., $M \subset I \subseteq R$, then there is $r \in I$ such that $r \notin M$. This implies that $r + M \neq M$. Since R/M is a field, there is the inverse $s + M$ of $r + M$. This implies that $(s + M)(r + M) = 1 + M$. This means that $sr = 1 + m$ for some $m \in M$. This equation becomes $1 = sr - m \in I$, i.e., $I = R$. So M is maximal. Conversely let's assume that M is maximal. Recall that R/M is a commutative ring with unity. It is enough to prove that every non-zero element $r + M$ of R/M has a multiplicative inverse. Since $r \notin M$, the ideal $< r > +M$ of R is R since M is maximal. This implies that $1 = rr_1 + m$ for some $r_1 \in R$. Then $1 + M = (rr_1 + m) + M = rr_1 + M = (r + M)(r_1 + M)$ so $r_1 + M$ is the multiplicative inverse of $r + M$ and this completes the proof. □

Example 15.4.6. *n is prime if and only if $n\mathbb{Z}$ is a maximal ideal of $\mathbb{Z}$.*

Proof. Let I be an ideal of $\mathbb{Z}$ such that it contains $p\mathbb{Z}$. If $I \neq p\mathbb{Z}$, then there is $a \in I$ such that $a \notin p\mathbb{Z}$. This implies that $(a, p) = 1$. So there are $s, t \in \mathbb{Z}$ such that $as + pt = 1$. For any $z \in \mathbb{Z}$, we have that $z = z(as + pt) = zas + zpt \in I$. This implies that $\mathbb{Z}$ is a subset of I so $I = \mathbb{Z}$ and this implies that $p\mathbb{Z}$ is a maximal ideal. Conversely, let us assume that p is not prime. There are two integers p_1 and p_2 such that $p = p_1p_2$ where $1 < p_1, p_2 < p$. This implies that the ideal $< p_1 >$ contains $p\mathbb{Z}$ and $< p_1 > \neq p\mathbb{Z}$. This implies that $p\mathbb{Z}$ is not maximal. Thus we have solved the example. □

This example implies that n is prime if and only if $\mathbb{Z}/n\mathbb{Z}$ is a field by Theorem

15.4.3.

Example 15.4.7. $17\mathbb{Z}$ *is a maximal ideal of* $\mathbb{Z}$. *This implies that* $\mathbb{Z}/17\mathbb{Z}$ *is a field with* 17 *elements.*

Chapter 15 Exercises.

1. Let $R = \mathbb{Z} + \mathbb{Z}[\sqrt{2}] = \{a + b\sqrt{2} | a, b \in \mathbb{Z}\}$.
   (a) Prove that R is a ring.
   (b) Prove that $4\mathbb{Z} + 4\mathbb{Z}[\sqrt{2}]$ is an additive subgroup of R.
   (c) Prove that $4\mathbb{Z} + 4\mathbb{Z}[\sqrt{2}]$ is a subring of R.
   (d) Prove that $4\mathbb{Z} + 4\mathbb{Z}[\sqrt{2}]$ is an ideal of R.
2. Let's consider $\mathbb{Z}/8\mathbb{Z}$.
   (a) Calculate $(3 + 8\mathbb{Z}) + (2 + 8\mathbb{Z})$.
   (b) Calculate $(7 + 8\mathbb{Z}) + (4 + 8\mathbb{Z})$.
   (c) Calculate $(2 + 8\mathbb{Z})(3 + 8\mathbb{Z})$.
   (d) Prove that $8\mathbb{Z}$ is not a prime ideal of $\mathbb{Z}$.
   (e) Prove that $8\mathbb{Z}$ is not a maximal ideal of $\mathbb{Z}$.
3. (a) Prove that the ideal $< 12 >$ of $\mathbb{Z}$ is not a prime ideal.
   (b) Prove that $< 12 >$ is not a maximal ideal.
4. (a) Prove that $5\mathbb{Z}$ is an ideal of $\mathbb{Z}$.
   (b) Identify all the elements of $\mathbb{Z}/5\mathbb{Z}$.

(c) Prove that $\mathbb{Z}/5\mathbb{Z}$ is a field.

5. Prove that $4\mathbb{Z}$ is not a maximal ideal of $\mathbb{Z}$ and find a maximal ideal of $\mathbb{Z}$ that contains $4\mathbb{Z}$.

6. Suppose I_i is an ideal of a ring R where $i \in J = \{1, \cdots, n\}$

   (a) Prove that $\cap_{i \in J} I_i$ is an ideal of R.

   (b) Prove that $I_1 + \cdots + I_n$ is an ideal of R

7. Let I be an ideal of a ring R with unity. If I contains the unity, then prove that $I = R$.

8. Let I be an ideal of a ring R with unity. If I contains a unit, then prove that $I = R$.

9. (a) Prove that $\mathbb{Z}/7\mathbb{Z}$ is a ring.

   (b) Prove that $\mathbb{Z}/7\mathbb{Z}$ is an integral domain.

   (c) Prove that $\mathbb{Z}/7\mathbb{Z}$ is a field.

   (d) Identify all the elements of $\mathbb{Z}/7\mathbb{Z}$.

   (e) Prove that $\mathbb{Z}/10\mathbb{Z}$ is a ring.

   (f) Prove that $\mathbb{Z}/10\mathbb{Z}$ is not an integral domain so it is not a field.

10. (a) Identify all the elements of $\mathbb{Z}_7$ and prove that it is a field.

    (b) Identify all the elements of $\mathbb{Z}_p$ and prove that it is a field where p is prime.

    (c) Identify all the elements of $\mathbb{Z}_8$ and prove that it is not a field.

11. For an element r of a ring R,

(a) Prove that rR is a subring R.

(b) Prove that rR is a right ideal of R.

(c) Prove that Rr is a left ideal of R.

12. For an ideal I of a ring R, prove that $I + I = I$ and $I - I = I$.

13. Suppose R is a ring.

(a) Prove that $S_1 = \left\{ \begin{pmatrix} a & b \\ 0 & 0 \end{pmatrix} | a, b \in R \right\}$ is a subring of $M_2(R)$.

(b) Prove that S_1 is a left ideal of $M_2(R)$.

(c) Prove that S_1 is not a field.

(d) Prove that $S_2 = \left\{ \begin{pmatrix} 0 & 0 \\ 0 & a \end{pmatrix} | a \in R \right\}$ is a left ideal but it is not a right ideal of $M_2(R)$.

(e) Prove that $S_3 = \left\{ \begin{pmatrix} 0 & 0 \\ a & b \end{pmatrix} | a, b \in R \right\}$ is a left ideal but it is not a right ideal of $M_2(R)$.

(f) Find a non-trivial ideal of $S_4 = \left\{ \begin{pmatrix} a & b \\ 0 & c \end{pmatrix} | a, b, c \in R \right\}$.

(g) Find a left ideal and a right ideal of $M_3(R)$.

14. Prove that $M_n(\mathbb{Q})$ is not an ideal of $M_n(\mathbb{R})$ or $M_n(\mathbb{C})$.

15. (a) Identify all elements of the ideal $< x + 1 >$ of $\mathbb{R}[x]$.

(b) Identify all elements of the ideal $< x^2 - 4x + 1 >$ of $\mathbb{R}[x]$.

16. (a) Show that $\mathbb{Z}/19\mathbb{Z}$ is a field.

(b) Find the additive inverse of $7 + 19\mathbb{Z}$ in $\mathbb{Z}/19\mathbb{Z}$

(c) Find the multiplicative inverse of $7 + 19\mathbb{Z}$ in $\mathbb{Z}/19\mathbb{Z}$

17. Find all maximal ideals of $\mathbb{Z}_{36}$.

18. (a) Prove that $< x+1 >$ is a prime ideal of $\mathbb{R}[x]$.

    (b) Prove that $< x+1 >$ is a maximal ideal of $\mathbb{R}[x]$.

19. Let $I =< x^2 + 6x + 12 > \lhd \mathbb{Q}[x]$

    (a) Calculate $(2x + 4 + I) + (5x - 7 + I)$ and $(x^2 + 4x - 1 + I)(x - 7 + I)$.

    (b) Identify all the elements of $\mathbb{Q}[x]/I$.

    (c) Is $\mathbb{Q}[x]/I$ a commutative ring with unity? Explain.

20. (a) Identify all the elements of $\mathbb{Q}[x]/< x^8 - 5 >$.

    (b) Is $\mathbb{Q}[x]/< x^8 - 5 >$ a field? Explain your reasoning.

21. (a) Identify all elements of $\mathbb{Z}_3[x]/< x^3 + 2x + 2 >$.

    (b) Show that $\mathbb{Z}_3[x]/< x^3 + 2x + 2 >$ is a field.

    (c) Find the multiplicative inverse of $x+1+ < x^3+2x+2 >$ in $\mathbb{Z}_3[x]/< x^3 + 2x + 2 >$.

22. (a) Identify all elements of $< 6 - i >$ in the **Gaussian integer ring** $\mathbb{Z}[i] = \{a + bi | a, b \in \mathbb{Z}, i = \sqrt{-1}\}$.

    (b) Identify all elements of $\mathbb{Z}[i]/ < 6 - i >$ and show that it is a field.

    (c) Let p be a prime such that $p = n^2 + 1$ for an integer n. Prove that $\mathbb{Z}[i]/ < n - i >$ is a field with p elements.

23. (a) Identify all prime ideals of $\mathbb{Z}$.

    (b) Identify all maximal ideals of $\mathbb{Z}$.

24. Suppose I_1 and I_2 are ideals of a ring R. Prove that $(I_1 : I_2) = \{a \in R | aI_2 \subset I_1\}$ is an ideal of R.

25. Suppose I is an ideal of a ring R and let $\sqrt{I} = \{a \in R | a^n \in I, 1 < n \in \mathbb{N}\}$.

    (a) Prove that $\sqrt{I}$ is an ideal of R. It is called **the radical ideal of** I.

    (b) Find $\sqrt{< 4 >}$ of $\mathbb{Z}$.

    (c) Find $\sqrt{< p >}$ of $\mathbb{Z}$ where p is prime.

26. Let I be an ideal of a ring R and R_1 be a subring of a ring R. Prove that if $I \subset R_1$, then I is an ideal of R_1.

27. Prove Proposition 15.2.1 part 3.

28. Prove that a principal ideal of a commutative ring with unity is actually an ideal.

29. Determine whether the following ideals of $\mathbb{R}[x]$ are maximal or not. Also determine whether they are prime or not.

    (a) $< x - 1 >$

    (b) $< x - a >$

    (c) $< x^2 + 3 >$

    (d) $< x^2 + a >$ where $a > 0$.

    (e) $< x^2 - 1 >$

30. (a) Is $\mathbb{R}[x]/< x^2 + 5 >$ an integral domain? A field? Explain.

    (b) Is $\mathbb{C}[x]/< x^2 + 5 >$ an integral domain? A field? Explain.

31. (a) Prove the additive identity of a ring is unique.

    (b) Prove that the unity of a unitary ring is unique.

32. Using Theorem 15.3.2, prove that $\mathbb{Q}[x]$, $\mathbb{R}[x]$, and $\mathbb{C}[x]$ are all PID's.

33. Let $f(x) = (x-3)(x^2+7x-2)$, $g(x) = (x-3)(x+4)$, and $h(x) = (x-3)(x^3-5)$. Recall that $< f(x) >= f(x)\mathbb{Q}[x] \lhd \mathbb{Q}[x]$. So, we can denote $f(x)\mathbb{Q}[x] + g(x)\mathbb{Q}[x] =: < f(x), g(x) >$.

    (a) Show that $< f(x), g(x) >=< x-3 >$.

    (b) Show that $< f(x), g(x), h(x) >=< x-3 >$.

34. (a) Identify the ideal of $\mathbb{Z}$ generated by 4 and 7, i.e. $< 4, 7 >$.

    (b) Identify the ideal of $\mathbb{Z}$ generated by 4 and 12.

    (c) Identify the ideal of $\mathbb{Z}$ generated by m and n where n is a divisor of m.

35. Let R be a commutative ring. $r \in R$ is called **nilpotent** is there is a natural number n such that $r^n = 0$.

    (a) Prove that if a and b are nilpotent , then so are ab and $a+b$.

    (b) Prove that if $a_1, \cdots, a_n$ are nilpotent, then so are $a_1 \cdots a_n$ and $a_1 + \cdots + a_n$.

    (c) If I is the set of all the nilpotent elements of R, then prove that I is an ideal of R. This set is called **the nil radical ideal** of R.

    (d) If a is nilpotent, then $1-a$ is invertible if R is unitary.

36. Let R be a commutative ring. For a non-zero element a of R, the set $ann(a) = \{r \in R | ar = 0\}$ is called the **annihilator** a in R.

    (a) Prove that $ann(a)$ is an additive subgroup of R.

    (b) Prove that $ann(a)$ is a subring of R.

    (c) Prove that $ann(a)$ is an ideal of R.

Chapter 16

Ring Homomorphisms

There are certain "known" rings: number rings, matrix rings, polynomial rings, etc. We also know many properties of these rings but there are many complicated but useful rings. How do we investigate these rings? One way is using a function. If we can find a function between a "known" ring and an "unknown" ring, and if this function preserves the ring operations, we can see the structure of a complicated ring through a well-known ring.

For given two rings R_1 and R_2, we define a map from the ring R_1 to the ring R_2 which preserves the addition and the multiplication and we call it **a ring homomorphism**. A group homomorphism preserves one operation[1] but a ring homomorphism preserves two operation. In this chapter we study such a map and its properties.

Let's take a look at a trivial example before we see a definition.

Example 16.0.8. *Let ι be the identity map from a ring R to itself. Then ι is a ring homomorphism.*

[1]Of course because there is only one operation.

Solution. For any $r_1, r_2 \in R$, $\iota(r_1+r_2) = r_1+r_2 = \iota(r_1)+\iota(r_2)$ and $\iota(r_1r_2) = r_1r_2 = \iota(r_1)\iota(r_2)$. This shows that the identity map ι is a ring homomorphism. □

16.1 Definition

Definition 16.1.1. *Let R_1 and R_2 be rings. A map θ from R_1 to R_2 is called* **a ring homomorphism** *if the following hold for $r_1, r_2 \in R_1$:*

$$\theta(r_1 + r_2) = \theta(r_1) + \theta(r_2),$$
$$\theta(r_1r_2) = \theta(r_1)\theta(r_2)$$

A ring homomorphism from a ring to itself is called **an endomorphism.** Also, if R_1 and R_2 in Definition 16.1.1 are fields, then the map θ is called **a field homomorphism**. Here is another obvious example.

Example 16.1.1. *Let $\mathbf{0}$ be the zero map from a ring R_1 to a ring R_2, i.e., $\mathbf{0}(r) = 0$ for $r \in R_1$. Then $\mathbf{0}$ is a ring homomorphism.*

Solution. For any $r_1, r_2 \in R_1$, $\mathbf{0}(r_1 + r_2) = 0 = \mathbf{0}(r_1) + \mathbf{0}(r_2)$ and $\mathbf{0}(r_1r_2) = 0 = \mathbf{0}(r_1)\mathbf{0}(r_2)$. This shows that $\mathbf{0}$ is a ring homomorphism. □

The following is an important homomorphism on $\mathbb{C}$.

Example 16.1.2. *Let θ be the map from $\mathbb{C}$ to itself such that $\theta(z) = \overline{z}$ where $\overline{z}$ is the conjugate of $z \in \mathbb{C}$, i.e., $\overline{a + bi} = a - bi$ for $a, b \in \mathbb{R}$. Then θ is a ring homomorphism.*

Solution. For any $a+bi, c+di \in \mathbb{C}$, we have that $\theta(a+bi+c+di) = \theta(a+c+bi+di) = a+c-bi-di = \theta(a+bi) = \theta(c+di)$. We also have that $\theta((a+bi)(c+di)) = \theta(ac-bd+(bc+ad)i) = ac-bd-(bc+ad)i = \theta((a+bi))\theta((c+di))$. Therefore, θ is a ring homomorphism. □

Recall that for any $\overline{a} \in \mathbb{Z}_n$ and $k \in \mathbb{Z}$, we have that $k\overline{a} = \overline{a} + \cdots + \overline{a}$ where $\overline{a} + \cdots + \overline{a}$ is the sum of k terms of $\overline{a}$.

Example 16.1.3. *Prove that the map $\theta : \mathbb{Z}_8 \to \mathbb{Z}_{12}$ defined by $\theta(x) = 9x$ mod 12 is a ring homomorphism.*

Solution. Note that

$$\begin{aligned}\theta(x+y) &= 9(x+y) \bmod 12 \\ &= (9x+9y) \bmod 12 \\ &= (9x \bmod 12 + 9y \bmod 12) \bmod 12 \\ &= \theta(x)+\theta(y).\end{aligned}$$

Also, note that

$$\begin{aligned}\theta(xy) &= 9xy \bmod 12 \\ &= 81xy \bmod 12 \\ &= (9x \bmod 12)(9y \bmod 12) \bmod 12 \\ &= \theta(x)\theta(y)\end{aligned}$$

since $81 = 9 \bmod 12$. Therefore, θ is a ring-homomorphism. □

Example 16.1.4. *Find all the ring homomorphisms from a ring $\mathbb{Z}_5$ to a ring $\mathbb{Z}_{10}$.*

Solution. Let θ be a homomorphism from $\mathbb{Z}_5$ to $\mathbb{Z}_{10}$. Since $\theta(\overline{1}) = \theta(\overline{1} \cdot \overline{1}) = \theta(\overline{1})\theta(\overline{1})$, $\theta(\overline{1})$ might be $\overline{0}, \overline{1}, \overline{5}$ (because $\overline{5}^2 \equiv 25 = \overline{5} \bmod 10$), or $\overline{6}$ (because $\overline{6}^2 \equiv 36 = \overline{6} \bmod 10$) in $\mathbb{Z}_{10}$. Note that $|\overline{1}| = 5$ in $\mathbb{Z}_5$ so $|\theta(\overline{1})| = 1$ or 5 and this implies that $\theta(\overline{1}) = \overline{0}, \overline{2}, \overline{4}, \overline{6}$, or $\overline{8}$. Since θ should satisfies both, only possible cases are $\theta(\overline{1}) = \overline{0}$ and $\theta(\overline{1}) = \overline{6}$. The 1st case is the zero map and the 2nd case makes that $\theta(\overline{0}) = \overline{0}$ (why?), $\theta(\overline{2}) = \overline{2}$, $\theta(\overline{3}) = \overline{8}$, and $\theta(\overline{4}) = \overline{4}$. It is easy to check that θ is a homomorphism. Therefore there are only two homomorphisms from the ring $\mathbb{Z}_5$ to the ring $\mathbb{Z}_{10}$. □

We want to look at more examples of all the homomorphisms from a ring to another ring. So it is a good place to make a good notation for that kind of sets.

Definition 16.1.2. *For given rings R_1 and R_2, $Hom_{ring}(R_1, R_2)$ is* **the set of all homomorphisms from** R_1 **to** R_2, *i.e.,*

$$Hom_{ring}(R_1, R_2) = \{\theta | \theta : R_1 \to R_2, \theta \text{ is a ring homomorphism}\}$$

Sometime, $Hom(R_1, R_2)$ is used for $Hom_{ring}(R_1, R_2)$. $End_{ring}(R_1)$ is defined in a similar way, i.e.

$$End_{ring}(R_1) = \{\theta | \theta : R_1 \to R_1, \theta \text{ is a ring endomorphism}\}$$

It is also simply written as $End(R_1)$.

Example 16.1.5. *Find $Hom_{ring}(\mathbb{Z}_6, \mathbb{Z}_7)$.*

Solution. Since all the multiplicative idempotents of $\mathbb{Z}_7$ are $\overline{0}$ and $\overline{1}$, we have the following possibilities $\theta(\overline{1}) = \overline{0}$ or $\theta(\overline{1}) = \overline{1}$. If $\theta(\overline{1}) = \overline{0}$ holds, then θ can be extended to the zero homomorphism. If $\theta(\overline{1}) = \overline{1}$ holds, then $\theta(x) = x$ for any $x \in \mathbb{Z}_6$. This leads that $\theta(\overline{1}) = \theta(\overline{2} + \overline{5}) = \theta(\overline{2}) + \theta(\overline{5}) = \overline{2} + \overline{5} = \overline{0}$, a contradiction, i.e. there is no homomorphism θ such that $\theta(\overline{1}) = \overline{1}$. Therefore, the zero map is the only ring homomorphism from the ring $\mathbb{Z}_6$ to the ring $\mathbb{Z}_7$.

□

Example 16.1.6. *Find* $End_{ring}(\mathbb{Z}_5)$.

Solution. Since all the idempotents of $\mathbb{Z}_5$ are $\overline{0}$ and $\overline{1}$, we have two possibilities $\theta(\overline{1}) = \overline{0}$ and $\theta(\overline{1}) = \overline{1}$. If $\theta(\overline{1}) = \overline{0}$, then it is easy to check that the homomorphism is the zero map (Why?). If $\theta(\overline{1}) = \overline{1}$, then it is easy to check that the homomorphism is the identity map (Why?). Therefore, these are only endomorphisms on $\mathbb{Z}_5$. □

Example 16.1.7. *Find* $End_{ring}(\mathbb{Z}_6)$.

Solution. Let θ be an endomorphism on $\mathbb{Z}_6$. Since $\theta(\overline{1})^2 = \theta(\overline{1})$, there are four possibilities: $\theta(\overline{1}) = \overline{0}$, $\theta(\overline{1}) = \overline{1}$, $\theta(\overline{1}) = \overline{3}$, and $\theta(\overline{1}) = \overline{4}$. If $\theta(\overline{1}) = \overline{0}$, then it is the zero map. If $\theta(\overline{1}) = \overline{1}$, then it is the identity map. If $\theta(\overline{1}) = \overline{3}$, then $\theta(\overline{0}) = \overline{0}$, $\theta(\overline{2}) = \overline{0}$, $\theta(\overline{3}) = \overline{3}$, $\theta(\overline{4}) = \overline{0}$, and $\theta(\overline{5}) = \overline{3}$. If $\theta(\overline{1}) = \overline{4}$, then $\theta(\overline{0}) = \overline{0}$, $\theta(\overline{2}) = \overline{2}$, $\theta(\overline{3}) = \overline{0}$, $\theta(\overline{4}) = \overline{4}$, and $\theta(\overline{5}) = \overline{2}$. Since these θ are ring homomorphisms, there are four different homomorphisms from $\mathbb{Z}_6$ to itself. □

16.2 Homomorphism and its Kernel

Now we want to look at a ring-theory version of the kernel of a homomorphism. Recall that the kernel of a group homomorphism is a normal subgroup and it makes all three isomorphism theorems possible.

Definition 16.2.1. *Let θ be a ring homomorphism from a ring R_1 to a ring R_2. Then* **the kernel** *θ is the set of elements of R_1 whose image is 0 and it is denoted by $ker(\theta)$, i.e.*

$$ker(\theta) = \{r_1 \in R_1 | \theta(r_1) = 0\}.$$

The set of all **images** *in R_2 is denoted by $im(\theta)$, i.e.,*

$$im(\theta) = \{\theta(r_1) | r_1 \in R_1)\}.$$

Note that R_2 is called **the range** of θ and the set of all images is not always the range of a function. If it is, the function is a surjection.

Example 16.2.1. *In Examples 16.0.8 and 16.1.1, we have that $ker(\iota) = \{0\}$ and $ker(\mathbf{0}) = R_1$.* □

In other words, the kernel of a zero function is everything and the kernel of the identity function is just a zero. We want to look at important properties of the kernel and the image of a homomorphism as we did in group theory, and these will be useful to develop a ring theory and extend it to a field theory later.

Theorem 16.2.1. *Let θ be a homomorphism from a ring R_1 to a ring R_2. Then*

1. *$ker(\theta)$ is an ideal of R_1 and*

2. *$im(\theta)$ is a subring of R_2.*

Proof.

1. For any $r_1, r_2 \in ker(\theta)$, $\theta(r_1 - r_2) = \theta(r_1) - \theta(r_2) = 0 - 0 = 0$. This implies that $r_1 - r_2 \in ker(\theta)$. We also have that $\theta(r_1 r_2) = \theta(r_1)\theta(r_2) = 0 \cdot 0 = 0$. This implies that $r_1 r_2 \in ker(\theta)$, i.e., $ker(\theta)$ is a subring of R_1. For any $r \in R_1$, we have that $\theta(rr_1) = \theta(r)\theta(r_1) = \theta(r)0 = 0$. So $r_1 r \in ker(\theta)$, i.e., $ker(\theta)$ is an ideal of R_1.

2. For any $r_1', r_2' \in im(\theta)$, there are $r_1, r_2 \in R_1$ such that $\theta(r_1) = r_1'$ and $\theta(r_2) = r_2'$. This implies that $\theta(r_1 - r_2) = \theta(r_1) - \theta(r_2) = r_1' - r_2' \in im(\theta)$. We also have that $\theta(r_1 r_2) = \theta(r_1)\theta(r_2) = r_1' r_2' \in im(\theta)$. This implies that $im(\theta)$ is a subring of R_2. $\square$

Corollary 16.2.1. *Let θ be a homomorphism from a ring R_1 to a ring R_2.*

1. *If I is an ideal of R_1, then $\theta(I)$ is an ideal of $\theta(R_1)$.*

2. *If S is a subring of R_2, then $\theta^{-1}(S)$ is a subring of R_1.*

3. *If θ is onto and I_2 is an ideal of R_2, then $\theta^{-1}(I_2)$ is an ideal of R_1.*

Proof.

1. By Theorem 16.2.1, $\theta(I)$ is a subring of $\theta(R_1)$. For any $\theta(r_1) \in \theta(R)$ and $\theta(i) \in \theta(I)$, since $r_1 i \in I$, we have that $\theta(r_1)\theta(i) = \theta(r_1 i) \in \theta(I)$ where $r_1 \in R_1$ and $i \in I$. This implies that $\theta(I)$ is an ideal of $\theta(R_1)$.

2. For $r_1, r_2 \in \theta^{-1}(S)$, we have that $\theta(r_1)$ and $\theta(r_2)$ are elements of S. This implies that $\theta(r_1)-\theta(r_2) = \theta(r_1-r_2) \in S$. So we have that $r_1-r_2 \in \theta^{-1}(S)$. This implies that $\theta^{-1}(S)$ is an additive subgroup of R_1. By $\theta(r_1)\theta(r_2) = \theta(r_1r_2) \in S$, we have that $r_1r_2 \in \theta^{-1}(R_4)$. This implies that $\theta^{-1}(S)$ is a subring of R_1.

3. By 2, we have that $\theta^{-1}(I_2)$ is a subring of R_1. For any $r_1 \in R_1$ and for any $i \in \theta^{-1}(I_2)$, we have that $\theta(i) \in I_2$. Then $\theta(r_1)\theta(i) = \theta(r_1 i) \in I$ and this makes that $r_1 i \in \theta^{-1}(I_2)$. So $\theta^{-1}(I_2)$ is an ideal of R_1. $\square$

Example 16.2.2. *If we define an endomorphism θ of $\mathbb{Z}_6$ by $\theta(\overline{1}) = \overline{3}$ as in Example 16.1.7, find $ker(\theta)$.*

Solution. From Example 16.1.7, we see that $\theta(\overline{0}) = \theta(\overline{2}) = \theta(\overline{4}) = \overline{0}$ so $ker(\theta) = \{\overline{0}, \overline{2}, \overline{4}\}$. $\square$

We want to classify several different homomorphisms and these words are convenient and universal. And then we will look at more properties to start the 1st isomorphism theorem.

Definition 16.2.2. *Let θ be a ring homomorphism. If θ is an injection (one to one), then it is called* **a monomorphism**. *If θ is a surjection (onto), then it is called* **an epimorphism**. *If θ is a bijection (one to one & onto), then it is called* **an isomorphism**. *If θ is an isomorphism from a ring to itself, then it is called* **an automorphism**. *If there is a ring isomorphism from a ring R_1 to a ring R_2, then we call that the rings R_1 and R_2 are* **isomorphic**, *and denoted by $R_1 \cong R_2$.*

Theorem 16.2.2. *For a ring homomorphism θ from a ring R_1 to a ring R_2,*

1. *θ is one to one if and only if $ker(\theta) = \{0\}$ and*

2. *θ is onto if and only if $im(\theta) = R_2$.*

Proof.

1. Let us assume that θ is one to one. Obviously $\theta(0) = 0$. Since θ is one to one, $\theta(r) = 0$ implies that $r = 0$. Thus $ker(\theta) = \{0\}$. Conversely, let us assume that $ker(\theta) = \{0\}$. If $\theta(r_1) = \theta(r_2)$ holds, then $\theta(r_1) - \theta(r_2) = 0$. This implies that $\theta(r_1 - r_2) = 0$. By assumption, we have that $r_1 - r_2 = 0$, i.e., $r_1 = r_2$. This implies that θ is one to one.

2. Let us assume that θ is onto. Then for any $r_1' \in R_2$, there is $r \in R_1$ such that $\theta(r_1) = r_1'$. Thus $\theta(R_1) = R_2$. Conversely, let us assume that $\theta(R_1) = R_2$. This implies that for any $r_1' \in R_2$, there is $r_1 \in R_1$ such that $\theta(r_1) = r_1'$. This implies that θ is onto. □

Example 16.2.3. *Let θ be a map from $\mathbb{Q}[\sqrt{3}]$ to itself such that*

$$\theta(a + b\sqrt{3}) = a - b\sqrt{3}$$

for $a + b\sqrt{3} \in \mathbb{Q}[\sqrt{3}]$.

1. *Prove that θ is a ring homomorphism.*

2. *Find the kernel $ker(\theta)$ of θ.*

3. *Is θ one-to-one?*

4. *Is θ onto?*

5. *Is θ an automorphism?*

Solution.

1. Obviously θ is an additive homomorphism. For $a+b\sqrt{3}, c+d\sqrt{3} \in \mathbb{Q}[\sqrt{3}]$, we have that $\theta((a+b\sqrt{3})(c+d\sqrt{3})) = \theta(ac+3bd+(ad+bc)\sqrt{3}) = ac+3bd-(ad+bc)\sqrt{3}$. On the other hand, we have that $\theta(a+b\sqrt{3})\theta(c+d\sqrt{3}) = (a-b\sqrt{3})(c-d\sqrt{3}) = ac+3bd-(ad+bc)\sqrt{3}$. Thus θ is a ring homomorphism.

2. Let $a+b\sqrt{3}$ be an element of $ker(\theta)$. This implies that $\theta(a+b\sqrt{3}) = a-b\sqrt{3} = 0$. This also implies that $a = b = 0$. We have that $ker(\theta) = \{0\}$.

3. By 2, θ is one to one.

4. For any $c+d\sqrt{3} \in \mathbb{Q}[\sqrt{3}]$, we have that $\theta(c-d\sqrt{3}) = c+d\sqrt{3}$. This shows that θ is onto.

5. By 1, 2, 3, and 4, θ is an automorphism. □

Example 16.2.4. *Prove that $\mathbb{Z}[i]/< 4-i >$ and $\mathbb{Z}_{17}$ are isomorphic.*

Solution. Note that $17 = (4-i)(4+i) \in < 4-i >$. Suppose $d \in < 4-i >$ where $1 \le d \le 16$. Then we have $(a+bi)(4-i) = d$ for some $a+bi \in \mathbb{Z}[i]$, and the real parts will give us $4a+b = d$ and the imaginary parts make $-a+4b = 0$. This implies that $b = \frac{1}{17}$, a contradiction. This means that d is not an element of $< 4-i >$ where $1 \le d \le 16$ and this implies that $\mathbb{Z}[i]/ < 4-i >$ has 17 elements and it is an Abelian group with the multiplication modulo 17. So, $\mathbb{Z}[i]/< 4-i >$ is a ring with 17 elements and every rings with 17 elements are isomorphic to $\mathbb{Z}_{17}$. Thus we have solved the example. □

Example 16.2.5. *Prove that $< 4-i >$ is a maximal ideal of the ring $\mathbb{Z}[i]$.*

Solution. Since 1 is the unity of $\mathbb{Z}[i]$, $\mathbb{Z}[i]/<4-i>$ has the unity $1+<4-i>$. By Example 16.2.4, $\mathbb{Z}[i]/<4-i>$ and $\mathbb{Z}_{17}$ are isomorphic. Since $\mathbb{Z}_{17}$ is a field, $\mathbb{Z}[i]/<4-i>$ is a field and this implies that $<4-i>$ is a maximal ideal of the ring. □

Proposition 16.2.1. *If R is an integral domain, then R^{-1} is a field and R is a subring of R^{-1}.*

Proof. We know that R^{-1} is a field by Theorem 13.4.1. If we define a map $\theta : R \to R^{-1}$ as for $r \in R$, $\theta(r) = \frac{r}{1} := r$, then it is easy to prove that θ is a ring homomorphism and one to one. So $\theta(R) = R$ is a subring of R^{-1} by Theorem 16.2.1. □

Lemma 16.2.1. *If θ_1 is a ring homomorphism from R_1 to R_2 and θ_2 is another ring homomorphism from R_2 to R_3, then the composition $\theta_2\theta_1$ is a ring homomorphism from R_1 to R_3.*

Proof. For $r_1, r_2 \in R_1$, we have that

$$\begin{aligned}(\theta_2\theta_1)(r_1 + r_2) &= (\theta_2)(\theta_1(r_1) + \theta_1(r_2)) \\ &= (\theta_2)(\theta_1(r_1)) + (\theta_2)(\theta_1(r_2)) \\ &= (\theta_2\theta_1)(r_1) + (\theta_2\theta_1)(r_2).\end{aligned}$$

This shows that $\theta_2\theta_1$ is a homomorphism that preserves the addition. We also have that

$$\begin{aligned}(\theta_2\theta_1)(r_1r_2) &= (\theta_2)(\theta_1(r_1)\theta_1(r_2))\\ &= (\theta_2)(\theta_1(r_1))(\theta_2)(\theta_1(r_2))\\ &= (\theta_2\theta_1)(r_1)(\theta_2\theta_1)(r_2).\end{aligned}$$

This shows that $\theta_2\theta_1$ is a homomorphism that preserves the multiplication. Therefore, $\theta_2\theta_1$ is a ring homomorphism. □

Theorem 16.2.3. *Let θ_1 be a ring homomorphism from R_1 to R_2 and θ_2 be another ring homomorphism from R_2 to R_3.*

1. *If θ_1 and θ_2 are one to one, then $\theta_2\theta_1$ is one to one.*

2. *If θ_1 and θ_2 are onto, then $\theta_2\theta_1$ is onto.*

3. *If θ_1 and θ_2 are isomorphisms, then $\theta_2\theta_1$ is an isomorphism.*

Proof. By Lemma 16.2.1, $\theta_2\theta_1$ is a homomorphism from R_1 to R_3. The rest of the proof is straightforward by Proposition 1.4.1, □

16.3 Isomorphism Theorems

Let I be an ideal of a ring R. If we define the map

$$\mathfrak{p} : R \to R/I \tag{16.1}$$

as $\mathfrak{p}(r) = r+I$ for any $r \in R$. Then for any $r_1, r_2 \in R$, we have that $\mathfrak{p}(r_1+r_2) = r_1 + r_2 + I = r_1 + I + r_2 + I = \mathfrak{p}(r_1) + \mathfrak{p}(r_2)$. We also have that $\mathfrak{p}(r_1r_2) =$

$r_1 r_2 + I = (r_1 + I)(r_2 + I) = \mathfrak{p}(r_1)\mathfrak{p}(r_2)$. Thus $\mathfrak{p}$ is a ring homomorphism and it is called **the canonical map.**

Theorem 16.3.1. (The 1st Isomorphism Theorem)

Let θ be a ring homomorphism from a ring R_1 to R_2. There is a ring isomorphism from $R_1/ker(\theta)$ to $im(\theta)$.

Proof. The proof of the theorem is almost the same as the proof of Theorem 5.2.2. □

Corollary 16.3.1. *Let θ be a ring homomorphism from R_1 to R_2. If θ is onto, then there is a ring isomorphism from $R_1/(ker(\theta))$ to R_2.*

Proof. Since $im(\theta) = R_2$, the proof of the corollary is straightforward by Theorem 16.3.1. □

Corollary 16.3.2. *Let I be an ideal of a ring R. I can be a kernel of a homomorphism.*

Proof. Let us define the map $\mathfrak{p}$ from R to R/I as follows:

$$\mathfrak{p}(r) = r + I$$

for $r \in R$. Then obviously $\mathfrak{p}$ is a ring homomorphism and its kernel is I. So the corollary holds by Theorem 16.3.1. □

Theorem 16.3.2. *Suppose I_1 and I_2 are ideals of a ring R.*

1. **(The 2nd Isomorphism Theorem)**

   $(I_1 + I_2)/I_1$ and $I_2/(I_1 \cap I_2)$ are isomorphic.

2. **(The 3rd Isomorphism Theorem)**

If I_1 is a subring of I_2, then R/I_2 and $(R/I_1)/(I_2/I_1)$ are isomorphic.

Proof. The proof of the theorem is straightforward by Theorem 5.2.3 by applying two operations. □

These isomorphism theorem will work nicely when we study and construct finite fields.

Chapter 16 Exercises.

1. Let θ be a ring homomorphism from a ring R_1 to a ring R_2.

   (a) Prove that if R_1 is a commutative ring, then $\theta(R_1)$ is a commutative ring.

   (b) Prove that if 1 is the unity of R_1, then $\theta(1)$ is the unity of $\theta(R_1)$.

   (c) Prove that if θ is one to one, then $\theta(1)$ is the unity of R_2.

2. Determine whether each of the following maps is a ring homomorphism or not.

   (a) $\theta : \mathbb{Z}_4 \to \mathbb{Z}_{10}$ defined by $\theta(x) = 5x$ for $x \in \mathbb{Z}_4$.

   (b) $\theta : \mathbb{Z}_4 \to \mathbb{Z}_{20}$ defined by $\theta(x) = 5x$ for $x \in \mathbb{Z}_4$.

   (c) $\theta : \mathbb{Z}_8 \to \mathbb{Z}_{12}$ defined by $\theta(x) = 4x$ for $x \in \mathbb{Z}_8$.

   (d) $\theta : \mathbb{Z}_9 \to \mathbb{Z}_{10}$ defined by $\theta(x) = 6x$ for $x \in \mathbb{Z}_9$.

   (e) $\theta : \mathbb{Z}_{11} \to \mathbb{Z}_6$ defined by $\theta(x) = 3x$ for $x \in \mathbb{Z}_{11}$.

3. Suppose $\theta : \mathbb{Z}_8 \to \mathbb{Z}_8$ and $\theta(1) = 2$.

(a) Prove that it can be extended to a ring endomorphism of $\mathbb{Z}_8$.

(b) Find $ker(\theta)$ and show that it is actually an ideal of $\mathbb{Z}_8$.

(c) Find $im(\theta) = \theta(\mathbb{Z}_8)$ and show that it is actually a subring of $\mathbb{Z}_8$.

4. Find all the ring homomorphisms from $\mathbb{Q}$ to itself, i.e., $Hom_{ring}(\mathbb{Q})$.

5. (a) Find $Hom_{ring}(\mathbb{Z})$.

   (b) Find $Hom_{ring}(\mathbb{Z} \oplus \mathbb{Z}, \mathbb{Z})$.

   (c) Find $Hom_{ring}(\mathbb{Z}, \mathbb{Z} \oplus \mathbb{Z})$.

6. Let $\theta : \mathbb{Z} \to \mathbb{Z}_n$ be a map such that for any $a \in \mathbb{Z}$, $\theta(a) = \overline{a}$. (This map is called **the evaluation map for** a.)

   (a) Prove that θ is a ring homomorphism.

   (b) Find $ker(\theta)$ and $im(\theta)$.

7. For a given $a \in \mathbb{R}$, let $\theta : \mathbb{R}[x] \to \mathbb{R}$ be a map defined by $\theta(f(x)) = f(a)$ for $f(x) \in \mathbb{R}[x]$.

   (a) Prove that θ is a ring homomorphism. (Hint: Use the Remainder Theorem).

   (b) Find $ker(\theta)$ and $im(\theta)$.

8. (a) If any, find a ring homomorphism θ from $\mathbb{Z}_4$ to $\mathbb{Z}_{10}$ such that θ is not the zero map.

   (b) Find $Hom(\mathbb{Z}_4, \mathbb{Z}_{10})$.

9. (a) Find $End_{ring}(\mathbb{Z}_7)$.

   (b) Find $End_{ring}(\mathbb{Z}_8)$.

   (c) Find $Hom_{ring}(\mathbb{Z}_5, \mathbb{Z}_8)$.

(d) Find $Hom_{ring}(\mathbb{Z}_8, \mathbb{Z}_{10})$.

10. Let θ be a map from $\mathbb{Z}_n$ $\mathbb{Z}_n$ such that $\theta(x) = -x$ for $x \in \mathbb{Z}_n$.

    (a) Prove or disprove that θ is a group homomorphism.

    (b) Prove or disprove that θ is a ring homomorphism.

11. Let θ be a map from $\mathbb{Z}_n$ to $\mathbb{Z}_m$ such that $\theta(x) = -x$ for any $x \in \mathbb{Z}_n$.

    (a) Prove or disprove that θ is an additive group homomorphism.

    (b) Prove or disprove that θ is a ring homomorphism.

    (c) Prove that $Hom_{ring}(\mathbb{Z}_n, \mathbb{Z}_n)$ is a subset of $Hom_{group}(\mathbb{Z}_n, \mathbb{Z}_n)$.

12. (a) Find $End_{ring}(\mathbb{Z}_{12})$ and $End_{group}(\mathbb{Z}_{12})$.

    (b) Find $End_{ring}(\mathbb{Z}_7)$ and $End_{group}(\mathbb{Z}_7)$.

    (c) Find $End_{ring}(\mathbb{Z}_p)$ and $Hom_{group}(\mathbb{Z}_p)$ where p is prime.

13. Let's define an addition $+$ and a multiplication $\cdot$ on $End(R)$ as follows:

$$(f+g)(x) = f(x) + g(x),$$
$$(f \cdot g)(x) = f(x)g(x)$$

    for $f, g \in End(R)$ and $x \in R$.

    (a) Prove that $End(R)$ is a ring.

    (b) Prove or disprove that $End(R)$ a commutative ring.

14. (a) Find a non-zero and non-identity homomorphism θ from the ring $\mathbb{Q}[\sqrt{2}]$ to itself.

    (b) Find a non-zero and non-identity homomorphism θ from the ring $\mathbb{Q}[\sqrt{5}]$ to itself.

15. Prove that if $(m, n) = 1$, then $\mathbb{Z}/(mn\mathbb{Z})$ and $(\mathbb{Z}/m\mathbb{Z}) \bigoplus (\mathbb{Z}/n\mathbb{Z})$ are isomorphic.

16. (a) Find a subring of $M_{2m}(\mathbb{R})$ that is isomorphic to $M_m(\mathbb{R}) \bigoplus M_m(\mathbb{R})$.

    (b) Find a subring of $M_{m+n}(\mathbb{R})$ that is isomorphic to $M_m(\mathbb{R}) \bigoplus M_n(\mathbb{R})$.

17. Let θ_1 be a ring homomorphism from a ring R_1 to a ring R_2 and let θ_2 be a ring homomorphism from a ring R_2 to a ring R_3. Prove that $ker((\theta_2\theta_1)) \subset ker(\theta_1)$.

18. (a) Find an ideal I of $\mathbb{Z}[i]$ such that the $\mathbb{Z}[i]/I$ is isomorphic to $\mathbb{Z}_{37}$.

    (b) Prove that I is a maximal ideal of $\mathbb{Z}[i]$.

    (c) Let $n^2 + 1$ be a prime number. Find an ideal I of $\mathbb{Z}[i]$ such that the $\mathbb{Z}[i]/I$ is isomorphic to $\mathbb{Z}_{n^2+1}$.

    (d) Prove that the ideal I in the part (c) is a maximal ideal of $\mathbb{Z}[i]$.

19. Suppose θ is a map from $\mathbb{Z}_4$ to $\mathbb{Z}_6$ with $\theta(\overline{1}) = \overline{3}$.

    (a) Prove that θ is a ring homomorphism.

    (b) Find $ker(\theta)$. Is θ one to one? Prove your claim.

    (c) Is θ onto? Prove your claim.

    (d) Is θ an isomorphism? Prove your claim.

    (e) If ψ is a map from $\mathbb{Z}_4$ to $\mathbb{Z}_6$ with $\psi(\overline{1}) = \overline{2}$, prove that ψ is not a ring homomorphism.

20. (a) Prove that two rings $\mathbb{Z}[\sqrt{2}]$ and $\mathbb{Z}[\sqrt{3}]$ are not isomorphic.

    (b) Prove that two rings $\mathbb{Z}[\sqrt{2}]$ and $\mathbb{Z}[\sqrt{p}]$ are not isomorphic where p is a prime such that $p \neq 2$.

Chapter 17

Factorizations of Polynomials

In chapter 1, we saw that a positive integer can be written as a product of primes through the Fundamental Theorem of Arithmetic. In this chapter, we will see a similar property in polynomials and it is

Every polynomial over a field can be factored as a product of irreducible polynomials.

We will start this chapter with the following definition and investigate several properties on the factorization of polynomials over a field. From now on, $\mathbb{F}$ will denote a field of any characteristic. It can be 0 or a prime number p.

17.1 Reducible and Irreducible Polynomials

Definition 17.1.1. *Suppose D is an integral domain and $f(x)$ is a non-zero, non-unit polynomial in $D[x]$. Then we say that $f(x)$ is* **irreducible over** *D if*

$f(x) = g(x)h(x)$ implies that $g(x)$ or $h(x)$ is unit in $D[x]$. Otherwise $f(x)$ is called **reducible over** D.

We will see that the irreducibility of a polynomial depends on the integral domain.

Example 17.1.1. *1. $5x - 10$ is irreducible over $\mathbb{Q}$.*

2. But it is reducible over $\mathbb{Z}$ since $5x - 10 = 5(x - 2)$, and both 5 and $x - 2$ are units.

3. $2x^2 + 1$ is reducible over $\mathbb{Z}_3$ since $2x^2 + 1 = (2x + 1)(x + 1)$, and both $2x + 1$ and $x + 1$ are units.

4. $2x^2 + 1$ is irreducible over $\mathbb{Z}_5$ since $2x^2 + 1$ does not have a linear factor over $\mathbb{Z}_5$.

Definition 17.1.2. *Let $f(x) = a_n x^n + \cdots + a_0 \in \mathbb{Z}[x]$ with $a_n \neq 0$.*

1. **The content of the polynomial** *$f(x)$ is the greatest common divisor of $a_n, \cdots, a_0$, and denoted by $C(f(x))$.*

2. *A polynomial $f(x)$ is called* **primitive** *if $C(f(x)) = 1$.*

3. *If $a_n = 1$, then $f(x)$ is called* **a monic polynomial**.

a_n is called **the leading coefficient of** $f(x)$ and denoted by $l(f(x))$.

Example 17.1.2. *1. For $f(x) = 6x^3 + 3x + 9 \in \mathbb{Z}[x]$, $C(f(x)) = 3$.*

2. $g(x) = 2x^2 + 8x + 9$ is primitive.

3. For a polynomial $a_nx^n + \cdots + a_0$ over $\mathbb{Z}$, if $a_n = 1$, then the polynomial is primitive (why?).

The following theorem says that a *product of primitive polynomials is primitive.*

Theorem 17.1.1. *If $f(x)$ and $g(x)$ are primitive, then $f(x)g(x)$ is primitive for any $f(x), g(x) \in \mathbb{Z}[x]$.*

Proof. Let $f(x) = c_mx^m + \cdots + c_0$, $g(x) = d_nx^n + \cdots + d_0$, and $h(x) = f(x)g(x)$ with appropriate coefficients. This implies that $h(x) = e_{m+n}x^{m+n} + \cdots + e_0$ where $e_k = \sum_{i+j=k} c_id_j$. Let us assume that $h(x)$ is not primitive. This implies that there is a prime p which divides e_i, $0 \leq i \leq m+n$. Since $f(x)$ and $g(x)$ are primitive, there exist largest non-negative integers r and s such that $p|c_j$, $j \leq r-1$, p does not divides c_r, and $p|d_k$, $k \leq s-1$, p does not divides d_s. Then we have that $e_{r+s} = c_0d_{r+s} + \cdots + c_{r-1}d_{s+1} + c_rd_s + c_{r+1}d_{s-1} + \cdots + c_{r+s}d_0$. This implies that $c_rd_s = e_{r+s} - c_0d_{r+s} + \cdots - c_{r-1}d_{s+1} - c_{r+1}d_{s-1} - \cdots - c_{r+s}d_0$. Since p divides $e_{r+s} - c_0d_{r+s} + \cdots - c_{r-1}d_{s+1} - c_{r+1}d_{s-1} - \cdots - c_{r+s}d_0$, we have that p divides c_rd_s. Since p is prime, we have that $p|c_r$ or $p|d_s$, a contradiction. Therefore, $f(x)g(x)$ is primitive. $\square$

The following lemma proves that a reducibility of a primitive polynomial is inherited from $\mathbb{Q}$ to $\mathbb{Z}$ if the polynomial is in $\mathbb{Z}[x]$.

Theorem 17.1.2. (Gauss' Lemma)

If a primitive polynomial $f(x) \in \mathbb{Z}[x]$ is reducible over $\mathbb{Q}$, then it is reducible over $\mathbb{Z}$ as well where $deg(f(x)) > 0$.

Proof. Suppose $f(x) \in \mathbb{Z}[x]$ is primitive and reducible over $\mathbb{Q}$. Then it can be factored into a product of two polynomials $g(x)$ and $h(x)$ of $\mathbb{Q}[x]$. Let $g(x) = \frac{a_n}{b_n}x^n + \cdots + \frac{a_0}{b_0}$ and $h(x) = \frac{a'_m}{b'_m}x^m + \cdots + \frac{a'_0}{b'_0}$ where $a_i, b_i(\neq 0), a'_i, b'_i(\neq 0) \in \mathbb{Z}$ for all i and $deg((g(x))), deg(h(x)) > 1$. Let b' be the least common multiple of the integers $b_n, \cdots, b_0$ and b'' be the least common multiple of the integers $b'_m, \cdots, b'_0$. We have that $f(x) = g(x)h(x) = \frac{b'}{b'}(\frac{a_n}{b_n}x^n + \cdots + \frac{a_0}{b_0})\frac{b''}{b''}(\frac{a'_m}{b'_m}x^m + \cdots + \frac{a'_0}{b'_0})$. This implies that $b'b''f(x) = (a'''_n x^n + \cdots + a'''_0)(b'''_m x^m + \cdots + b'''_0) = abg_1(x)h_1(x)$ such that $ag_1(x) = a'''_n x^n + \cdots + a'''_0$ and $bh_1(x) = b'''_m x^m + \cdots + b'''_0$ where $g_1(x)$ and $h_1(x)$ are primitive. We have that $b'b''f(x) = abg_1(x)h_1(x)$. Then since $g_1(x)h_1(x)$ is primitive By Theorem 17.1.1 and $f(x)$ is primitive by assumption, we derive that $b'b'' = ab$. This implies that it is reducible in $\mathbb{Z}[x]$ and this proves the theorem. □

Corollary 17.1.1. *If a primitive polynomial $f(x) \in \mathbb{Z}[x]$ is irreducible over $\mathbb{Z}$, then it is irreducible over $\mathbb{Q}$ as well where $deg(f(x)) > 0$.*

17.2 Tests for Irreducibility of Polynomials over $\mathbb{Z}$

We have known an easier term than "reducible" polynomial: factorable. If a polynomial has a small degree like 2 or 3, it is not hard to factorize it. But if a polynomial has a high degree, for example $3x^4 - 3x^2 - 9x + 1$, then it is not that easy or simple even though we can apply the Rational Root Theorem. In this section, we'd like to present several tests to check whether a polynomial in $\mathbb{Z}[x]$ is irreducible. If it is, then we don't need to try to factor it. Moreover, it will be convenient to build a finite field using a polynomial if it is irreducible.

Theorem 17.2.1. (Eisenstein's Irreducibility Criterion)

Let $f(x) = a_nx^n + \cdots + a_0 \in \mathbb{Z}[x]$. If there is a prime p such that

1. $p \nmid a_n$
2. $p|a_{n-1}, \cdots, p|a_0$
3. $p^2 \nmid a_0$

then $f(x)$ is irreducible over $\mathbb{Q}$.

Proof. Without loss of generality, let's assume that $f(x)$ is primitive since we can always factor out the content $C(f(x))$ of $f(x)$. Let us assume that $f(x) = g(x)h(x)$ where $g(x)$ and $h(x)$ are non-constant polynomials in $\mathbb{Q}[x]$ i.e., $f(x) = (b_rx^r + \cdots + b_0)(c_sx^s + \cdots + c_0)$ with appropriate coefficients. This implies that $a_h = \sum_{i+j=h} b_ic_j$. Since $a_0 = b_0c_0$, $p|b_0c_0$, and $p^2 \not| b_0c_0$, we have that $p|b_0$ or $p|c_0$. Let us assume that $p|b_0$ and $p \not| c_0$. If $p|b_v$, $0 \le v \le r$, then $f(x)$ is divisible by p. By this contradiction, we can find the first minimal integer k, $0 \le k < r$ such that $p \not| b_k$ and $p|b_{k-1}$. By $a_k = b_kc_0 + \cdots + b_0c_k$, we have that $b_kc_0 = a_k - b_{k-1}c_1 - \cdots - b_1c_{k-1} - b_0c_k$. This implies that $p|b_kc_0$. Since p is prime, we have that $p|b_k$ or $p|c_0$, a contradiction. Therefore, $f(x)$ is irreducible in $\mathbb{Q}[x]$. □

Corollary 17.2.1. *Let $f(x) = a_nx^n + \cdots + a_0 \in \mathbb{Z}[x]$. If there is a prime p such that*

1. $p \nmid a_0$
2. $p|a_n, \cdots, p|a_1$
3. $p^2 \nmid a_n$

then $f(x)$ is irreducible over $\mathbb{Q}$.

Proof. This is the reversed statement of Eisenstein's Criterion. □

Example 17.2.1. *x^3-4x^2-2x+2 is irreducible over $\mathbb{Z}$ by Eisenstein's Criterion using $p=2$.*

Example 17.2.2. *$3x^3-6x^2+9x+5$ and $3x^3+9x+5$ are irreducible over $\mathbb{Q}$ by Corollary 17.2.1 using $p=3$.*

Example 17.2.3. *$x^{10}-7 \in \mathbb{Z}[x]$ is irreducible over $\mathbb{Q}$ by Eisenstein's Criterion using $p=7$.*

In general, x^n-p and px^n-1 are irreducible for any natural number n and any prime p.

Example 17.2.4. *x^8-10 is irreducible over $\mathbb{Q}$ by Eisenstein's Criterion using $p=5$ (or 2).*

In general, x^n-pq and pqx^n-1 are irreducible for any natural number n and any primes p,q such that $p \nmid q$.

Example 17.2.5. *$f(x)=x^{p-1}+\cdots+x+1$ is irreducible over $\mathbb{Q}$ for any prime number p.*

Proof. If $g(x)\in\mathbb{Q}[x]$, $g(x)=r(x)s(x)$ if and only if $g(x+a)=r(x+a)s(x+a)$ where $r(x)$ and $s(x)$ are non-constant polynomials and $a\in\mathbb{Q}$. This means that $g(x)$ is reducible (resp. irreducible) if and only if $g(x+a)$ is reducible (resp. irreducible). Since $f(x) = \frac{x^p-1}{x-1}$, we have that $f(x+1) = \frac{(x+1)^p-1}{(x+1)-1} = \frac{x^p+px^{p-1}+\cdots+px+1-1}{x} = x^{p-1} + px^{p-2} + \cdots + p$. Then, $f(x+1)$ is irreducible over $\mathbb{Q}$ by Eisenstein's Criterion and so is $f(x)$. □

Recall that $\mathbb{Z}_p = \{\overline{0}, \overline{1}, \cdots \overline{p-1}\}$. Then for a given polynomial $f(x) = a_n x^n + \cdots + a_0 \in \mathbb{Z}[x]$, we can define $\overline{f(x)} = \overline{a}_n x^n + \cdots + \overline{a}_0 \in \mathbb{Z}_p[x]$ where $a_i \equiv \overline{a}_i$ mod p. For example, for a prime number $p = 5$, we can have $\overline{6x^3 + 8x^2 - 4x - 2} = \overline{1}x^3 + \overline{3}x^2 + \overline{1}x + \overline{3} \in \mathbb{Z}_5[x]$. We can simply write it as $x^3 + \overline{3}x^2 + x + \overline{3} \in \mathbb{Z}_5[x]$. Now we are ready to state the following useful theorem to check irreducibility of a polynomial.

Theorem 17.2.2. (Mod p Irreducibility Criterion)
Suppose $f(x) \in \mathbb{Z}$ is a non-constant polynomial and let $\overline{f(x)} \in \mathbb{Z}_p$ be the image of $f(x)$ under the mod p map for a prime number p. Then, if $\overline{f(x)}$ is irreducible over $\mathbb{Z}_p$ and $deg(f(x)) = deg(\overline{f(x)})$, $f(x)$ is irreducible over $\mathbb{Q}$.

Proof. Let's assume that $f(x) \in \mathbb{Z}[x]$ is reducible over $\mathbb{Q}$ with $deg(f(x)) = n$. By Theorem 17.1.2, $f(x)$ is reducible over $\mathbb{Z}$. This implies that $f(x) = g(x)h(x)$ for $g(x), h(x) \in \mathbb{Z}[x]$. Note that $deg(f(x)) > deg(g(x)) > 0$ and $deg(f(x)) > deg(h(x)) > 0$. This implies that $\overline{f(x)} = \overline{g_1(x)h_1(x)} = \overline{g_1(x)}\,\overline{h_1(x)}$ in $\mathbb{Z}_p[x]$ with $deg(g(x)) \geq deg(\overline{g_1(x)})$ and $deg(h(x)) \geq deg(\overline{h_1(x)})$. Since $deg(f(x)) = deg(\overline{f(x)}) = n$, we have that $deg(g(x)) = deg(\overline{g_1(x)})$ and $deg(h(x)) = deg(\overline{h_1(x)})$. So $\overline{f(x)}$ is reducible over $\mathbb{Z}_p$, a contradiction. This implies that $f(x)$ is irreducible over $\mathbb{Z}$ and so is over $\mathbb{Q}$. Thus we have proven the theorem. □

Example 17.2.6. *Show that $x^3 + 2x^2 + 1$ is irreducible over $\mathbb{Z}$.*

Solution. Let's use the *mod* 3 test. The image of $x^3 + 2x^2 + 1 \in \mathbb{Z}[x]$ under *mod* 3 map is $x^3 + \overline{2}x^2 + \overline{1} \in \mathbb{Z}_3[x]$. If $x^3 + \overline{2}x^2 + \overline{1}$ is reducible over $\mathbb{Z}$, it has a linear factor over $\mathbb{Z}_3$. Then, there must be a zero of $x^3 + \overline{2}x^2 + \overline{1}$ among $\overline{0}$, $\overline{1}$, and $\overline{2}$ but any of them can not be a zero of $x^3 + \overline{2}x^2 + \overline{1}$. This contradiction shows that $x^3 + 2x^2 + 1$ is irreducible over $\mathbb{Z}$ by the *mod* 3 test. $\square$

17.3 Irreducible Polynomials and Fields

The following theorem explains how to construct a new field out of a given field using an irreducible polynomial and it will work for any characteristic.

Theorem 17.3.1. *Let $f(x)$ be a polynomial over $\mathbb{F}$. Then, $f(x)$ is irreducible if and only if $\mathbb{F}[x]/< f(x) >$ is a field.*

Proof. Let's assume that $f(x)$ is irreducible over $\mathbb{F}$. Since $\mathbb{F}[x]/< f(x) >$ is a commutative ring with unity, it is enough to prove that for every non-zero element $g(x)+ < f(x) >$ has the multiplicative inverse. Since $g(x) \notin < f(x) >$, $g(x) \neq f(x)r(x)$ for some $r(x) \in \mathbb{F}[x]$ and since $f(x)$ is irreducible, $(g(x), f(x)) = 1$. This implies that there exist $s(x), t(x) \in \mathbb{F}[x]$ such that $g(x)s(x) + t(x)f(x) = 1$. Then, $g(x)s(x) + t(x)f(x)+ < f(x) >= 1+ < f(x) >$ and this makes $g(x)s(x)+ < f(x) >= 1+ < f(x) >$. Then, $(g(x)+ < f(x) >)(s(x)+ < f(x) >) = 1+ < f(x) >$ and this means that $g(x)+ < f(x) >$ is invertible. Thus $\mathbb{F}[x]/< f(x) >$ is a field. Conversely let's assume that $f(x)$ is not irreducible. Suppose that $f(x) = g(x)h(x)$ for some non-constant polynomials $g(x)$ and $h(x)$. Then, $(g(x)+ < f(x) >)(h(x)+ < f(x) >) = g(x)h(x)+ < f(x) >=< f(x) >$. This implies that $g(x)+ < f(x) >$ and $h(x)+ < f(x) >$ are zero divisors of $\mathbb{F}[x]/< f(x) >$ since $g(x), h(x) \notin < f(x) >$, a contradiction because $\mathbb{F}[x]/< f(x) >$ is a field. So $f(x)$ is irreducible and this completes the theorem. $\square$

Example 17.3.1. *By Example (17.2.6), $x^3 + 2x^2 + 1$ is irreducible over $\mathbb{Z}$ and $\mathbb{Z}_3$. Then it should be irreducible over $\mathbb{Q}$. Therefore $\mathbb{Q}[x]/< x^3 + 2x^2 + 1 >$ and $\mathbb{Z}_3[x]/< x^3 + \overline{2}x^2 + \overline{1} >$ are fields.*

The following is another equivalent statement for a irreducible polynomial.

Corollary 17.3.1. *Let $f(x)$ be a polynomial over $\mathbb{F}$. Then, $f(x)$ is irreducible if and only if $< f(x) >$ is a maximal ideal.*

Proof. By Theorem 15.4.3, $< f(x) >$ is maximal if and only if $\mathbb{F}[x]/< f(x) >$ is field. By Theorem 17.3.1, $\mathbb{F}[x]/< f(x) >$ is field if and only if $f(x)$ is irreducible. This implies that $f(x)$ is irreducible if and only if $< f(x) >$ is maximal. $\square$

So, we can have these equivalent conditions:

$$\begin{aligned} f(x) \text{ is irreducible over } \mathbb{F} &\iff < f(x) > \text{ is maximal in } \mathbb{F}[x] \\ &\iff \mathbb{F}[x]/ < f(x) > \text{ is a field} \end{aligned}$$

Example 17.3.2. *Show that $\mathbb{Q}[x]/< x^3 + 2x - 6x + 2 >$ is a field and identify all the elements of this field.*

Solution. Since $2 \nmid 1, 2|2, 2|(-6), 2|2$, and $2^2 \nmid 2$, $x^3 + 2x - 6x + 2$ is irreducible by Eisenstein's Criterion. So $\mathbb{Q}[x]/< x^3 + 2x - 6x + 2 >$ is a field. Suppose that $f(x)+ < x^3 + 2x - 6x + 2 > \in \mathbb{Q}[x]/ < x^3 + 2x - 6x + 2 >$ with $deg(f(x)) \geq 3$. Then by the Division Algorithm, there exist $q(x), r(x) \in \mathbb{Q}[x]$ such that $f(x) = (x^3 + 2x - 6x + 2)q(x) + r(x)$ where $r(x) = 0$ or $0 \leq deg(r(x)) < 3$. This implies that $f(x)+ < x^3 + 2x - 6x + 2 > = (x^3 + 2x - 6x + 2)q(x) + r(x)+ < x^3 + 2x -$

$6x+2 >= r(x)+ < x^3+2x-6x+2 >$, which means that all elements in the field $\mathbb{Q}[x]/< x^3+2x-6x+2 >$ have the form $ax^2+bx+c+ < x^3+2x-6x+2 >$ where $a,b,c \in \mathbb{Q}$. □

Example 17.3.2 is an infinite field and the following example is a finite field.

Example 17.3.3. *Show that $\mathbb{Z}_2[x]/ < x^2+x+\overline{1} >$ is a field and identify all elements of this field. Also find the order of the field and the (multiplicative) inverse of $x+\overline{1}+ < x^2+x+\overline{1} >$.*

Solution. Let $I =< x^2+x+\overline{1} >$. Since $\mathbb{Z}_2$ is a field, $\mathbb{Z}_2[x]/I$ is a commutative ring with unity. Since 0 and 1 cannot be zeros of $x^2+x+\overline{1}$, $x^2+x+\overline{1}$ is irreducible over $\mathbb{Z}_2$ and it makes $\mathbb{Z}_2[x]/I$ a field. By Example 17.3.2, for any $l \in \mathbb{Z}_2[x]/I$, $l = ax+b+I$ for $a,b \in \mathbb{Z}_2$. So the order of $\mathbb{Z}_2[x]/I$ is 4. In fact, $\mathbb{Z}_2[x]/I = \{x+\overline{1}+I, x+I, \overline{1}+I, I\}$. The inverse of $x+\overline{1}+I$ is $x+I$ since $(x+\overline{1}+I)(x+I) = (x+\overline{1})x+I = (x^2+x)+I = (x^2+x)+(x^2+x+\overline{1})+I = \overline{1}+I$. □

Example 17.3.4. *Prove that $\mathbb{R}[x]/< x^2+1 >$ is a field and isomorphic to $\mathbb{C}$.*

Proof. The zeros of x^2+1 are i and $-i$. So x^2+1 is irreducible over $\mathbb{R}$ and it makes $\mathbb{R}[x]/< x^2+1 >$ a field. Note that $\mathbb{R}[x]/< x^2+1 > = \{ax+b+ < x^2+1 > | a,b \in \mathbb{R}\}$. Let's define a map $\theta : \mathbb{R}[x]/< x^2+1 > \to \mathbb{C}$ by

$$\theta(ax+b+ < x^2+1 >) = b+ai$$

for $ax+b+ < x^2+1 > \in \mathbb{R}[x]/< x^2+1 >$. Then θ is obviously a bijection and we will that it is a field homomorphism. Suppose $a_1x+b_1+ < x^2+1 >$ and $a_2x+b_2+ < x^2+1 >$ are two elements in $\mathbb{R}[x]/< x^2+1 >$. Then $\theta(a_1x+b_1+ <$

$x^2+1> +a_2x+b_2+ < x^2+1 >) = b_1+b_2+(a_1+a_2)i$. On the other hand, $\theta(a_1x+b_1+ < x^2+1 >)+\theta(a_2x+b_2+ < x^2+1 >) = b_1+b_2+(a_1+a_2)i$. So θ is an additive homomorphism. It also preserves the multiplication because of these two computations:

$$\begin{aligned} & \theta((a_1x+b_1+ < x^2+1 >)(a_2x+b_2+ < x^2+1 >)) \\ = & \theta(a_1a_2x^2+(a_1b_2+a_2b_1)x+b_1b_2+ < x^2+1 >)) \\ = & \theta((a_1b_2+a_2b_1)x+(\text{-}a_1a_2+b_1b_2)+ < x^2+1 >) \\ = & \text{-}a_1a_2+b_1b_2+(a_1b_2+a_2b_1)i \end{aligned}$$

$$\begin{aligned} & \theta(a_1x+b_1+ < x^2+1 >)\theta(a_2x+b_2+ < x^2+1 >) \\ = & (b_1+a_1i)(b_2+a_2i) = (b_1b_2-a_1a_2)+(a_1b_2+b_1a_2)i \end{aligned}$$

So θ is a field isomorphism and it proves the example. $\square$

Example 17.3.5. *Prove that the ideal $< x^2+3 >$ is a maximal ideal of $\mathbb{R}[x]$ and the field $\mathbb{R}[x]/ < x^2+3 >$ is isomorphic to the field $\mathbb{C}$.*

Solution. Since the zeros of x^2+3 are $\pm i\sqrt{3}$, it is an irreducible polynomial over $\mathbb{R}$ and this makes $< x^2+3 >$ a maximal ideal of $\mathbb{R}[x]$. So $\mathbb{R}[x]/< x^2+3 >$ is a field. Let's define a map θ from $\mathbb{R}[x]/< x^2+3 >$ to $\mathbb{C}$ as follows:

$$\theta(ax+b+ < x^2+3 >) = a\sqrt{3}i+b$$

for $ax+b+ < x^2+3 >\in \mathbb{R}[x]/< x^2+3 >$. Then it is easy to check that θ is a field isomorphism. $\square$

Therefore, we can conclude that $\mathbb{R}[x]/ < x^2+1 >\cong \mathbb{C} \cong \mathbb{R}[x]/ < x^2+3 >$.

As matter of fact, $\mathbb{R}[x]/ < f(x) >\cong \mathbb{C}$ if $f(x)$ is irreducible over $\mathbb{R}$ and its degree is 2.

17.4 Prime Numbers and Prime Polynomials

Let $\mathbb{F}$ be a field of any characteristic. A a non-constant polynomial $p(x)$ over $\mathbb{F}$ is called **prime** if $p(x)|r(x)s(x)$ implies $p(x)|r(x)$ or $p(x)|s(x)$. $f(x)$ and $g(x)$ are called **relatively prime** if a common divisor of them is an element of $\mathbb{F}$. Now we start to see why the word "prime" is used for numbers and polynomials.

Proposition 17.4.1. *If $p(x) \in \mathbb{F}[x]$ is a prime polynomial, then $< p(x) > \lhd \mathbb{F}[x]$ is a prime ideal.*

Proof. Suppose that $p(x) \in \mathbb{F}[x]$ is a prime polynomial and $f(x)g(x) \in< p(x) >$. Then $f(x)g(x) = r(x)p(x)$ for some $r(x) \in \mathbb{F}[x]$. This implies that $p(x)$ divides $f(x)g(x)$. Since $p(x)$ is prime, $p(x)$ divides $f(x)$ or $g(x)$. Then $f(x) = f_1(x)p(x)$ or $g(x) = g_1(x)p(x)$. This means that $f(x) \in< p(x) >$ or $g(x) \in< p(x) >$. So $< p(x) >$ is a prime ideal and this proves the proposition. □

An irreducible polynomial is a generalized form of a prime number. The following property reflect the fact precisely.

Corollary 17.4.1. *If $f(x)$ is irreducible over a field $\mathbb{F}$ and $f(x)|g(x)h(x)$, then $f(x)|g(x)$ or $f(x)|h(x)$ where $g(x), h(x) \in \mathbb{F}[x]$.*

Proof. Since $f(x)|g(x)h(x)$, there exists $r(x) \in \mathbb{F}[x]$ such that $f(x)r(x) = g(x)h(x)$. This means that $f(x)r(x)+ < f(x) >= g(x)h(x)+ < f(x) >$ so $<$

$f(x) >= \big(g(x)+ < f(x) > \big)\big(h(x)+ < f(x) > \big)$. Then, since $f(x)$ is irreducible, , $\mathbb{F}[x]/< f(x) >$ is a field by Theorem 17.3.1 and since a field is an integral domain, we can derive that $< f(x) >= g(x)+ < f(x) >$ or $< f(x) >= h(x)+ < f(x) >$. This implies that $g(x) \in < f(x) >$ or $h(x) \in < f(x) >$. Then, there exist $g_1(x), g_2(x) \in \mathbb{F}[x]$ such that $g(x) = f(x)g_1(x)$ or $h(x) = f(x)g_2(x)$. Therefore, we can conclude that $f(x)|g(x)$ or $f(x)|h(x)$. □

Once again, like a factorization of a natural number using prime numbers, a factorization of a polynomial using prime polynomials is unique.

Theorem 17.4.1. (Unique Factorization Theorem)
For a non-zero polynomial $f(x)$ over a field $\mathbb{F}$, $f(x)$ is unit or $f(x) = p_1(x)\cdots p_s(x)$ where $p_1(x), \cdots, p_s(x)$ are prime polynomials and $s \geq 1$. If

$$p_1(x)\cdots p_s(x) = q_1(x)\cdots q_t(x) \tag{17.1}$$

are two prime factorizations of $f(x)$, then $s = t$ and for an appropriate ordering, $p_i(x) = u_i q_i(x)$, $1 \leq i \leq s$ u_i is unit.

Proof. Let us prove the first part of the proposition by induction on $deg(f(x))$. If $deg(f(x))$ is zero or one, then there is noting to prove (why?). Let us assume that by induction, a polynomial $g(x)$ has a factorization when $deg(g(x))$ is less than $deg(f(x))$ where $g(x) \in \mathbb{F}[x]$. If $f(x)$ is prime, then there is nothing to prove. Let us assume that $f(x)$ is not prime. This implies that $f(x) = p(x)q(x)$ where $p(x)$ is prime and $p(x), q(x) \in \mathbb{F}[x]$. Since $deg(p(x)) \geq 1$, we have that $deg(q(x)) < deg(f(x))$. By induction, we have that $q(x) = q_1(x)\cdots q_r(x)$. This implies that $f(x) = p(x)q_1(x)\cdots q_r(x)$. So $f(x)$ has a required factorization of the theorem. Let us assume that (17.1) holds and prove the theorem by induction on s of (17.1). If $s = 1$, then there is noting to prove. Let us assume that the theorem is true when the number of prime factors of $f(x)$ is less than s. By (17.1), $p_1(x)|q_1(x)\cdots q_t(x)$ where $q_1(x),\cdots,q_t(x)$ are primes. By Corollary 17.4.1, we can assume that $p_1(x)$ divides $q_1(x)$. This implies that $q_1(x) = u_1p_1(x)$ where $u_1 \in \mathbb{F}$. We have that $p_1(x)\cdots p_s(x) = u_1p_1(x)q_2(x)\cdots q_t(x)$. By eliminating $p_1(x)$, we have that $p_2(x)\cdots p_s(x) = u_1q_2(x)\cdots q_t(x)$. This implies that by induction, $s = t$ and $p_i(x) = u_iq_i(x)$ where u_i, $2 \le i \le s$, is unit. Thus we have proven the theorem. □

$x^2 - 3$ is irreducible over $\mathbb{Q}$ but it is reducible over $\mathbb{R}$. $x^2 + 3$ is irreducible over $\mathbb{R}$ but it is reducible over $\mathbb{C}$. So we can observe this: if we extend our field, an irreducible polynomial could become reducible. We also ask a question like "Is there an irreducible polynomial over $\mathbb{C}$? If there is one, could we extend $\mathbb{C}$ so that that irreducible polynomial becomes reducible?" The following definition is helpful to answer these type of questions.

Definition 17.4.1. *A field $\mathbb{F}$ is called* **algebraically closed** *if every non-constant polynomial $f(x)$ of $\mathbb{F}[x]$ has a zero in $\mathbb{F}$.*

If there is an irreducible polynomial, we could extend the given field for that polynomial to be reducible. If we can find an irreducible polynomial over this extended field, we could extend it again for this polynomial to be reducible. If you keep extending these fields, there must be a field that no polynomial is irreducible. This field is **algebraically closed** and we know one good example.

The following example is called **the Fundamental Theorem of Algebra** and it was first proved by Gauss in 1799. He also made two other proofs later

(1816, 1849)[1]

Example 17.4.1. (Fundamental Theorem of Algebra) $\mathbb{C}$ *is an algebraically closed field,*

This means that every polynomial of $\mathbb{C}[x]$ is a product of linear polynomials and those polynomials are in $\mathbb{C}[x]$. Actual proof is omitted.[2] The following theorem explains Theorem 17.4.1 over an algebraically closed field.

Theorem 17.4.2. *If $f(x) \in \mathbb{F}[x]$ and $\mathbb{F}$ is an algebraically closed field, then $f(x)$ can be factored as follows:*

$$f(x) = (x - a_1)^{d_1} \cdots (x - a_m)^{d_m} \tag{17.2}$$

where $a_i \neq a_j \in \mathbb{F}$, $1 \leq i \neq j \leq m$, and $d_1, \cdots, d_m \in \mathbb{N}$.

Proof. Theorem 17.4.1 says that $f(x)$ can be factored into prime polynomials over $\mathbb{F}$. Since $\mathbb{F}$ is algebraically closed, only prime polynomials are linear polynomials and it explains (17.2). □

Chapter 17 Exercises.

1. (a) Prove that $3x^2 + 15x + 6$ is reducible over $\mathbb{Z}$.

   (b) Prove that $3x^2 + 15x + 6$ is irreducible over $\mathbb{Q}$.

   (c) Prove that $3x^2 + 15x + 6$ is reducible over $\mathbb{R}$.

2. (a) Prove that $2x^2 - 8x + 20$ is reducible over $\mathbb{Z}$.

[1] These days it is called "classical Fundamental Theorem of Algebra ".
[2] It is even not "algebraic".

(b) Prove that $3x^2 + 15x + 21$ is irreducible over $\mathbb{Q}$.

(c) Prove that $3x^2 + 15x + 21$ is reducible over $\mathbb{C}$.

3. Prove that $x^5 + 3x + 1 \in \mathbb{Z}[x]$ is irreducible over $\mathbb{Z}$. (Hint: Use the mod p-test)

4. Factor the following polynomials over the indicated fields completely.

   (a) $x^2 - 6x - 7$ over $\mathbb{R}$

   (b) $2x^2 - 5x - 12$ over $\mathbb{R}$

   (c) $x^3 + x^2 - 4x - 4$ over $\mathbb{R}$

   (d) $x^3 - x^2 - 5x - 3$ over $\mathbb{R}$

   (e) $x^4 - 16$ over $\mathbb{Z}$

   (f) $x^4 - 16$ over $\mathbb{R}$

   (g) $x^4 - 16$ over $\mathbb{C}$

5. Consider $x^2 + 5 \in \mathbb{R}[x]$.

   (a) Identify all the elements of the ideal $< x^2 + 5 > \lhd \mathbb{R}[x]$.

   (b) Identify all the elements of $\mathbb{R}[x]/< x^2 + 5 >$.

   (c) Prove that $\mathbb{R}[x]/< x^2 + 5 >$ is a ring.

   (d) Prove that $\mathbb{R}[x]/< x^2 + 5 >$ is a field.

   (e) Find the multiplicative inverse of $x+3+ < x^2+5 >$ in $\mathbb{R}[x]/ < x^2+5 >$.

6. Suppose a, b, and p are integers. If p is prime and p divides ab, then prove that p divides a or b.

7. (a) Prove that $x + 2$ is not a unit but it is prime in $\mathbb{R}[x]$.

   (b) Prove that $x + 2$ is irreducible over $\mathbb{R}$.

(c) Prove that $6x+12$ is irreducible over $\mathbb{R}$.

(d) Prove that $6x+12$ is reducible over $\mathbb{Z}$.

8. (a) Prove that $5x+10$ is reducible over $\mathbb{Z}$.

   (b) Prove that $5x+10$ is irreducible over $\mathbb{Q}$.

9. (a) Prove that x^2+6 is irreducible over $\mathbb{Z}$.

   (b) Prove that x^2+6 is irreducible over $\mathbb{Q}$.

   (c) Prove that x^2+6 is reducible over $\mathbb{C}$.

10. Prove that the following polynomials are irreducible over $\mathbb{Q}$.

    (a) $2x^4-3x^2-9x+3$

    (b) $3x^4-3x^2-9x+1$

    (c) x^7-5

    (d) x^3+2x+2

    (e) x^3+2x^2+2x+2

    (f) $7x^4+5x^3+10x^2-15x+35$

11. Prove that for a commutative ring R, if $u \in R$ is unit, then for $a, b \in R$, $<a>=<ub>$.

12. (a) Prove that the ideal $<x^2+5>$ is maximal ideal of $\mathbb{R}[x]$ and $\mathbb{R}/<x^2+5>$ is isomorphic to $\mathbb{C}$.

    (b) Prove that $<x^2+p>$ is a maximal ideal of $\mathbb{R}[x]$ and $\mathbb{R}/<x^2+p>$ is isomorphic to $\mathbb{C}$ where p is prime.

    (c) If d is a square free positive integer, then prove that $\mathbb{R}[x]/<x^2+d>$ is isomorphic to $\mathbb{C}$.

13. (a) Prove that $3x^3 + 4x^2 + 6x + 10$ is primitive.

    (b) Prove that $3x^3 + 4x^2 + 6x + 10$ is irreducible over $\mathbb{Q}$.

    (c) Prove that $\mathbb{Q}[x]/< 3x^3 + 4x^2 + 6x + 10 >$ is a field.

14. (a) Prove that $\mathbb{Z}_3[i]$ is a field where $i = \sqrt{-1}$.

    (b) Prove that $x^2 + 1$ is irreducible over $\mathbb{Z}_3$.

    (c) Prove that $\mathbb{Z}_3[x]/< 4x^2 + 1 >$ is a field.

    (d) Prove that $\mathbb{Z}_3[i]$ and $\mathbb{Z}_3[x]/< 4x^2 + 1 >$ are isomorphic.

15. Prove that $x^5 + x^4 + x^3 + x^2 + x + 1$ is reducible without using Example 17.2.5 directly.

16. (a) Construct a field with 49 elements.

    (b) Construct a field with 125 elements.

    (c) Construct a field with 625 elements.

17. (a) Prove that $\frac{x^5-1}{x-1}$ is irreducible over $\mathbb{Z}$ without using Example 17.2.5 directly.

    (b) Prove that $\frac{x^8-1}{x-1}$ is reducible over $\mathbb{Z}$.

    (c) Prove that $\frac{x^p-1}{x-1}$ is reducible over $\mathbb{Z}$ if p is not prime.

18. Let $p(x)$ be a irreducible in $\mathbb{F}[x]$ where $\mathbb{F}$ is a field. If $p(x) = f(x)g(x)$ and $< p(x) >=< f(x) >$ where $f(x), g(x) \in \mathbb{F}[x]$, prove that $g(x)$ is unit.

19. (a) If any, find all irreducible elements of $\mathbb{Z}$.

    (b) If any, find all irreducible elements of $\mathbb{Q}[x]$.

    (c) If any, find all irreducible elements of $\mathbb{Q}$.

20. (a) Prove that $\mathbb{Z}[\sqrt{-5}]$ is a ring.

(b) Prove that $1+\sqrt{-5}$ is irreducible.

(c) Prove that $1+\sqrt{-5}$ is prime.

21. Consider $f(x) = 3x^3 + 5x^2 + 2x + 1 \in \mathbb{Z}[x]$.

(a) Prove that $f(x)$ is irreducible over $\mathbb{Z}$.

(b) Prove that $\mathbb{Q}[x]/< 3x^3 + 5x^2 + 2x + 1 >$ is a field.

(c) Prove that $\mathbb{Z}_2[x]/< x^3 + x^2 + \overline{1} >$ is a field.

(d) What is the order of $\mathbb{Z}_2[x]/< \overline{3}x^3 + \overline{5}x^2 + \overline{2}x + \overline{1} >$?

22. Let $p(x)$ be a non-constant polynomial over a field $\mathbb{F}$. If $< p(x) >$ is a prime ideal of $\mathbb{F}[x]$, then prove that $p(x)$ is a prime element of $\mathbb{F}$.[3]

23. (a) Factor $x^3 - 3x^2 - 2x + 6$ over $\mathbb{Z}$ using irreducible factors. Make sure the factors are irreducible.

(b) Factor $x^3 - 3x^2 - 2x + 6$ over $\mathbb{R}$ using irreducible factors.

24. (a) Prove that 5 is a prime element of $\mathbb{Z}$.

(b) Prove that 5 is a prime element of $\mathbb{Z}[x]$.

(c) Prove that $x^2 + 4$ is a prime element of $\mathbb{Z}[x]$.

(d) Prove that $6x - 12$ is not a prime element of $\mathbb{Z}[x]$.

(e) Prove that $x^2 - 6x - 7$ is not a prime element of $\mathbb{Z}[x]$.

[3]Now we can say that $p(x)$ is a prime polynomial if and only if $< p(x) >$ is a prime ideal by Proposition 17.4.1.

Chapter 18

Euclidean Domains

In Chapter 14, we have studied the Division Algorithm on polynomial rings (see Theorem 14.2.1). In this chapter, we'd like to extend polynomial rings so that the Division Algorithm is still applicable. When we divide a number by a divisor, the number must be "bigger" then the divisor. Otherwise, the quotient is 0. So we need a ring with a "proper" order so that we can compare a dividend and a divisor in some sense. Good examples of this type of order are the degrees of polynomials because we compare their degrees first when we divide one polynomial by another.

We will generalize this concept to the definition of **Euclidean domain**. We will also prove that every Euclidean domain is a principle ideal domain (**PID**) using the Well Ordering Principal.

18.1 Definition and Examples

Definition 18.1.1. *An integral domain R is called a* **Euclidean domain** *if there is a map $d : R - \{0\} \to \mathbb{N}$ such that for all $a, b \in R$*

1. $d(a) \leq d(ab)$ *and*

2. *there are* $q, r \in R$ *such that* $a = bq + r$ *where* $r = 0$ *or* $d(r) < d(b)$.

Note that we do not assign a value of 0.[1] We call this map d a **Euclidean map** or a **norm**[2].

Example 18.1.1. $\mathbb{Z}$ *is a Euclidean domain since the absolute value function is obviously a Euclidean map.*

Example 18.1.2. $\mathbb{R}[x]$ *is a Euclidean domain since the degree function is another obvious example of a Euclidean map using the Division Algorithm of* $\mathbb{R}[x]$.

Example 18.1.3. *The Gaussian integers* $\mathbb{Z}[i]$ *is a Euclidean domain.*

Solution. For any $a + bi \in \mathbb{Z}[i]$, let us define the map $d : \mathbb{Z}[i] \to \mathbb{N}$ as follows:

$$d(a + bi) = a^2 + b^2.$$

Then we have the following for any $a_1 + b_1 i, a_2 + b_2 i \in \mathbb{Z}[i]$:

1. $d((a_1 + b_1 i)(a_2 + b_2 i)) = d((a_1 + b_1 i))d((a_2 + b_2 i)) \geq d((a_1 + b_1 i))$

2. If $a_2 + b_2 i \neq 0$, then $\frac{a_1+b_1 i}{a_2+b_2 i} \in \mathbb{Q}[i]$. Let us put $\frac{a_1+b_1 i}{a_2+b_2 i} = u + vi$. If $u + vi \in \mathbb{Z}[i]$, then

$$a_1 + b_1 i = (a_2 + b_2 i)(u + vi), \tag{18.1}$$

i.e., we take $u + vi$ as the quotient and 0 as the remainder. So we have solved the example. If $u + vi \notin \mathbb{Z}[i]$, then we take $m, n \in \mathbb{Z}$ so that

[1] We don't want to use 0 for a divisor.

[2] This term "norm" is usually used in Linear Algebra

$|m-u| \leq \frac{1}{2}$ and $|n-v| \leq \frac{1}{2}$. Then we have that $u+vi = m-m+u+(n-n+v)i = m+ni+(u-m+(v-n))i$. So by (18.1), we have that $a_1+b_1i = (a_2+b_2i)(u+vi) = (m+ni)(a_2+b_2i)+(u-m+(v-n)i)(a_2+b_2i)$. This implies that $(u-m+(v-n)i)(a_2+b_2i) \in \mathbb{Z}[i]$ and we have that $d((u-m+(v-n)i)(a_2+b_2i)) = d(u-m+(v-n)i)d(a_2+b_2i) = ((u-m)^2+(v-n)^2)d(a_2+b_2i) \leq (\frac{1}{4}+\frac{1}{4})d(a_2+b_2i) < d(a_2+b_2i)$.

This implies that d is a Euclidean map and it makes $\mathbb{Z}[i]$ a Euclidean domain.
□

Theorem 18.1.1. (ED $\Longrightarrow$ PID) *If R is a Euclidean domain, then it is a principle integral domain.*

Proof. Suppose I be an ideal of a Euclidean domain R and let $d : R-\{0\} \to \mathbb{N}$ be a Euclidean map of R. If I is the zero ideal, then there is nothing to prove so let's assume that I is a non-zero ideal. Then there must be an element $l \in I$ such that $d(l) \leq d(x)$ for all $x \in I$ by the Well Ordering Principal. Let $< l >$ be the ideal of R generated by l. Obviously $< l > \subset I$. Note that for any $x \in I$, there are $q, r \in R$ such that $x = lq + r$ with $r = 0$ or $d(r) < d(l)$. If $d(r) < d(l)$, then $r = x - lq \in I$ and it implies that $r \in I$. This contradicts the choice of l. So we have that $r = 0$, i.e., $x = lq \in < l >$ and this makes $I \subset < l >$. Since an arbitrary ideal $I = < l >$, R is a principal integral domain. □

Theorem 18.1.1 says that "if R is a Euclidean domain, then for any $I \lhd R$, there exists an element $r \in R$ such that $< r > = I$" and it is why the usual division works in $\mathbb{Z}$ which is a PID.

18.2 Divisions In Rings

Before we consider a division on a Euclidean Domain, we need to re-visit several terms including divisibility and irreducibility.[3] We need to define a division on a ring without defining it directly because a ring requires an addition and a multiplication, not a division. It will be defined from multiplication and we'd like to keep the same or similar properties of prime numbers of $\mathbb{Z}$ in rings.

Definition 18.2.1. *Let a be a non-zero element of a commutative ring R. We say that a* **divides** *$b \in R$ if there is an element $c \in R$ such that $b = ac$ and we denote it as $a|b$. So, $a \nmid b$ means that a does not divide b. A non-zero and non-unit element a is called* **irreducible** *if $a = bc$ implies that b or c is a unit.*

This division in a commutative ring is the same as one in $\mathbb{Z}$ or a polynomial ring. We can see it through the following examples and definition.

Example 18.2.1. 5 *divides* 30 *in the ring* $\mathbb{Z}$ *since* $30 = 6 \cdot 5$ *so we can write or* $5|30$. *Obviously* $7 \nmid 30$.

Definition 18.2.2. *For a, b of a commutative ring R, $d \in R$ is called* **a common divisor of** *a* **and** *b if $d|a$ and $d|b$.*

Example 18.2.2. 7 *is a common divisor of* 35 *and* 42*, and* 5 *is not a common divisor of them.*

---

[3]They were defined on $\mathbb{Z}$ in Chapter 1 and on polynomial rings in Chapter 14.

Definition 18.2.3. *$d \in R$ is called* **a greatest common divisor of** a **and** b *of a commutative ring R if*

1. *d is a common divisor of a and b, and*

2. *$d_1|a$ and $d_1|b$ implieas that $d_1|d$ where $d_1 \in R$.*

Sometimes this d is denoted by **gcd(a,b)** *or simply* **(a,b)**.

Lemma 18.2.1. *Let R be a Euclidean domain. For $a, b \in R$, there exists a greatest common divisor (a, b) of a and b and there exist $s, t \in R$ such that $as + bt = (a, b)$.*

Proof. This proof is similar to the proofs of Theorem 1.1.2 and Theorem 14.2.2. Let a and b be elements in a Euclidean domain R and let d be a Euclidean map on R. Also let $S = \{as + bt|s, t \in R\}$ and $d(S) = \{d(as + bt)|s, t \in R\}$. Then there is $v \in S$ such that $d(v)$ is the minimum element of $d(S)$ by the Well-Ordering Principle. Note that a and b are elements of S. First we show that v divides a. Suppose that $a = vq + r$ with $r = 0$ or $d(r) < d(v)$ for some q and r of R. If $r = 0$, then v divides a. Assume that $r \neq 0$. Since we have that $r = a - vq = a - (as_1 + bt_1) = a(1 - v) + b(-t_1) \in S$, r is a non-zero element of S such that $d(r) < d(v)$, a contradiction to the choice of v. This implies that v divides a. Similarly we are able to prove that v is a divisor of b. Now we will show that v is the greatest common divisor of a and b. Suppose w is a common divisor of a and b. This implies that $a = ww_1$ and $b = ww_2$ for some $w_1, w_2 \in R$. This implies that $v = as_1 + bt_1 = ww_1s_1 + ww_2t_1 = w(w_1s_1 + w_2t_1)$. So we have that w divides v and this makes v be the greatest common divisor of a and b. Thus we have proven the lemma. □

The division we define here is little bit different from the one in $\mathbb{Z}$ and the following is such a property.

Lemma 18.2.2. *Let R be an integral domain. For $a, b \in R$, $a|b$ and $b|a$ if and only if $a = bu$ for some unit u of R.*

Proof. If $a|b$ and $b|a$, then $b = ax_1$ and $a = bx_2$ for some $x_1, x_2 \in R$. We have that $b = ax_1 = bx_2x_1$. This implies that $1 = x_1x_2$ and this shows that x_1 and x_2 are units. Conversely if $a = ub$ for some unit $u \in R$, then $u^{-1}a = b$. This shows that $b|a$ and $a|b$ respectively. □

This relation in Lemma 18.2.2 is a useful tool to define "a prime" (again).

Definition 18.2.4. *Two elements a and b of a commutative ring R are called* **associates** *if $a = ub$ for some unit $u \in R$. An element $p \in R$ is called* **prime** *if $p|ab$ implies that $p|a$ or $p|b$.*

Example 18.2.3. *For the ring $\mathbb{Z}$, 3 and -3 are associate, and 6 and 7 are not associate. 11 is prime (why?).*

If we think about a prime number, it is "irreducible" because there is no other factors other than 1 and itself, which are trivial factors. It is true not only in $\mathbb{Z}$ but also in an integral domain.

Proposition 18.2.1. *Let R be an integral domain and p be a non-zero element of R. If p is prime, then it is irreducible.*

Proof. Let p be a non-zero element of an integral domain R. Suppose $p = ab$ for some $a, b \in R$. Since p divides itself, p divides ab. This implies that $p|a$ or $p|b$. Without loss of generality, let's assume that $p|a$. Then $a = pu$ for some $u \in R$ and this makes that $p = ab = pub$. This implies that $p = pub$ and $p(1 - ub) = 0$. Since p is prime, it is a non-zero element. This means that $1 - ub = 0$ since R is an integral domain and this implies that b is unit. Therefore p is irreducible. □

Example 18.2.4. *All prime numbers in $\mathbb{Z}$ are prime. It is also irreducible since $\mathbb{Z}$ is an integral domain.*

Definition 18.2.5. *Two elements a and b of a Euclidean domain R are called* **relatively prime** *if $gcd(a, b)$ is a unit of R.*

Lemma 18.2.3. *Let a, b, and c are elements in a Euclidean domain. If $a|(bc)$ and $gcd(a, b) = 1$, then $a|c$.*

Proof. Since $gcd(a, b) = 1$, there exist $s, t \in R$ such that $as + bt = 1$ by Lemma 18.2.1. This makes that $asc + btc = c$ and since $a|(bc)$ and $a|(asc)$, $a|c$. □

Proposition 18.2.2. *Suppose p is a prime element of a Euclidean domain R. If p divides a product $a_1 \cdots a_n$, then p divides at least one a_i, $1 \leq i \leq n$.*

Proof. Let us prove the proposition by induction on n of the element $a_1 \cdots a_n$. If $n = 1$, then there is nothing to prove. Let us assume that the proposition holds for any element $a_1 \cdots a_k$ of R where $k > 1$. Suppose that p divides $a_1 \cdots a_{k+1}$.

If p divides a_{k+1}, then there is nothing to prove. Let us assume that p does not divide a_{k+1}. This makes $(p, a_{k+1}) = 1$ and by Lemma 18.2.1, there exist $s, t \in R$ such that $ps + a_{k+1}t = 1$. This implies that $psa_1 \cdots a_k + a_1 \cdots a_k a_{k+1} t = a_1 \cdots a_k$. This implies that p divides $a_1 \cdots a_k$. So by induction, there is a_h, $1 \leq h \leq k$, such that p divides a_h. Thus we have proven the proposition. □

Corollary 18.2.1. *If p is a prime element of a Euclidean domain R and p divides ab, then $p|a$ or $p|b$ where $a, b \in R$.*

Theorem 18.2.1. *Let R be a Euclidean domain and a be a non-zero and non-unit in R. If $a = p_1 \cdots p_n = p'_1 \cdots p'_m$ where $p_1, p_2, \cdots, p'_{m-1}, p'_m$ are primes of R, then $n = m$ and each p_i is associate to some p'_j. (This means that each p'_j is associate to some p_i.)*

Proof. By assumption, we have that $p_1 \cdots p_n = p'_1 \cdots p'_m$. Since p_1 is prime, we have that $p_1 | (p'_1 \cdots p'_m)$. By Proposition 18.2.2, $p_1 | p'_j$ for some j. This implies that $p'_j = u_1 p_1$ where u_1 is unit. This implies that $p_2 \cdots p_n = u_1 p'_1 \cdots p'_{j-1} p'_{j+1} \cdots p'_m$. Consecutively we can cancel all $p_2, \cdots, p_n$, so $n \leq m$. Similarly we can prove that $m \leq n$ and this makes $m = n$. We have also proven that $p'_j = u_k p_k$ where u_k is unit for some k and this proves the theorem. □

Theorem 18.2.2. *Let R be a Euclidean domain. For a non-zero element p of R, the ideal $< p >$ of R is maximal if and only if p is prime.*

Proof. Let us assume that p is not prime. This means that $p = ab$ holds for some non-units $a, b \in R$ and this implies that $p \in < a >$. Let $I = < p >$, the ideal of R generated by p. At this point $< a > \supseteq I$. If $< a > = I$, then $a \in < p > = I$. This implies that $a = ps$ so $a = ps = abs$. Then by cancellation we have that $bs = 1$ and it makes that b is a unit, a contradiction. So, $< a > \supsetneq I$. Now assume that $< a >= R$. Then $1 \in < a >$ and this implies that $1 = ax$ for some $x \in R$. This makes that a is unit and we have another contradiction. So $R \supsetneq < a > \supsetneq I$ and this implies that I is not maximal. Thus, we prove that if $< p >$ is maximal, p is prime. Conversely let us assume that p is prime. Suppose M is an ideal of R such that $I \subset M \subset R$. Since R is a Euclidean domain, $M =< a >$ for some $a \in R$. We have that $< p > \subset < a > \subset R$. This implies that $p \in < a >$. We have that $p = au$ for some $u \in R$. Since p is prime, a is unit or u is unit. If a is unit, then $M =< a >= R$. Let u be an unit. $p = au$. We have that $I \subset M$. Since u is unit, we have that $pu^{-1} = a$. This implies that $a \in < p >$. So we have that $< a > \subset < p >$ and this implies that $M \subset I$. Since we have derived that $M = I$, I is maximal and therefore we have proven the theorem. □

Example 18.2.5. *Since $x + 5$ is prime in $\mathbb{R}[x]$, $< x + 5 >$ is a maximal ideal of $\mathbb{R}[x]$.*

Example 18.2.6. *Since 5 is prime in $\mathbb{Z}$, $< 5 >$ is a maximal ideal of $\mathbb{Z}$.*

Example 18.2.7. *Since 6 is not prime in $\mathbb{Z}$, $< 6 >$ is not a maximal ideal of $\mathbb{Z}$. In fact, $< 6 > \subsetneq < 2 > \lhd \mathbb{Z}$. Note that $< 2 >$ a maximal ideal since 2 is prime.*

Corollary 18.2.2. *Let R be a Euclidean domain and p be a non-zero element of R. Then $R/< p >$ is field if and only if p is prime.*

Chapter 18 Exercises.

1. Let's consider $\mathbb{Z}$.

   (a) Is 11 prime? irreducible? 12 prime? Justify your answers.

   (b) Is 12 irreducible?

   (c) Find all primes of $\mathbb{Z}$.

   (d) Find all irreducibles of $\mathbb{Z}$.

2. Let a, b, and c are elements in a ring R.

   (a) Prove that if $a|b$ and $b|c$, then $a|c$.

   (b) Prove that if $a|b$ and $a|c$, then $a|(b+c)$ and $a|(b-c)$.

   (c) Prove that $a|ax$ for any $x \in R$.

   (d) Prove or disprove the following statement: $(a, b) = 1$ if and only if $(a, bc) = 1$. Note that a counter-example will disprove the statement.

3. (a) In the ring $\mathbb{Z}$, find all the elements that are associates to 7.

   (b) In the ring $\mathbb{Z}$, show that 4 and 9 are not associates

4. Prove the following rings are Euclidean domains.

   (a) $\mathbb{Z}$

   (b) $\mathbb{Q}[x]$

   (c) $\mathbb{R}[x]$

   (d) $\mathbb{C}[x]$

5. (a) Prove that $2\mathbb{Z}[i]$ is a subring of $\mathbb{Z}[i]$.

   (b) Prove that $2\mathbb{Z}[i]$ is an ideal of $\mathbb{Z}[i]$.

(c) Prove or disprove that $2\mathbb{Z}[i]$ is a Euclidean domain.

6. (a) Find $a+bi, u+vi \in \mathbb{Z}[i]$ such that $8+4i = (3-2i)(a+bi)+u+vi$ and $d(u+vi) < d(3-2i)$ where d is the norm of $\mathbb{Z}[i]$.

   (b) Find $a+bi, u+vi \in \mathbb{Z}[i]$ such that $10+7i = (1-2i)(a+bi)+u+vi$ and $d(u+vi) < d(3-2i)$.

7. Let R be a Euclidean domain, prove that if u is unit of R, then $d(u) = 1$ where d is the Euclidean map.

8. Let R be a Euclidean domain.

   (a) Prove that if r is a prime element of R, then r is irreducible.

   (b) Prove that if r is an irreducible element of R, then r is prime.

   (c) If r is irreducible, prove that the ideal $< r >$ of R generated by r is maximal.

Chapter 19

Factorizations in an Integral Domain

We have studied factorizations of polynomials in Chapter 17. Since a polynomial ring is an integral domain, it is natural to generalize a factorization on polynomials to a factorization on elements of an integral domain and that is what we will see in this chapter.

19.1 Definitions and Examples

Since our main interest is integral domains, we'd like to re-visit Definition 18.2.4 of commutative rings for integral domains.

Definition 19.1.1. *Let a and b be elements of an integral domain D.*

1. *a and b are called* **associate** *if $a = bu$ for a unit $u \in D$.*
2. *A non-zero, non-unit a is called* **irreducible** *if $a = bc$ implies that b or c is a unit of D.*

3. p is called **prime** *if $p|ab$ implies that $p|a$ or $p|b$ in D.*

Example 19.1.1. *All the units of $\mathbb{Z}[i]$ are 1, -1, i, and $-i$ (why?).*

Example 19.1.2. *A prime number of $\mathbb{Z}$ is prime and irreducible.*

Example 19.1.3. *7 is associate to -7 and 6 is not associate to 12 in $\mathbb{Z}$.*

Lemma 19.1.1. *If D is an integral domain, then a prime element of D is irreducible.*

Proof. Suppose a is prime and $a = bc$ for $b, c \in D$. Since a is prime and $a|bc$, we have that $a|b$ or $a|c$. If $a|b$, then $b = ar$ for $r \in D$. This implies that $a = bc = arc$ so $a(1 - rc) = 0$. Since $a \neq 0$, $1 = rc$ so c is unit. If $a|c$, then similarly we can prove that b is unit. This implies that a is irreducible. □

The other way is not always true in an integral domain but it is true in a PID.

Theorem 19.1.1. *If D is a principal ideal domain, then a non-zero element $a \in D$ is prime if and only if a is irreducible and a is not a unit.*

Proof. If $a \in D$ is prime, then it is irreducible by Lemma 19.1.1. Conversely, suppose that $a \in D$ is irreducible. Let us assume that $a|bc$ holds. Let I be the ideal of D generated by a and b, i.e., $I = \{ax + by|x, y \in D\}$. Since D is a principal ideal domain, there is $d \in D$ such that $I =< d >$ and this implies that $a = du$ for $u \in D$. Since a is irreducible, d or u is unit. If d is unit, then

$I = D$ and this implies that there are $x_1, y_1 \in D$ such that $1 = ax_1 + by_1$. Then we can derive that $c = acx_1 + bcy_1$. Since $a|a$ and $a|bc$, we have that $a|c$. Now, let us assume that u is unit. Then $< a >=< d >$ and since $b \in< d >$, we have that $b \in< a >$. This means that $b = au_1$ for $u_1 \in D$ and this makes $a|b$. So a is prime and this completes the proof of the theorem. □

The following is a counter example of an irreducible element that is not prime in an integral domain. Let d be a square free integer and consider $\mathbb{Z}[\sqrt{d}] = \{a + b\sqrt{d} | a, b \in \mathbb{Z}\}$. We can define a norm N on $\mathbb{Z}[\sqrt{d}]$ as follows:

$$N : \mathbb{Z}[\sqrt{d}] \to \mathbb{N} \tag{19.1}$$

with $N(a + b\sqrt{d}) = |a^2 - b^2 d|$ for $a + b\sqrt{d} \in \mathbb{Z}[\sqrt{d}]$. Then the following hold:

1. $x = 0$ if and only if $N(x) = 0$ where $x \in \mathbb{Z}[\sqrt{d}]$.
2. $N(xy) = N(x)N(y)$ for $x, y \in \mathbb{Z}[\sqrt{d}]$.
3. x is unit if and only if $N(x) = 1$ for $x \in \mathbb{Z}[\sqrt{d}]$.
4. If $N(x)$ is prime, then x is irreducible for $x \in \mathbb{Z}[\sqrt{d}]$.

Using this N, we can consider the following examples.

Example 19.1.4. *Prove that* $2 - \sqrt{-7}$ *of* $\mathbb{Z}[\sqrt{-7}]$ *is irreducible.*

Solution. Let's say that $2 - \sqrt{-7} = xy$ for $x, y \in \mathbb{Z}[\sqrt{-7}]$. Since $N(2 - \sqrt{-7}) = N(xy) = N(x)N(y) = 11$, $N(x)$ or $N(y)$ is 1. This implies that x or y is unit. Thus $2 - \sqrt{-7}$ is irreducible. □

Example 19.1.5. *Prove that* $1 - \sqrt{-3}$ *of* $\mathbb{Z}[\sqrt{-3}]$ *is irreducible but it is not prime.*

Solution. Note that $N(1-\sqrt{-3}) = 4$. Let's assume that $1-\sqrt{-3} = xy$. Since $4 = N(1-\sqrt{-3}) = N(xy) = N(x)N(y)$, $N(x) = 1, 2$, or 4. If $N(x) = 1$, then x is unit. If $N(x) = 4$, then y is unit. So let's assume that $N(x) = 2$ and set $x = a + b\sqrt{-3}$. Then $N(x) = a^2 + 3b^2 = 2$ but there are no integers that holds $a^2 + 3b^2 = 2$, a contraction. This proves that $1-\sqrt{-3}$ is irreducible. Now, we will show that $1-\sqrt{-3}$ is not prime. Note that $1-\sqrt{-3}$ divides 4 since $(1-\sqrt{-3})(1+\sqrt{-3}) = 4 = 2 \cdot 2$. If $1-\sqrt{-3}$ is prime, then $1-\sqrt{-3}$ divides 2. This implies that $2 = (1-\sqrt{-3})(a + b\sqrt{-3})$. Since $2 = a - 3b + (a+b)\sqrt{-3}$, we can derive that $b = -\frac{1}{2} \notin \mathbb{Z}$, a contradiction. This shows that $1-\sqrt{-3}$ is not prime. $\square$

Examples 19.1.4 and 19.1.5 show us that the concepts of irreducible and prime are different and it depends on the ground ring.

19.2 ED, PID, & UFD

ED, PID, & UFD use different concepts: distance, ideal, and factorization. But somehow they are related and we'd like to investigate their relations in this section.

Definition 19.2.1. *An integral domain D is a* **unique factorization domain (UFD)** *if the following hold:*

1. *All non-zero and non-unit elements of D are products of irreducibles of D.*
2. *By taking appropriate order of the factorization of irreducibles, the product is unique up to associates of the factors.*

A simplest example of a UFD is $\mathbb{Z}$. In $\mathbb{Z}$, prime numbers are irreducible, and any number can be factorized using prime numbers only. Moreover, this factorization is unique.

This UFD is probably a member of largest collections of rings that don't require a division[1] so an ED or a PID is a UFD. Before we show them, we'll take a look at a couple of properties of PID's.

Theorem 19.2.1. *If D is a principal ideal domain and $I_1 \subset I_2 \subset \cdots$ is a series of ideals of D, then there is $n \in \mathbb{N}$ such that $I_n = I_{n+1} = \cdots$.*

Proof. Let us put $I = \cup_{i\in\mathbb{N}} I_i$. Then I is an ideal of D,[2] a PID, so there is $d \in D$ such that $< d >= I$. Since $d \in I_n$ for some $n \in \mathbb{N}$, we have that $< d > \subset I_n \subset I =< d >$ and this implies that $I_n = I$. □

Proposition 19.2.1. *If a is irreducible in a PID D and $a = bc$, then $< a >=< b >$ or $< a >=< c >$.*

Proof. By $a = bc$, b or c is a unit since a is irreducible. Without loss of generality, we can assume that b is a unit. Since $a = bc$, $< a > \subset < c >$ and since b is a unit, $b^{-1}a = c$. This implies that $< c > \subset < a >$. So $< a >=< c >$ and this completes the proof. □

Theorem 19.2.2. (PID$\Longrightarrow$UFD)

If D is a principal ideal domain, then D is a unique factorization domain.

Proof. Let a_0 be a non-zero and non-unit element of D. Let us show that a_0 has a factor which is irreducible. If a_0 is irreducible, then there is nothing

[1] At least not directly. If a division is defined directly, it is a division ring and a commutative division ring is a field.

[2] See Exercise 6.

to prove. Let us assume that a_0 is not irreducible and $a_0 = a_1 b_1$ for non-units $a_1, b_1 \in D$. If a_1 is irreducible, then we have proven the statement. If a_1 is not irreducible, $a_1 = a_2 b_2$ for non-units $a_2, b_2 \in D$ where $a_2 \neq 0$. If a_2 is irreducible, then we have proven the statement. Otherwise, we can assume that a_2 is not irreducible. So inductively we can assume that $a_n = a_{n+1} b_{n+1}$ where a_{n+1} and b_{n+1} are not units. This implies that

$$< a_0 > \subset < a_1 > \subset < a_2 > \subset \cdots \subset < a_n > \subset < a_{n+1} > \subset \cdots \tag{19.2}$$

is a strictly increasing chain of ideals of R. By Theorem 19.2.1, there exists $k \in \mathbb{N}$ such that $< a_k > = < a_{k+1} > = \cdots$. This implies that a_k is an irreducible factor of a_0. So a_0 has an irreducible factor. So we have that $a_0 = p_1 c_1$ where p_1 is irreducible and c_1 is not a unit. If c_1 is not unit, $c_1 = p_2 c_2$ where p_2 is irreducible and c_2 is not unit. Whether c_2 is irreducible or not, inductively as (19.2), we have that that $< a_0 > \subset < c_1 > \subset < c_2 > \subset \cdots$. By Theorem 19.2.1, there is r such that $< c_r > = < c_{r+1} > = \cdots$. Then c_r is irreducible, i.e., it is an irreducible factor of a_0. This implies that $a_0 = p_1 \cdots p_r c_r$. So a_0 is a product of irreducibles. Let us assume that a_0 can be written as follows:

$$a_0 = p_1 \cdots p_r = q_1 \cdots q_s \tag{19.3}$$

where $p_1, \cdots, p_r, q_1, \cdots, q_s$ are irreducibles. Let us prove the theorem by induction on r. If $r = 1$, then $a = p_1 = q_1$, then we have proven the theorem. Let us assume that every element that can be written as a product of less than r irreducibles has a unique factorization up to associates well orderly. Let us assume that a_0 can be written as (19.3). Since p_1 divides $p_1 \cdots p_r$, p_1 divides $q_1 \cdots q_s$. This implies that p_1 divides q_i for $1 \leq i \leq s$. Let us put $i = 1$. This implies that $q_1 = u_1 p_1$ where u_1 is unit. We have that $p_1 \cdots p_r = u_1 p_1 q_2 \cdots q_s$ and this implies that $p_2 \cdots p_r = u_1 q_2 \cdots q_s$, i.e., two factorizations are the same unique up to associates by induction. Therefore, D is a UFD. □

Corollary 19.2.1. (ED⟹UFD)

If D is an Euclidean domain, then D is a unique factorization domain.

Proof. If D is an Euclidean domain, it is a PID by Theorem 18.1.1 and then it is a UFD by Theorem 19.2.2. □

Example 19.1.5 show that $\mathbb{Z}[\sqrt{-3}]$ is not a UFD. So it is not a PID nor an ED. Now we have the following relation:

$$\text{ED} \Longrightarrow \text{PID} \Longrightarrow \text{UFD}$$

Chapter 19 Exercises.

1. Prove the properties 1 through 4 of the norm (19.1) in $\mathbb{Z}[d]$.

2. Let a and b are elements of a commutative ring R with unity.

   (a) Let $a \sim b$ if and only if a is associate to b. Prove that $\sim$ is an equivalence relation.

   (b) If $a \sim b$, prove that $< a >=< b >$.

3. Show that $\mathbb{Z}$ is an ED.[3] So, it is a PID and therefore a UFD.

4. (a) Find an associate of $3 - 6i$ in $\mathbb{Z}[i]$. Let's call it a.

[3]See Example 18.1.1.

(b) Prove that $< a >=< 3 - 6i >$ in $\mathbb{Z}[i]$.

5. (a) Identify all the irreducible elements of $\mathbb{Z}$.

   (b) Identify all the primes of $\mathbb{Q}[x]$.

   (c) If any, find all the irreducibles of $\mathbb{R}$.

6. (a) If D is a PID and $I_1 \subset I_2 \subset \cdots$ is a series of ideals of D, then prove that $\cup_{i \in N} I_i$ is an ideal of D.

   (b) Is it true if D is an arbitrary ring, not a PID?

   (c) Let I_1 and I_2 be ideals of a ring R. Is $I \cup I_2$ an ideal of R? Justify your claim.

7. (a) Prove that $1 - \sqrt{-3} \in \mathbb{Z}[\sqrt{-3}]$ is irreducible in $\mathbb{Z}[\sqrt{-3}]$.

   (b) Prove that $1 - \sqrt{-3} \in \mathbb{Z}[\sqrt{-3}]$ is not a prime in $\mathbb{Z}[\sqrt{-3}]$.

8. Prove that 3 is irreducible in $\mathbb{Z}[\sqrt{5}]$.

9. Prove that $2\mathbb{Z}[x]$ is not a PID.

10. (a) Let R be a commutative ring with unity and $a, b \in R$. If b is irreducible in R and $b \in< a >$, then prove that $< a >=< b >$ or $< a >= R$.

    (b) Let R be an integral domain and $a, b \in R$. If $< a >=< b >$, then prove that a and b are associates.

Chapter 20

Vector Spaces

A vector space is a very important algebraic structure to analyze many problem in physical sciences, social sciences, engineering, economics, etc. In this chapter, $\mathbb{F}$ will denote a field of any characteristic, zero or p. Recall that a field $\mathbb{F}$ contains 0 and 1 as the additive and multiplicative identity respectively. We will start this chapter with definitions of a vector space and its subspace.

20.1 Definition and Examples

Definition 20.1.1. *A* **vector space** $\mathbf{V}$ *over* $\mathbb{F}$ *is a set with two binary operations: an addition* $+$ *between* $\mathbf{v}_1, \mathbf{v}_2 \in \mathbf{V}$ *denoted by* $\mathbf{v}_1 + \mathbf{v}_2 \in \mathbf{V}$*, and a scalar multiplication between* $\mathbf{v}_1 \in \mathbf{V}$ *and* $a \in \mathbb{F}$ *denoted by* $a\mathbf{v}_1 \in \mathbf{V}$ *satisfying the following properties:*

1. $\mathbf{v}_1 + \mathbf{v}_2 = \mathbf{v}_2 + \mathbf{v}_1$ **(commutativity)**

2. $\mathbf{v}_1 + (\mathbf{v}_2 + \mathbf{v}_3) = (\mathbf{v}_1 + \mathbf{v}_2) + \mathbf{v}_3$ **(associativity for addition)**

3. *There is* $\mathbf{0} \in \mathbf{V}$ *such that* $\mathbf{v}_1 + \mathbf{0} = \mathbf{0} + \mathbf{v}_1$ **(additive identity)**

4. *There is* $-\mathbf{v}_1 \in \mathbf{V}$ *such that* $\mathbf{v}_1 + (-\mathbf{v}_1) = (-\mathbf{v}_1) + \mathbf{v}_1 = 0$ **(additive inverse)**

5. $a(\mathbf{v}_1 + \mathbf{v}_2) = a\mathbf{v}_1 + a\mathbf{v}_2$ **(distributive property using a scalar)**

6. $(a + b)\mathbf{v}_1 = a\mathbf{v}_1 + b\mathbf{v}_1$ **(distributive property using a vector)**

7. $(ab)\mathbf{v}_1 = a(b\mathbf{v}_1)$ **(associativity for multiplication)**

8. $1\mathbf{v}_1 = \mathbf{v}_1$ **(multiplicative identity)**

for $\mathbf{v}_1, \mathbf{v}_2, \mathbf{v}_3 \in \mathbf{V}$ *and* $a, b, 1 \in \mathbb{F}$. *An element of* $\mathbf{V}$ *is called* **a vector** *and an elements of* $\mathbb{F}$ *is called* **a scalar**.

By Definition 20.1.1, a vector space $\mathbf{V}$ is an additive Abelian group, not a ring.

Example 20.1.1. $\mathbb{R}^n = \{(a_1, \cdots, a_n) | a_u \in \mathbb{R}, 1 \leq u \leq n\}$ *is a vector space if we define the addition and multiplication as follows:*

$$\begin{aligned} (a_1, \cdots, a_n) + (b_1, \cdots, b_n) &= (a_1 + b_1, \cdots, a_n + b_n) \\ c(a_1, \cdots, a_n) &= (ca_1, \cdots, ca_n) \end{aligned}$$

for any $\mathbf{v}_1 = (a_1, \cdots, a_n), \mathbf{v}_2 = (b_1, \cdots, b_n) \in \mathbb{R}^n$ *and* $c \in \mathbb{R}$.

Solution. The operations defined above are well-defined since each coordinate in $\mathbf{v}_1 + \mathbf{v}_2$ is well-defined (closed) on $\mathbb{R}^n$ and each coordinate on $c\mathbf{v}_1$ is uniquely determined in $\mathbb{R}^n$. Let $\mathbf{v}_1 = (a_1, \cdots, a_n), \mathbf{v}_2 = (b_1, \cdots, b_n), \mathbf{v}_3 = (c_1, \cdots, c_n)$ be elements in $\mathbb{R}^n$, and $a, b \in \mathbb{R}$. Then

1. $\mathbf{v}_1 + \mathbf{v}_2 = (a_1 + b_1, \cdots, a_n + b_n) = (b_1 + a_1, \cdots, b_n + a_n) = \mathbf{v}_2 + \mathbf{v}_1$ by the commutativity in $\mathbb{R}$.

2. $\mathbf{v}_1 + (\mathbf{v}_2 + \mathbf{v}_3) = \mathbf{v}_1 + (b_1 + c_1, \cdots, b_n + c_n) = ((a_1 + b_1) + c_1, \cdots, (a_n + b_n) + c_n) = (\mathbf{v}_1 + \mathbf{v}_2) + \mathbf{v}_3$ by the associativity in $\mathbb{R}$.

3. $(0, \cdots, 0) \in \mathbb{R}^n$ works as $\mathbf{0}$

4. $(-a_1, \cdots, -a_n) \in \mathbb{R}^n$ works as -$\mathbf{v}_1$

5. $a(\mathbf{v}_1 + \mathbf{v}_2) = (aa_1 + ab_1, \cdots, aa_n + ab_n) = a\mathbf{v}_1 + a\mathbf{v}_2$.

6. $(a + b)\mathbf{v}_1 = (aa_1 + ba_1, \cdots, aa_n + ba_n) = a\mathbf{v}_1 + b\mathbf{v}_1$.

7. $(ab)\mathbf{v}_1 = (a(ba_1), \cdots, a(ba_n)) = a(b\mathbf{v}_1)$.

8. Obvious.

Thus $\mathbb{R}^n$ is a vector space over $\mathbb{R}$. □

Example 20.1.2. $\mathbb{F}^n = \{(a_1, \cdots, a_n) | a_u \in \mathbb{F}, 1 \leq u \leq n\}$ *is not a vector space if we define the addition and multiplication as follows:*

$$\begin{aligned} (a_1, \cdots, a_n) + (b_1, \cdots, b_n) &= (a_1 + b_1, \cdots, a_n + b_n) \\ c(a_1, \cdots, a_n) &= (ca_1, a_2, \cdots, a_n), \end{aligned}$$

where $(a_1, \cdots, a_n), (b_1, \cdots, b_n) \in \mathbb{F}^n$ *and* $c \in \mathbb{F}$.

Solution. For $(a_1, \cdots, a_n) \in \mathbb{F}^n$, we have that $(a_1, \cdots, a_n) - (a_1, \cdots, a_n) = (0, \cdots, 0)$. On the other hand, we have that

$$\begin{aligned} (a_1, \cdots, a_n) - (a_1, \cdots, a_n) &= (a_1, \cdots, a_n) + (\text{-}1)(a_1, a_2, \cdots, a_n) \\ &= (a_1, \cdots, a_n) + (\text{-}a_1, a_2, \cdots, a_n) \\ &= (0, a_2, \cdots, a_n) \end{aligned}$$

So we have a contradiction if $a_2 \neq 0$ for example. Thus it is not a vector space over $\mathbb{F}$. $\square$

Example 20.1.3. $\mathbb{F}[x]$ *is a vector space over* $\mathbb{F}$.

Solution. Recall that $\mathbb{F}[x]$ is the set of all polynomials over $\mathbb{F}$ and we have shown that $\mathbb{F}[x]$ is a ring in Chapter 13. So it satisfies the properties 1-4. For any two polynomials $f(x) = \sum_{i=0}^{m} a_i x^i, g(x) = \sum_{i=0}^{n} b_i x^i \in \mathbb{F}[x]$ and $a, b \in \mathbb{F}$ with $m \leq n$, we have that $a(f(x) + g(x)) = \sum_{i=0}^{n}(aa_i + ab_i)x^i = af(x) + ag(x)$ where $a_i = 0$ for $m + 1 \leq i \leq n$. We also have that $(ab)f(x) = \sum_{i=0}^{m}(ab)a_i x^i = \sum_{i=0}^{m} a(ba_i)x^i = a(bf(x))$ by the associativity in $\mathbb{F}$. Obviously, $1f(x) = f(x)$ holds. Thus $\mathbb{F}[x]$ is a vector space over $\mathbb{F}$. $\square$

Example 20.1.4. *Let* $\mathbb{F}_n[x]$ *be a subset of* $\mathbb{F}[x]$ *consisting of all polynomials whose degrees are less than or equal to* n. *Prove that* $\mathbb{F}_n[x]$ *is a vector space over* $\mathbb{F}$.

Solution. The proof is similar to the proof of Example 20.1.3. $\square$

Lemma 20.1.1. (Additive Cancellation) *Let* $\mathbf{V}$ *be a vector space and* $\mathbf{v}_1, \mathbf{v}_2, \mathbf{v}_3 \in \mathbf{V}$. *If* $\mathbf{v}_1 + \mathbf{v}_2 = \mathbf{v}_1 + \mathbf{v}_3$, *then* $\mathbf{v}_2 = \mathbf{v}_3$.

Proof. Suppose that $\mathbf{v}_1 + \mathbf{v}_2 = \mathbf{v}_1 + \mathbf{v}_3$. Since $\mathbf{v}_1 \in \mathbf{V}$, there exists -$\mathbf{v}_1$ such that -$\mathbf{v}_1 + \mathbf{v}_1 = \mathbf{0}$. Then since $(\text{-}\mathbf{v}_1) + (\mathbf{v}_1 + \mathbf{v}_2) = (\text{-}\mathbf{v}_1) + (\mathbf{v}_1 + \mathbf{v}_3)$ implies that $(\text{-}\mathbf{v}_1 + \mathbf{v}_1) + \mathbf{v}_2 = (\text{-}\mathbf{v}_1 + \mathbf{v}_1) + \mathbf{v}_3$ by the associativity, we derive that $\mathbf{v}_2 = \mathbf{v}_3$. $\square$

Proposition 20.1.1. *Let* $\mathbf{v}$ *be a vector in vector space* $\mathbf{V}$. *Then*

1. $0\mathbf{v} = \mathbf{0}$

2. $(\text{-}1)\mathbf{v} = \text{-}\mathbf{v}$.

Proof.

1. Note that $0 \in \mathbb{F}$ is a zero scalar and $\mathbf{0} \in \mathbf{V}$ is a zero vector. Since $0\mathbf{v} = (0+0)\mathbf{v} = 0\mathbf{v}+0\mathbf{v}$, $0\mathbf{v} = 0\mathbf{v}+0\mathbf{v}$. This implies that $\mathbf{0}+0\mathbf{v} = 0\mathbf{v}+0\mathbf{v}$ since $\mathbf{0}$ is the additive identity. So we derive that $\mathbf{0} = 0\mathbf{v}$ by the additive cancellation.

2. Since $(1-1)\mathbf{v} = \mathbf{0}$. we derive that $1\mathbf{v} + (\text{-}1)\mathbf{v} = \mathbf{0}$ or $\mathbf{v} + (\text{-}1)\mathbf{v} = \mathbf{0}$. Also, since $\mathbf{v} - \mathbf{v} = \mathbf{0}$, we conclude that $(\text{-}1)\mathbf{v} = \text{-}\mathbf{v}$. □

For example, $0\mathbf{0} = \mathbf{0}$ by Proposition 20.1.1. Now let's prove the uniqueness of a zero vector and an additive inverse of a vector.

Theorem 20.1.1. *Let* $\mathbf{V}$ *be a vector space and* $\mathbf{v} \in \mathbf{V}$. *Then* $\mathbf{0}$ *and* $\text{-}\mathbf{v}$ *are unique in* $\mathbf{V}$

Proof. Suppose that $\mathbf{0}$ and $\mathbf{0}'$ are two zero vectors in $\mathbf{V}$. Then $\mathbf{0} = \mathbf{0} + \mathbf{0}'$ since $\mathbf{0}'$ is an additive identity. On the other hand, $\mathbf{0} + \mathbf{0}' = \mathbf{0}'$ since $\mathbf{0}$ is an additive identity. So the zero vector of $\mathbf{V}$ is unique. Now, suppose that $\mathbf{v}_1$ and $\mathbf{v}_2$ are two additive inverses of $\mathbf{v}$. This means that $\mathbf{v} + \mathbf{v}_1 = \mathbf{0}$ and $\mathbf{v} + \mathbf{v}_2 = \mathbf{0}$, and this makes $\mathbf{v} + \mathbf{v}_1 = \mathbf{v} + \mathbf{v}_2$. This implies that $\mathbf{v}_1 = \mathbf{v}_2$ by Lemma 20.1.1. So, there exists only one additive inverse of $\mathbf{v}$. □

Chapter 20 Section 1 Exercises.

1. (a) Prove that $\mathbb{R}$ is a vector space over $\mathbb{Q}$.

(b) Prove that $\mathbb{R}$ is a vector space over itself.

2. (a) Prove that $\mathbb{C}$ is a vector space over $\mathbb{Q}$.

   (b) Prove that $\mathbb{C}$ is a vector space over $\mathbb{R}$.

3. Let $\mathbb{F}$ be a subfield of $\mathbb{E}$, then prove that $\mathbb{E}$ is a vector space over $\mathbb{F}$ but generally $\mathbb{F}$ is not a vector space over $\mathbb{E}$.

4. (a) Prove that $\mathbb{Z}_7$ is a vector space over itself.

   (b) Prove that $\mathbb{Z}_p$ is a vector space over itself where p is prime.

   (c) Prove that a field is a vector space over itself.

5. (a) Prove that $M_n(\mathbb{F})$ is a vector space over $\mathbb{F}$ where $M_n(\mathbb{F})$ is the set of all $n \times n$ matrices over a field $\mathbb{F}$

   (b) Prove that $U_n(\mathbb{F})$ is a vector space over $\mathbb{F}$ where $U_n(\mathbb{F})$ is the set of all $n \times n$ upper triangular matrices over $\mathbb{F}$

   (c) Prove that $L_n(\mathbb{F})$ is a vector space over $\mathbb{F}$ where $L_n(\mathbb{F})$ is the set of all $n \times n$ lower triangular matrices over $\mathbb{F}$

   (d) Prove that $D_n(\mathbb{F})$ is a vector space over $\mathbb{F}$ where $D_n(\mathbb{F})$ is the set of all $n \times n$ diagonal matrices over $\mathbb{F}$.

6. (a) Let S be a non-empty set and $Map(S, \mathbb{F})$ be the set of all functions from S into $\mathbb{F}$ and define an addition and a scalar multiplication of $Map(S, \mathbb{F})$ as follows:

$$\begin{aligned}(f+g)(s) &= f(s)+g(s)\\ (cf)(s) &= cf(s)\end{aligned}$$

for any $f, g \in Map(S, \mathbb{F})$, $s \in S$, and $c \in \mathbb{F}$. Prove that $Map(S, \mathbb{F})$ is a vector space over $\mathbb{F}$.

(b) Let S_1 be a non-empty subset of S and $Map(S, \mathbb{F})_{S_1}$ be the subset of $Map(S, \mathbb{F})$ such that $\{f \in Map(S, \mathbb{F}) | f(S_1) = 0\}$. Prove that $Map(S, \mathbb{F})_{S_1}$ is a vector space over $\mathbb{F}$.

7. (a) Identify all the elements of $\mathbb{F}_3[x]$ where $\mathbb{F} = \mathbb{Z}_3$.

   (b) Identify all the elements of $\mathbb{F}_4[x]$ where $\mathbb{F} = \mathbb{Z}_5$.

20.2 Subspaces and Spanning Sets

$\mathbb{R}^m$ has a subset that is isomorphic to $\mathbb{R}^n$ if $m \geq n$[1] and since $\mathbb{R}^n$ is a vector space over $\mathbb{R}$, we can also consider the subset as a vector space over $\mathbb{R}$. In this section, we will study this type of subset of a vector space and called it a **subspace.**

Definition 20.2.1. *Let* $\mathbf{V}$ *be a vector space over a field* $\mathbb{F}$. *A subset* $\mathbf{S}$ *of* $\mathbf{V}$ *is a (vector)* **subspace** *of* $\mathbf{V}$ *if* $\mathbf{S}$ *is a subset of* $\mathbf{V}$ *and it is a vector space over* $\mathbb{F}$ *under the operations inherited from* $\mathbf{V}$.

If $\mathbf{v} = \sum_{1 \leq s \leq r} a_s \mathbf{v}_s$ for $a_s \in \mathbb{F}$, we say that $\mathbf{v}$ is **a linear combination** of $\{\mathbf{v}_1, \cdots, \mathbf{v}_r\} \subset \mathbf{V}$ over $\mathbb{F}$.

Lemma 20.2.1. (Subspace Test 1) *A non-empty subset* $\mathbf{S}$ *of* $\mathbf{V}$ *is a subspace of* $\mathbf{V}$ *if and only if* $\mathbf{u} + \mathbf{v} \in \mathbf{S}$ *and* $a\mathbf{v} \in \mathbf{S}$ *hold for any* $\mathbf{u}, \mathbf{v} \in S$ *and* $a \in \mathbb{F}$.

[1]See Example 20.2.1.

Proof. If $\mathbf{S}$ is a subspace of $\mathbf{V}$, obviously $\mathbf{u}+\mathbf{v}$ and $a\mathbf{u} \in \mathbf{S}$ for any $\mathbf{u}, \mathbf{v} \in \mathbf{S}$ and $a \in \mathbb{F}$ by the definition of a vector space. Let's assume $\mathbf{u}+\mathbf{v}$ and $a\mathbf{u} \in \mathbf{S}$ for $\mathbf{u}, \mathbf{v} \in \mathbf{S}$ and $a \in \mathbb{F}$. Since $\mathbf{S}$ is non-empty, for any $\mathbf{u} \in \mathbf{S}$, -$\mathbf{v} \in \mathbf{S}$. This also implies that the zero vector is an element of $\mathbf{S}$. Since $\mathbf{S}$ is a subset of $\mathbf{V}$, other axioms of Definition 20.1.1 are true automatically. Therefore, $\mathbf{S}$ is a subspace of $\mathbf{V}$. □

We can combine the necessary conditions of Lemma 20.2.1.

Lemma 20.2.2. (Subspace Test 2) *A non-empty subset* $\mathbf{S}$ *of* $\mathbf{V}$ *is a subspace of* $\mathbf{V}$ *if and only if* $a\mathbf{u}+b\mathbf{v} \in \mathbf{S}$ *holds for any* $\mathbf{u}, \mathbf{v} \in \mathbf{S}$ *and* $a, b \in \mathbb{F}$.

Proof. If $\mathbf{S}$ is a subspace of $\mathbf{V}$, then for any $\mathbf{u}, \mathbf{v} \in S$ and $a, b \in \mathbb{F}$, $a\mathbf{u}+b\mathbf{v} \in \mathbf{S}$. Conversely let us assume that for any $\mathbf{u}, \mathbf{v} \in \mathbf{S}$ and $a, b \in \mathbb{F}$, $a\mathbf{u}+b\mathbf{v} \in \mathbf{S}$. If $a = b = 1$, then $\mathbf{u}+\mathbf{v} \in S$ since $\mathbf{S}$ is closed under addition. Also, if we set $b = 0$, then $a\mathbf{u} \in \mathbf{S}$. By Lemma 20.2.1, $\mathbf{S}$ is a subspace of $\mathbf{V}$. □

Example 20.2.1. *If* $n \geq m$, *then* $\mathbb{F}^n$ *has a subspace that is isomorphic to* $\mathbb{F}^m$.

Solution Note that each vector in $\mathbb{F}^n$ has n coordinates. Since $n \geq m$, pick any m coordinates of a vector in $\mathbb{F}^n$ and replace the rest of $n-m$ coordinates with 0. If we do the same procedure for all vectors in $\mathbb{F}^n$, then the set of all of this type of vectors is a subspace of $\mathbb{F}^n$ and it is isomorphic to $\mathbb{F}^m$. □

We can consider the subspace described in Example 20.2.1 as an internal subspace of $\mathbb{F}^n$ and $\mathbb{F}^m$ as an external subspace of $\mathbb{F}^n$. So, it is fair to say that $\mathbb{F}^m$ a subspace of $\mathbb{F}^n$ if there is no ambiguity.

Example 20.2.2. *If* $n \geq m$, *then* $\mathbb{F}_m[x]$ *is a subspace of* $\mathbb{F}_n[x]$.

Definition 20.2.2. *A subset* $\mathbf{S}$ *of a vector space* $\mathbf{V}$ *is called* **a spanning set of V** *if every vector* $\mathbf{v} \in \mathbf{V}$ *is a linear combination of finitely many non-zero elements of* $\mathbf{S}$*, i.e.,*

$$\mathbf{V} = \{\mathbf{v} \in \mathbf{V} | \mathbf{v} = \sum_{i=1}^{n} a_i \mathbf{v}_i \text{ where } \mathbf{v}_i \in \mathbf{S}, a_i \in \mathbb{F}\}$$

and we denote it as $span(\mathbf{S}) = \mathbf{V}$.

We also say that $\mathbb{V}$ **is spanned by S**.

Proposition 20.2.1. *For any subset* $\mathbf{S}S$ *of* $\mathbf{V}$*,* $span(\mathbf{S})$ *is a subspace of* $\mathbf{V}$.

Proof. Let $\mathbf{S}S = \{\mathbf{v}_i | i \in J\}$ where J is an index set. Suppose that $\sum_{i\in J} a_i \mathbf{v}_i, \sum_{i\in J} b_i \mathbf{v}_i \in span(\mathbf{S}S)$ and $c_1, c_2 \in \mathbb{F}$.. Then we have that $c_1 \sum_{i\in J} a_i \mathbf{v}_i + c_2 \sum_{i\in J} b_i \mathbf{v}_i = \sum_{i\in J}(c_1 a_i + c_2 b_i)\mathbf{v}_i \in span(\mathbf{S}S)$ where $a_i, b_i \in \mathbb{F}$ for $i \in J$. This implies that $span(\mathbf{S}S)$ is a subspace of $\mathbf{V}$ by Lemma 20.2.2. □

Let $\mathbf{V}_1$ and $\mathbf{V}_2$ be subspaces of a vector space $\mathbf{V}$. The sum $\mathbf{V}_1 + \mathbf{V}_2$ is a subset of $\mathbf{V}$ such that

$$\mathbf{V}_1 + \mathbf{V}_2 = \{\mathbf{v}_1 + \mathbf{v}_2 | \mathbf{v}_1 \in \mathbf{V}_1, \mathbf{v}_2 \in \mathbf{V}_1\}. \tag{20.1}$$

It is easy to prove that $\mathbf{V}_1 + \mathbf{V}_2$ is a subspace of $\mathbf{V}$. **The direct sum** $\mathbf{V}_1 \oplus \mathbf{V}_2$ of $\mathbf{V}_1$ and $\mathbf{V}_2$ is defined as the sum $\mathbf{V}_1 + \mathbf{V}_2$ with $\mathbf{V}_1 \cap \mathbf{V}_2 = \{\mathbf{0}\}$. Extensively we can define the direct sum of vector spaces $\mathbf{V}_i$, $1 \leq i \leq n$ as $\oplus_{1\leq i\leq n} \mathbf{V}_i = \mathbf{V}_1 \oplus \cdots \oplus \mathbf{V}_n$ with $\mathbf{V}_i \cap \mathbf{V}_j = \{\mathbf{0}\}$ if $i \neq j$.

Theorem 20.2.1. *Let* $\mathbf{V}_1$ *and* $\mathbf{V}_2$ *be subspaces of a vector space* $\mathbf{V}$*. Then* $\mathbf{V} = \mathbf{V}_1 \oplus \mathbf{V}_2$ *if and only if for every vector* $\mathbf{v}$ *of* $\mathbf{V}$*,* $\mathbf{v}$ *can be uniquely written* *as* $\mathbf{v}_1 + \mathbf{v}_2$ *where* $\mathbf{v}_1 \in \mathbf{V}_1$ *and* $\mathbf{v}_2 \in \mathbf{V}_2$.

Proof. Let's assume that $\mathbf{V} = \mathbf{V}_1 \oplus \mathbf{V}_2$ and $\mathbf{v} \in \mathbf{V}$. Let's say $\mathbf{v} = \mathbf{v}_1 + \mathbf{v}_2$ and $\mathbf{v} = \mathbf{v}_1' + \mathbf{v}_2'$ where $\mathbf{v}_1, \mathbf{v}_1' \in \mathbf{V}_1$ and $\mathbf{v}_2, \mathbf{v}_2' \in \mathbf{V}_2$. Then we have $\mathbf{0} = \mathbf{v}_1 - \mathbf{v}_1' = \mathbf{v}_2' - \mathbf{v}_2$, and this implies that $\mathbf{v}_1 = \mathbf{v}_1'$ and $\mathbf{v}_2' = \mathbf{v}_2$. Thus the expression is unique. Conversely, suppose that for every vector $\mathbf{v}$ of $\mathbf{V}$, $\mathbf{v}$ can be written uniquely as $\mathbf{v}_1 + \mathbf{v}_2$ where $\mathbf{v}_1 \in \mathbf{V}_1$ and $\mathbf{v}_2 \in \mathbf{V}_2$. This implies that $\mathbf{V} = \mathbf{V}_1 + \mathbf{V}_2$. Assume that there is a non-zero vector $\mathbf{v}_3 \in \mathbf{V}_1 \cap \mathbf{V}_2$. Note that $\mathbf{0} = \mathbf{0} + \mathbf{0} = \mathbf{v}_3 - \mathbf{v}_3$. This means that $\mathbf{0} \in \mathbf{V}$ can be written as $\mathbf{0} + \mathbf{0}$ and $\mathbf{v}_3 - \mathbf{v}_3$. a contradiction to the uniqueness. This implies that $\mathbf{v}_3 = \mathbf{0}$ and therefore the sum is direct. □

Example 20.2.3. *Let*

$$\mathbf{V}_1 = \left\{ \begin{pmatrix} a & b \\ 0 & 0 \end{pmatrix} | a, b \in \mathbb{F} \right\} \quad \textit{and} \quad \mathbf{V}_2 = \left\{ \begin{pmatrix} 0 & a \\ b & c \end{pmatrix} | a, b, c \in \mathbb{F} \right\}$$

Then, $\mathbf{V}_1$ *and* $\mathbf{V}_2$ *are subspaces of* $M_2(\mathbb{F})$ *such that* $M_2(\mathbb{F}) = \mathbf{V}_1 + \mathbf{V}_2$ *but* $M_2(\mathbb{F}) \neq \mathbf{V}_1 \oplus \mathbf{V}_2$.

Example 20.2.4. *Let*

$$\mathbf{V}_1 = \left\{ \begin{pmatrix} a & b \\ 0 & 0 \end{pmatrix} | a, b \in \mathbb{F} \right\} \quad \textit{and} \quad \mathbf{V}_2 = \left\{ \begin{pmatrix} 0 & 0 \\ a & b \end{pmatrix} | a, b \in \mathbb{F} \right\}$$

Then, $\mathbf{V}_1$ *and* $\mathbf{V}_2$ *are subspaces of* $M_2(\mathbb{F})$ *such that* $M_2(\mathbb{F}) = \mathbf{V}_1 \oplus \mathbf{V}_2$.

Example 20.2.5. *Find* $span(\{(1, 0), (0, 2)\})$ *in* $\mathbb{R}^2$.

Solution. For any $(x, y) \in \mathbb{R}^2$, $(x, y) = x(1, 0) + \frac{y}{2}(0, 2)$. This implies that $span(\{(1,0),(0,2)\}) = \mathbb{R}^2$. $\square$

Note that $span(\{(1,0),(0,2)\}) = span(\{(1,0)\}) \oplus span(\{(0,2)\}) = \mathbb{R}^2$.

Chapter 20 Section 2 Exercises.

1. Let $\mathbf{V}_i$, $1 \leq i \leq n$, be vector spaces over $\mathbb{F}$.

   (a) Prove that $\mathbf{V}_i \cap \mathbf{V}_j$, $1 \leq i, j \leq n$, is a vector subspace of $\mathbf{V}_i$ or $\mathbf{V}_j$.

   (b) Prove that $\cap_{1 \leq i \leq n} \mathbf{V}_i$ is a vector subspace of $\mathbf{V}_j$, $1 \leq j \leq n$.

   (c) Prove or disprove that $\mathbf{V}_i \cup \mathbf{V}_j$ is a vector space over $\mathbb{F}$.

2. Let $\mathbf{S}$ be a subset of a vector space $\mathbf{V}$. For a subset $\mathbf{S}_1$ of $\mathbf{V}$, prove that if $\mathbf{S} \subset\subset S_1$, then $span(\mathbf{S})$ is a subspace of $span(\mathbf{S}_1)$.

3. Prove that $D_n(\mathbb{F})$ is a subspace of $M_n(\mathbb{F})$.[2]

4. (a) Prove that $U_n(\mathbb{F})$ and $L_n(\mathbb{F}))$ are subspaces of $M_n(\mathbb{F})$.[3]

   (b) Prove that $U_n^+(\mathbb{F})$ (resp. $L_n^-(\mathbb{F})$) is a subspace of $U_n(\mathbb{F})$ (resp. $L_n(\mathbb{F})$).

   (c) Prove that $U_n^+(\mathbb{F}) \cap U_n(\mathbb{F})$ is a subspace of $M_n(\mathbb{F})$.

5. Let $Map_0(\mathbf{S}, \mathbb{F})$ be the subset of $Map(\mathbf{S}, \mathbb{F})$ such that $Map_0(\mathbf{S}, \mathbb{F}) = \{f | f : \mathbf{S} \to \mathbb{F}, f(\mathrm{s}) = 0 \text{ for all } \mathrm{s} \in \mathbf{S}\}$. Prove that $Map_0(\mathbf{S}, \mathbb{F})$ is a subspace of $Map(\mathbf{S}, \mathbb{F})$.

6. (a) Prove that $\mathbb{F}_3[x]$ is a subspace of $\mathbb{F}_4[x]$.

   (b) If $m \leq n$, prove that $\mathbb{F}_m[x]$ is a subspace of $\mathbb{F}_n[x]$.

[2]Check Exercise 5d of Section 20.1 for the definition.

(c) For $m \leq n$, prove that $M_{n,m}(\mathbb{F}) =$ is a subspace of $M_n(\mathbb{F})$ where $M_{n,m}(\mathbb{F})$ is the set of $m \times n$ matrices over $\mathbb{F}$ with the form

$$\begin{pmatrix} a_{11} & \cdots & a_{1m} & 0 & \cdots & 0 \\ a_{21} & \cdots & a_{2m} & 0 & \cdots & 0 \\ & \cdots & & & & \\ a_{m1} & \cdots & a_{mm} & 0 & \cdots & 0 \\ 0 & \cdots & 0 & 0 & \cdots & 0 \\ & \cdots & & & & \\ 0 & \cdots & 0 & 0 & \cdots & 0 \end{pmatrix}$$

7. Let $\mathbf{S} = \{(1,0,1,1),(0,1,1,2),(0,2,1,1),(1,0,0,0)\}$. Show that $\mathbf{S} = \mathbb{R}^4$.

8. Find three vector subspaces $\mathbf{V}_1$, $\mathbf{V}_2$, and $\mathbf{V}_3$ of $M_3(\mathbb{R})$ such that $M_3(\mathbb{R}) = \mathbf{V}_1 \oplus \mathbf{V}_2 \oplus \mathbf{V}_3$.

20.3 Basis of a Vector Space

Even though a vector space can have infinitely many elements, a spanning set doesn't have to have infinitely many elements to span the vector space. If the size of a spanning set is minimal, we call it a **basis**. Since it is a spanning set, every element of the vector space can be written as a linear sum of elements of the set. Also, this expression is unique because the size is minimal. To find such a basis of a vector space, we start this section with basic definitions and concepts.

Definition 20.3.1. *Let* $\mathbf{V}$ *be a vector space over a field* $\mathbb{F}$. *A subset* $\{\mathbf{v}_1, \cdots, \mathbf{v}_r\}$ *of* $\mathbf{V}$ *is called* **linearly dependent** *if there are scalars* $a_1, \cdots, a_r$ *such that* $\sum_{1 \le u \le r} a_u \mathbf{v}_u = \mathbf{0}$ *where at least one of* $a_1, \cdots, a_r$ *is not zero. If it is not linearly dependent, it is called* **linearly independent**.

Example 20.3.1. *If* $\mathbf{v}$ *is a non-zero vector of a vector space* $\mathbf{V}$, *then* $\{\mathbf{v}\}$ *is a linearly independent set.*

Example 20.3.2. *Show that* $\{(1, 0, 0), (0, 2, 0), (0, 0, 3)\}$ *is a linearly independent subset of* $\mathbb{F}^3$.

Solution. Let's set $a(1, 0, 0) + b(0, 2, 0) + c(0, 0, 3) = (0, 0, 0)$ for $a, b, c \in \mathbb{F}$. Then we have $(a, 2b, c) = (0, 0, 0)$ and this makes that $a = b = c = 0$. So it is linearly independent since 0 is the only solutions for a, b, and c. □

If we make a linear combination of all vectors in a set and set it $\mathbf{0}$, there is always a set of scalars that makes the linear combination $\mathbf{0}$: zero scalars. If zeros' are only possible scalars, the set is linearly independent. If a non-zero scalar can be a part of the linear combination, the set is linearly dependent. We state this useful property as a theorem.

Theorem 20.3.1. *Let* $S = \{\mathbf{v}_1, \cdots, \mathbf{v}_r\}$ *be a linearly independent subset of a vector space* $\mathbf{V}$. *If* $\sum_{1 \le u \le r} a_u \mathbf{v}_u = 0$, *then* a_u, $1 \le u \le r$, *are zeros.*

Proposition 20.3.1. *Let* S *be a non-empty subset of a vector space* $\mathbf{V}$ *consisting of* $\mathbf{v}_1, \cdots, \mathbf{v}_r$. *The set* S *is linearly dependent if and only if there is* $\mathbf{v}_s \in S$ *such that* $\mathbf{v}_s$ *is a linear combination of other vectors in* S.

Proof. If $dim(\mathbf{V}) = 1$, then the proof of proposition is trivial. So we can assume that $dim(\mathbf{V}) > 1$. Suppose that S is linearly dependent and let $\sum_{1\leq u\leq r} a_u \mathbf{v}_u = 0$ where $a_i \in \mathbb{F}$. Then there must be $a_s \neq 0$, $1 \leq s \leq n$. This implies that $\mathbf{v}_s = -\frac{a_1}{a_s}\mathbf{v}_1 - \cdots - \frac{a_r}{a_s}\mathbf{v}_r$. Conversely let us assume that there is $\mathbf{v}_s \in S$ such that $\mathbf{v}_s = b_1\mathbf{v}_1 + \cdots + b_r\mathbf{v}_r$ for $b_1, \cdots, b_r \in \mathbb{F}$. This implies that $b_1\mathbf{v}_1 + \cdots + b_r\mathbf{v}_r - \mathbf{v}_s = 0$ and this means that S is linearly dependent. □

Example 20.3.3. *The subset* $\{(1,2,0),(-1,1,1),(2,1,2)\}$ *of* $\mathbb{R}^3$ *is linearly independent.*

Solution. Suppose that such that $a(1,2,0)+b(-1,1,1)+c(2,1,2) = (0,0,0)$. Then since $(a-b+2c, 2a+b+c, b+2c) = (0,0,0)$, we have that $a-b+2c = 0$, $2a+b+c = 0$, and $b+2c = 0$. The solution of the system of linear equations is $(0,0,0)$. i.e. $(0,0,0)$ is the only solution. Therefore $\{(1,2,0),(-1,1,1),(2,1,2)\}$ of $\mathbb{R}^3$ is linearly independent. □

Example 20.3.4. *Since* $2\cdot 1 + 0\cdot(x+1) + 2(x^2-3) + (\text{-}2)(2x^2-4) = 0$, *the subset* $\{1, x+1, x^2-3, 2x^2-4\}$ *of* $\mathbb{F}_2[x]$ *is linearly dependent.*

Example 20.3.5. *Since* $(1,2,3,-4)+(-2,1,4,3) = (-1,3,7,-1)$, *the subset* $\{(1,2,3,-4), (-2,1,4,3), (-1,3,7,-1)\}$ *of* $\mathbb{R}^4$ *is linearly dependent.*

Example 20.3.6. *Since* $x^2+3x-1-(x^2-4)-\frac{5}{3}(2x+3)+\frac{1}{3}x = 0$, *the subset* $\{x^2+3x+1, x^2-4, 2x+3, x\}$ *of* $\mathbb{F}[x]$ *is linearly dependent.*

Example 20.3.7. *Since* $\begin{vmatrix} 1 & 2 & 1 \\ 2 & 1 & 3 \\ 1 & 2 & 2 \end{vmatrix} \neq 0$, *we have that* $\{(1,2,1),(2,1,3),(1,2,2)\}$ *is a linearly independent set of* $\mathbb{R}^3$.

Proposition 20.3.2. *Let* S *be a linearly independent subset of a vector space* $\mathbf{V}$ *and* $\mathbf{v}$ *be a non-zero vector of* $\mathbf{V}$. *If* $\mathbf{v} = \sum_{i=1}^{n} a_i\mathbf{v}_i = \sum_{i=1}^{n} b_i\mathbf{v}_i$, *then* $a_i = b_i$, $1 \le i \le n$, *where* $\mathbf{v}_i \in S$ *and* $a_i \in \mathbb{F}$.

Proof. Suppose $\mathbf{v} \in \mathbf{V}$ is a non-zero vector satisfying $\mathbf{v} = \sum_{i=1}^{n} a_i\mathbf{v}_i = \sum_{i=1}^{n} b_i\mathbf{v}_i$. Then we can derive that $\sum_{i=1}^{n}(a_i - b_i)\mathbf{v}_i$ and this makes $a_i - b_i = 0$ for $1 \le i \le n$ since $\{\mathbf{v}_1, \cdots, \mathbf{v}_n\}$ is linearly independent. □

This linear independency makes a linear combination unique and it will be a part of a definition of a basis of a vector space.

Definition 20.3.2. *A subset* B *of a vector space* $\mathbf{V}$ *is called* **a basis** *of* $\mathbf{V}$ *if* B *is linearly independent and span*$(B) = \mathbf{V}$. *The cardinality* $|B|$ *is called* **the dimension of** $\mathbf{V}$ *and denoted by dim*$(\mathbf{V})$. *If dim*$(\mathbf{V})$ *is a non-zero integer, then* $\mathbf{V}$ *is called* **a finite dimensional vector space.**

In other words, a basis is a linearly independent spanning set for a vector space.

Example 20.3.8. *Show that* $S = \{(1,0,0),(0,1,0),(0,0,1),(1,0,1)\}$ *is not a basis of* $\mathbb{F}^3$.

Soltion. Let $(a_1, a_2, a_3) \in \mathbb{F}^3$. Then $a_1(1,0,0) + a_2(0,1,0) + a_3(0,0,1) + 0 \cdot (1,0,1) = (a_1, a_2, a_3)$ and this implies that S is a spanning set of $\mathbb{F}^3$, i.e., $\mathbb{F}^3 =$

span$(\{(1,0,0),(0,1,0),(0,0,1),(1,0,1)\})$ but S is not a linearly independent since $(1,0,0)+(0,0,1)=(1,0,1)$. Thus S is not a basis of $\mathbb{F}$. It is just a spanning set of $\mathbb{F}$. $\square$

Example 20.3.9. *Since* $B=\{(1,0,0,0),(0,1,0,0),(0,0,1,0),(0,0,0,1)\}$ *is linearly independent spanning set of* $\mathbb{F}^4$*, it is a basis of* $\mathbb{F}^4$*.*

According to Proposition 20.3.2, for a given basis $B=\{\mathbf{v}_1,\cdots,\mathbf{v}_n\}$ of a vector space $\mathbf{V}$, every element $\mathbf{v}$ can be uniquely written as follows:

$$\mathbf{v}=\sum_{1\leq u\leq n} a_i\mathbf{v}_i$$

where $a_i\in\mathbb{F}$. This makes the dimension of a vector space well-defined and a rigorous proof will be stated later. This a_i is called **the component** (or **coefficient**) of $\mathbf{v}_i$ for $\mathbf{v}$.

Lemma 20.3.1. *Let* $\mathbf{V}$ *be a finite dimensional vector space. Then there is a linearly independent subset* S *of* $\mathbf{V}$ *such that* $|S|\geq|T|$ *for any linearly independent set* T *of* $\mathbf{V}$*. Recall that* $|S|$ *is the cardinality of* S*.*

Proof. Since *span*$(\mathbf{V})=\mathbf{V}$, there is a subset S of $\mathbf{V}$ *span*$(S)=\mathbf{V}$. Since $\mathbf{V}$ is a finite dimensional vector space, we can assume that $|S|<\infty$. If S is a linearly dependent set, then we can remove an element $\mathbf{v}_n$ from S so that $S_n=S-\{\mathbf{v}_n\}$ and *span*$(S_n)=\mathbf{V}$. If S_n is a linearly dependent set, then we can remove an element $\mathbf{v}_{n-1}$ from S_n so that $S_{n-1}=S_n-\{\mathbf{v}_{n-1}\}$ and *span*$(S_{n-1})=\mathbf{V}$. Continuing these processes, we can find a linearly independent maximal set S_k so that *span*$(S_k)=\mathbf{V}$ and *span*$(S_{k-1})\neq\mathbf{V}$. Let us prove the lemma by induction on the number k of S_k. Let us put $S_k=\{\mathbf{v}_1,\cdots,\mathbf{v}_k\}$.

Let T_0 be a linearly independent set of $\mathbf{V}$ where $T_0 = \{\mathbf{w}_1, \cdots, \mathbf{w}_m\}$. Since $\{\mathbf{v}_1, \cdots, \mathbf{v}_k, \mathbf{w}_m\}$ is linearly dependent, there are scalars $a_1, \cdots, a_{k+1}$ so that $a_1\mathbf{v}_1 + \cdots + a_k\mathbf{v}_k + a_{k+1}\mathbf{w}_1 = 0$ such that $a_1, a_{k+1} \neq 0$. This implies that

$$\mathbf{v}_1 = -\frac{a_2}{a_1}\mathbf{v}_2 - \cdots - \frac{a_{k+1}}{a_1}\mathbf{w}_1.$$

Since $span(S_k) = \mathbf{V}$, we have that $span(\{\mathbf{v}_2, \cdots, \mathbf{v}_k, \mathbf{w}_1\}) = \mathbf{V}$. If $\{\mathbf{v}_2, \cdots, \mathbf{v}_k, \mathbf{w}_1\}$ is linearly dependent, then there is a proper subset T_1 of $\{\mathbf{v}_2, \cdots, \mathbf{v}_k, \mathbf{w}_1\}$ so that $span(T_1) = \mathbf{V}$. By induction, we have a contradiction. So $\{\mathbf{v}_2, \cdots, \mathbf{v}_k, \mathbf{w}_1\}$ is linearly independent. Similarly we have a linearly independent set $T_2 = \{\mathbf{v}_3, \cdots, \mathbf{v}_k, \mathbf{w}_1, \mathbf{w}_2\}$ which spans $\mathbf{V}$. Continuing these processes, we have that $T_k = \{\mathbf{v}_{m+1}, \cdots, \mathbf{v}_k, \mathbf{w}_1, \mathbf{w}_2, \cdots, \mathbf{w}_m\}$. So $|S| \geq |T_k|$ holds. Therefore we have proven the lemma. □

Theorem 20.3.2. *If B_1 and B_2 are bases of a finite dimensional vector space $\mathbf{V}$, then $|B_1| = |B_2|$.*

Proof. Since B_1 and B_2 are bases of $\mathbf{V}$. Since they are linearly independent sets, if we put $S = B_2$ and $S_1 = B_1$ of Lemma 20.3.1, we have that $|B_1| \leq |B_2|$. Similarly we are able to prove that $|B_2| \leq |B_1|$. Therefore $|B_1| = |B_2|$. □

Theorem 20.3.2 gives a uniqueness of a dimension of a vector space, i.e., we know that any two bases of a vector space $\mathbf{V}$ have the same number of elements and this unique number is called the *dimension* of the vector space. A spanning set of a vector space is not necessarily a basis but it at least includes a basis of the vector space and the following proposition proves it.

Proposition 20.3.3. *Let S be a subset of a finite dimensional vector space $\mathbf{V}$. If $span(S) = \mathbf{V}$, then there is a subset B of S such that B is a basis of $\mathbf{V}$.*

Proof. By Lemma 20.3.1, we can find a minimal spanning set S_m such that $S_m \subset S$ and $span(S_m) = \mathbf{V}$. We only need to show that S_m is linearly independent. Let $S_m = \{\mathbf{v}_1, \cdots, \mathbf{v}_k\}$. If S_m is not linearly independent, there are scalars $a_1, \cdots, a_k$ such that $\sum_{1 \leq i \leq k} a_i \mathbf{v}_i = 0$. If we say that a_k is not zero, then $\mathbf{v}_k = \sum_{1 \leq i \leq k-1} \frac{a_i}{a_k} \mathbf{v}_i$. This implies that $span(S_m - \{a_k\}) = \mathbf{V}$. This contradicts that S_m is a minimal spanning set so S_m is linearly independent. $\square$

Example 20.3.10. $S = \{(1,0,0), (0,1,0), (0,1,1), (1,0,1)\}$ *is a subset of* $\mathbb{F}^3$ *but not a basis. If we take* $S_1 = \{(1,0,0), (0,1,0), (0,1,1)\}$, *then* $S_1 \subseteq S$ *and* S_1 *is a basis of* $\mathbb{F}^3$.

Example 20.3.11. $S = \{(1,0,1), (1,1,2)\}$ *is not a basis of* $\mathbb{F}^3$ *but if we add an element* $(0,0,1)$ *to the set* S, *then* $S \cup \{(0,0,1)\}$ *becomes a basis of* $\mathbb{F}^3$.

Example 20.3.12. *A basis* $\mathcal{E}_n = \{\mathbf{e}_1 = (1, 0, \cdots, 0), \cdots, \mathbf{e}_n = (0, \cdots, 0, 1)\}$ *of* $\mathbb{F}^n$ *is called* **the standard basis of $\mathbb{F}^n$**.

If $\mathbb{F}$ is a field of characteristic zero, there are infinitely many different bases of $\mathbb{F}_m[x]$. Note that $dim(\mathbb{F}_m[x]) = m + 1$.

Example 20.3.13. *Prove that* $\{1, x, \cdots, x^m\}$ *is a basis of* $\mathbb{F}_m[x]$.

Solution For any $f(x) \in \mathbb{F}_m[x]$, we have that $f(x) = \sum_{0 \leq i \leq m} a_i x^i$ where $a_i \in \mathbb{F}$. This implies that $\mathbb{F}_m[x] = span(\{1, x, \cdots, x^m\})$. Next we show that

$\{1, x, \cdots, x^m\}$ is linearly independent. Suppose that

$$\sum_{0 \leq i \leq m} a'_i x^i = 0 \tag{20.2}$$

where $a'_i \in \mathbb{F}$. If we apply $(\frac{d}{dx})^m$ to (20.2), we have that a'_m is zero. So we can conclude that all a'_i are zeros and this shows that $\{1, x, \cdots, x^m\}$ is linearly independent. □

We often consider the basis in Example 20.3.13 as the standard basis for $\mathbb{F}_m[x]$.

Now we conclude this chapter by proving that the dimension of a vector space is the minimum number of linearly independent vectors in the vector space.

Proposition 20.3.4. *Let* $\mathbf{V}$ *be an* n*-dimensional vector space and* S *be a subset of* $\mathbf{V}$. *If* $|S| = m > n$, *then* S *is linearly dependent.*

Proof. Let S be a subset of $\mathbf{V}$ with $|S| = m > n = dim(\mathbf{V})$. If S contains $\mathbf{0}$, then S is linearly dependent. So let's consider the case that S does not contain $\mathbf{0}$. Let's assume that S is linearly independent. If $span(S) = \mathbf{V}$, then S is a basis, a contradiction since $|S| = m > n = dim(\mathbf{V})$. So $span(S)$ is a proper subspace of $\mathbf{V}$. Now we need to add more non-zero vectors to S to span $\mathbf{V}$, i.e. $span(S \cup \{\mathbf{v}_1, \cdots, \mathbf{v}_r\}) = \mathbf{V}$. This means the dimension of $\mathbf{V}$ is greater than m and $m > n = dim(\mathbf{V})$, another contradiction. Therefore S is linearly dependent. □

Even though all the vector spaces we consider in this book are finite dimensional, there are infinite dimensional vector spaces. For example, $\mathbb{F}_m[x]$ is has $m + 1$ dimension but $\mathbb{F}[x]$ is an infinite dimensional vector space over $\mathbb{F}[x]$. We only consider finite dimensional vector spaces in this book.

Chapter 20 Section 3 Exercises.

1. Show that the claim in Example 20.3.7 is valid.

2. Show that a subset S of a vector space is linearly dependent if $S \ni \mathbf{o}$.

3. Find a non-standard basis of $\mathbb{F}^n$.

4. (a) Let S_1 be a maximal linearly independent set of a vector space $\mathbf{V}$. If S is a subset of S_1, then prove that S is linearly independent.

   (b) Prove that if S is a spanning set of $\mathbf{V}$ and $S \subset S_1$, then $\mathbf{V} = span(S_1)$.

   (c) Let S be a linearly dependent subset of $\mathbf{V}$. For $S_1 \subset \mathbf{V}$, if S is a subset of S_1, then prove that S_1 is linearly dependent.

5. For a subset S of $\mathbf{V}$, prove that if S is linearly independent and S_1 is subset of S, then S_1 is linearly independent.

6. (a) Prove that $\{1, x, x^2\}$ is a linearly independent subset of $\mathbb{F}[x]$.

   (b) Prove that $\{1, x, \cdots, x^n, \cdots\}$ is a linearly independent subset of $\mathbb{F}[x]$.

   (c) Prove that $span(\{1, x, \cdots, x^n, \cdots\}) = \mathbb{F}[x]$.

   (d) Prove that $\{1, x, \cdots, x^n, \cdots\}$ is a basis of $\mathbb{F}[x]$. What is $dim(\mathbb{F}[x])$?

   (e) Find a basis of the vector space $\mathbb{F}[x_1, \cdots, x_n]$. What is $dim(\mathbb{F}[x_1, \cdots, x_n])$?

7. Prove that $\{x, 3x+1, 2x^2 - x, x^3 + x^2\}$ is a basis of $\mathbb{F}_4[x]$.

8. (a) Is $B_1 = \left\{ \begin{pmatrix} 1 & 1 \\ 0 & 1 \end{pmatrix}, \begin{pmatrix} 1 & 0 \\ -1 & 0 \end{pmatrix}, \begin{pmatrix} 1 & 0 \\ 0 & -1 \end{pmatrix}, \begin{pmatrix} 0 & 0 \\ 0 & 1 \end{pmatrix} \right\}$ a linearly independent subset of $M_2(\mathbb{F})$?

(b) Is $span(B_1) = M_2(\mathbb{F})$? Prove your claim.

(c) Is $B_2 = \left\{ \begin{pmatrix} 1 & 0 \\ 0 & 1 \end{pmatrix}, \begin{pmatrix} 1 & 1 \\ 0 & 0 \end{pmatrix}, \begin{pmatrix} 0 & 0 \\ 1 & 0 \end{pmatrix}, \begin{pmatrix} 0 & 0 \\ 0 & 1 \end{pmatrix} \right\}$ a basis of $M_2(\mathbb{F})$?

(d) If $B_3 = \left\{ \begin{pmatrix} 1 & 1 \\ 0 & 1 \end{pmatrix}, \begin{pmatrix} 1 & 0 \\ -1 & 0 \end{pmatrix}, \begin{pmatrix} 1 & 0 \\ 0 & -1 \end{pmatrix} \right\}$, find a basis B of $M_2(\mathbb{F})$ such that $B_3 \subset B$.

9. For a given set $\{(1,2,3),(3,1,0)\}$ of $\mathbb{F}^3$, find a basis B of $\mathbb{F}^3$ such that $\{(1,2,3),(3,1,0)\} \subset B$.

10. A matrix A is called **symmetric** if $A^t = A$.[4]

    (a) Let Sym_2 be the set of all 2×2 symmetric matrices over $\mathbb{R}$. Find all the elements of Sym_2, prove that S_2 is a vector space, and find its dimension.

    (b) Find a basis of Sym_3 of all 3×3 symmetric matrices over $\mathbb{R}$ and its dimension.

    (c) Find a basis of Sym_n of all $n \times n$ symmetric matrices over $\mathbb{R}$ and its dimension.

11. (a) Find a basis of the vector space $U_n(\mathbb{F})$ (resp. $L_n(\mathbb{F})$) and its dimension.[5]

    (b) Find a basis of the vector space $U_n^+(\mathbb{F})$ (resp. $L_n^-(\mathbb{F})$) and its dimension.

    (c) Find two different bases of the vector space $D_n(\mathbb{F})$. Prove your assertion.

12. (a) Prove that all the solutions of $3x_1 + 4x_2 = 0$ is a subspace of $\mathbb{R}^2$.

    (b) Prove that all the solutions of $3x_1 + 4x_2 + 5x_4 = 0$ is a subspace of $\mathbb{R}^3$.

[4] If $A = (a_{ij})_{m\times n}$, then $A^t = (a_{ji})_{n\times m}$.
[5] Check Exercise 5d of Section 20.1 for the definition.

(c) Prove that all the solutions of $\sum_{1\leq i\leq n} a_i x_i = 0$ is a subspace of $\mathbb{R}^n$. This solution set is called **the hyperplane of** $\mathbb{R}^n$.

13. For a non-zero vector $\mathbf{v}$ of a vector space $\mathbf{V}$, prove that there is a basis B of $\mathbf{V}$ such that $\mathbf{v} \in B$.

14. Let S be a subset of an n-dimensional vector space $\mathbf{V}$. Prove that if $span(S) = \mathbf{V}$, then $|S| \geq n$.

Chapter 21

Basic Field Theory: Prime Fields

A field is a ring of a special type. It is a commutative ring with division so every non-zero elements are units. Examples of fields include $\mathbb{C}$, $\mathbb{R}$, $\mathbb{Q}$, $\mathbb{Z}_p$ for a prime p, etc., and we'd like to classify them first.

Starting this chapter, our main attention will be field theory. Recall that a field is a commutative division ring. We will use $\mathbb{F}$ for a field from now on.

21.1 Prime Fields

Example 21.1.1. (Function field)
Let $\mathbb{F}(x) = \left\{ \dfrac{f(x)}{g(x)} | f(x), g(x) \in \mathbb{F}[x], g(x) \neq 0 \right\}$. *Then* $\mathbb{F}(x)$ *is a field.*

Proposition 21.1.1. *If* $\mathbb{F}_i$, $1 \leq i \leq n$, *is a subfield of* $\mathbb{F}$, *then intersection* $\cap_{1 \leq i \leq n} \mathbb{F}_i$ *of* $\mathbb{F}_i$, $1 \leq i \leq n$, *is a subfield of* $\mathbb{F}$.

Proof. The proof of the lemma is similar to the proof of Proposition 15.2.1. □.

As a result of Proposition 21.1.1, we have the following definition.

Definition 21.1.1. *The intersection of all subfields of* $\mathbb{F}$ *is called* **the prime field of** $\mathbb{F}$.

Two important prime fields are $\mathbb{Q}$ and $\mathbb{Z}_p$ for a prime number p, and these are more important than they look.

Recall Corollary 13.3.1: the characteristic of a field is either 0 or a prime p. For example, $\mathbb{Q}$ is a field of characteristic zero and $\mathbb{Z}_p$ is a field of characteristic p where p is prime. Therefore, there are two different types of prime fields: infinite prime fields with characteristic 0 and finite prime fields with prime characteristic. The following propositions classify and specify these two different prime fields.

Proposition 21.1.2. *If* $\mathbb{F}$ *is a field of characteristic* p*, then* p *is prime.*

Proof. By Theorem 13.3.2, an integral domain has characteristic 0 or a prime p. Therefore, the characteristic of $\mathbb{F}$ is a prime number since a field is an integral domain by Proposition 13.2.1. □

Proposition 21.1.3. *The prime field of* $\mathbb{F}$ *is isomorphic to* $\mathbb{Q}$ *or* $\mathbb{Z}_p$ *where* p *is prime.*

Proof. Suppose that $\mathbb{F}$ is a field of characteristic zero and let $\{\mathbb{F}_i\}_{i\in I}$ be the set of all subfields of $\mathbb{F}$ where I is an index set. Then, $\cap_{i\in I}\mathbb{F}_i$ is a subfield of $\mathbb{F}$

by Proposition 21.1.1 so 0 and $1 \in \cap_{i \in I} \mathbb{F}_i$. This implies that $\cap_{i \in I} \mathbb{F}_i$ contains a subring that is isomorphic to $\mathbb{Z}$. Furthermore, by adding multiplicative inverses (= reciprocals) we can make a subring that is isomorphic to $\mathbb{Q}$.[1] So the prime subfield of $\mathbb{F}$ is isomorphic to $\mathbb{Q}$ because $\mathbb{Q}$ does not have any nontrivial subfield. A proof for a field of characteristic p similar. $\square$

Example 21.1.2. *Find a prime subfield of* $\mathbb{Z}_5[x]/< x^2 + x + 1 >$.[2]

Solution. The subfield $\{a+ < x^2 + x + 1 > | a \in \mathbb{Z}_5\}$ of $\mathbb{Z}_5[x]/< x^2 + x + 1 >$ is the prime subfield. Note that it is isomorphic to $\mathbb{Z}_5$. $\square$

21.2 Finite and Infinite Fields.

Once again, finite fields and prime characteristics are interchangeable as infinite fields and 0 characteristics are interchangeable. We start with properties of a finite field using its characteristic.

Proposition 21.2.1. *Let* $\mathbb{F}$ *be a field of characteristic* p. *Then*

1. *For any* $x \in \mathbb{F}$, $px = 0$.
2. *Let* $\theta : \mathbb{F} \to \mathbb{F}$ *be a map defined by*

$$\theta(x) = x^{p^k}, \tag{21.1}$$

for $x \in \mathbb{F}$, *then* θ *is a field homomorphism for* $k \geq 1$.

Proof.

[1]It is called **localization**. Check Section 13.4 for a more general case: integral domains.
[2]Why is $\mathbb{Z}_5[x]/< x^2 + x + 1 >$ a field?

1. Since $\mathbb{F}$ is a field, there is the unity $1 \in \mathbb{F}$ and $p \cdot 1 = 0$. Then, for any $a \in \mathbb{F}$, $a = a \cdot 1$ so $pa = p(a1) = (pa)1 = (ap)1 = a(p1) = a0 = 0$.

2. We prove (21.1) by induction on k. If $k = 1$, then (21.1) becomes $\theta(x) = x^p$. Then, $\theta(x+y) = (x+y)^p = C(p,0)x^p + \cdots + C(p,i)x^{p-i}y^i + \cdots + C(p,p)y^p$ where $C(p,i) = \dfrac{p!}{(p-i)!i!}$. Since p is the characteristic of $\mathbb{F}$, we have that $\theta(x+y) = x^p + y^p$. So (21.1) holds for $k = 1$. Assume that (21.1) holds for k and let's consider the case of $k+1$. Since $\theta(x) = x^{p^{k+1}}$ by (21.1), $\theta(x+y) = (x+y)^{p^{k+1}} = (x+y)^{p^k p} = (x^{p^k} + y^{p^k})^p = x^{pp^k} + y^{pp^k} = x^{p^{k+1}} + y^{p^{k+1}}$. This implies that θ is an additive homomorphism. We also have that $\theta(xy) = (xy)^{p^{k+1}} = x^{p^{k+1}} y^{p^{k+1}} = \theta(x)\theta(y)$. This implies that θ is a multiplicative homomorphism. Therefore, θ is a field homomorphism.

$\square$

The following is for arbitrary fields, finite or infinite.

Proposition 21.2.2. *A homomorphism θ from a field $\mathbb{F}$ to a field $\mathbb{E}$ is either injective or the zero map.*

Proof. Suppose that θ is a homomorphism from $\mathbb{F}$ to $\mathbb{E}$. Assume that θ is not injective. Then, there are two distinct elements x_1 and x_2 in $\mathbb{F}$ such that $\theta(x_1) = \theta(x_2)$. This implies that $\theta(x_1 - x_2) = 0$. Since $x_1 - x_2$ is not zero, there is an inverse $\dfrac{1}{x_1 - x_2}$ in $\mathbb{F}$. Since $\theta\left(x_1 - x_2 \dfrac{1}{x_1 - x_2}\right) = \theta\left(x_1 - x_2\right)\theta\left(\dfrac{1}{x_1 - x_2}\right) = 0$, $\theta(1) = 0$. Then, $\theta(x) = \theta(x1) = \theta(x)\theta(1) = \theta(x)0 = 0$ for any $x \in \mathbb{F}$ so θ is the zero map and this proves the the proposition. $\square$

Example 21.2.1. *Find all field homomorphisms from $\mathbb{Z}_3$ to $\mathbb{Z}_p$ where p is prime and $p > 3$.*

Solution. Let θ be a field homomorphism from $\mathbb{Z}_3$ to $\mathbb{Z}_p$. Assume that θ is not the zero map. Then $\theta(\overline{1}) = \overline{1}$ and this implies $\theta(\overline{2}) = \overline{2}$ so we have $\theta((\overline{2})(\overline{2})) = \theta(\overline{4}) = \theta(\overline{1}) = \overline{1}$. On the other hand, we have $\theta(\overline{2})\theta(\overline{2}) = (\overline{2})(\overline{2}) = \overline{4} \neq \overline{1}$ since $p > 3$, a contradiction. Therefore the zero map is the only homomorphism from $\mathbb{Z}_3$ to $\mathbb{Z}_p$. $\square$

By Example 21.2.1, we can see that two finite fields, $\mathbb{Z}_p$ and $\mathbb{Z}_q$ are not (field) isomorphic if p and q are two distinct prime numbers. The following example is for infinite fields.

Example 21.2.2. *Prove that the fields $\mathbb{Q}(\sqrt{3})$ and $\mathbb{Q}(\sqrt{2})$ are not isomorphic.*

Solution. Let's assume that there is a field isomorphism θ from $\mathbb{Q}(\sqrt{3})$ to $\mathbb{Q}(\sqrt{2})$. Since $\theta(1) = 1$, we have that $\theta(\mathbb{Q}) \subset \mathbb{Q}$. Let $\theta(\sqrt{3}) = a + b\sqrt{2}$ for some $a, b \in \mathbb{Q}$. Then, since $\theta(\sqrt{3}\sqrt{3}) = \theta(3)$, we have that $a^2 + 2b^2 + 2ab\sqrt{2} = 3$ and this implies that $2ab = 0$. This makes $a = 0$ or $b = 0$. If $a = 0$, then we have that $2b^2 = 3$, a contradiction. If $b = 0$, then we have that $\theta(\mathbb{Q}(\sqrt{3})) \subset \mathbb{Q}$, another contradiction. Thus they are not isomorphic. $\square$

Chapter 21 Exercises.

1. Prove that if n is not prime, then $\mathbb{Z}_n$ is not a field but a ring.

2. Is there a field consisting of 4 elements? Justify your answer.

3. Suppose an addition $+$ and a multiplication $\cdot$ are defined in $\mathbb{R}^2 =$ by

$$(a, b) + (c, d) = (a + c, b + d)$$

$$(a, b) \cdot (c, d) = (ac - bd, ad + bc)$$

(a) Prove that $(\mathbb{R}^2, +, \cdot)$ is a field.[3]

(b) Prove that the fields $(\mathbb{R}^2, +, \cdot)$ and $\mathbb{C}$ are isomorphic.

4. Prove that $\left\{ \frac{f(x)}{g(x)} | f(x), g(x) \in \mathbb{F}[x], g(x) \neq 0 \right\}$ is a field under the usual addition and the multiplication.

5. (a) Find all the field homomorphisms from $\mathbb{Z}_5$ to itself, i.e., endomorphisms on $\mathbb{Z}_5$.

   (b) Find all the field endomorphisms on $\mathbb{Z}_3$.

   (c) Find all the field endomorphisms on $\mathbb{Z}_p$ where p is prime.

6. Find all the field homomorphisms from $\mathbb{Z}_5$ to $\mathbb{Q}$.

7. Prove that any ring homomorphism from a field to a ring is either injective or the zero map.

8. Suppose $\mathbb{F}$ is a field and let $\mathbb{F}^* = \mathbb{F}/\{0\}$. Prove that $\mathbb{F}^*$ is a multiplicative group and show that $\mathbb{F}^*$ a ring (so it is not a field).

9. Show that $\mathbb{Q}[\sqrt{2}]$ a subfield of $\mathbb{R}$.

10. (a) Prove that $\mathbb{Q}[\sqrt{-3}]$ is a field.[4]

    (b) Prove that $\mathbb{Q}[\sqrt{-p}]$ is a field where p is a prime number.

11. Let $\mathbb{E}$ and $\mathbb{F}$ be fields. If θ is a non-zero field homomorphism from $\mathbb{E}$ to $\mathbb{F}$, then prove each of the following:

    (a) $\theta(0) = 0$

[3] $(\mathbb{R}^2, +, \cdot)$ means that $\mathbb{R}^2$ is an algebraic structure with two operations, $+$ and $\cdot$.
[4] Show that $\mathbb{Q}[\sqrt{-3}] = \mathbb{Q}(\sqrt{-3})$.

(b) $\theta(1) = 1$

(c) Prove that $ker(\theta)$ is a subfield of $\mathbb{E}$.

12. Let $\mathbb{F}_1$ and $\mathbb{F}_2$ are fields. Prove that $\mathbb{F}_1 \cap \mathbb{F}_2$ is a subfield of $\mathbb{F}_1$ and $\mathbb{F}_2$.

13. (a) Find the prime subfield of $\mathbb{Z}_5/< x^3 + x^2 + 1 >$.

(b) Find a non-prime field of characteristic p.

(c) Find a non-prime field of characteristic zero.

14. (a) Prove that $\mathbb{Z}_3[i] = \{a + bi | a, b \in \mathbb{Z}_3, i^2 = -1 \equiv 2 \mod 3\}$ is $\mathbb{Z}_3(i)$. (So, it is a field.)

(b) What is the characteristic of $\mathbb{Z}_3[i]$?

15. (a) Prove that the fields $\mathbb{Q}(\sqrt{2})$ and $\mathbb{Q}(\sqrt{5})$ are not isomorphic.

(b) Prove that the fields $\mathbb{Q}(\sqrt{2}, \cdots, \sqrt{p})$ and $\mathbb{Q}(\sqrt{5})$ are not isomorphic where p is a prime.

Chapter 22

Field Extensions Using a Zero of a Polynomial

As $\mathbb{R}$ is a subfield of $\mathbb{C}$, an arbitrary field $\mathbb{F}$ can be extended to a larger field so that $\mathbb{F}$ becomes a subfield and it is our main interest of this chapter. Recall that if $\mathbb{F}$ is a field, $\mathbb{F}[x]$ is a polynomial ring. If $p(x) \in \mathbb{F}[x]$ is irreducible, $\mathbb{F}[x]/< p(x) >$ is a field and has[1] $\mathbb{F}$ as its subfield. Moreover $p(x)$ has a zero in the field $\mathbb{F}[x]/< p(x) >$.

22.1 Irreducible Polynomials and Extension Fields

We start with a simple definition of an extension field.

Definition 22.1.1. *A field $\mathbb{E}$ is called an* **extension field** *of a field $\mathbb{F}$ if $\mathbb{F} \subset \mathbb{E}$ and it has the same operations of $\mathbb{F}$.*

[1]Physically $\mathbb{F}[x]/< p(x) >$ doesn't have $\mathbb{F}$ but $\mathbb{F}[x]/< p(x) >$ has a subfield that is isomorphic to $\mathbb{F}$

Example 22.1.1. $\mathbb{R}$ *is an extension field of* $\mathbb{Q}$.

Example 22.1.2. $\mathbb{C}$ *is an extension field of* $\mathbb{R}$. *Also,* $\mathbb{C}$ *is an extension field of* $\mathbb{Q}$.

Example 22.1.3. $\mathbb{Q}(\sqrt{p})$ *is an extension field of* $\mathbb{Q}$ *for* $p \in \mathbb{R}$. *It is also true if* $p \in \mathbb{C}$.[2].

For a given polynomial $f(x)$ over a field $\mathbb{F}$, it is an interesting question whether $f(x)$ has a zero or not in $\mathbb{F}$. If there is no zero of $f(x)$ in $\mathbb{F}$, there must[3] be a zero if $\mathbb{F}$ is extended to a larger field, an extension field. If $\mathbb{F}$ doesn't have a zero of $f(x)$, $f(x)$ might[4] be irreducible over $\mathbb{F}$ so we consider irreducible polynomials, in particular over the number field $\mathbb{Q}$.

Besides the Eisenstein criterion and the mod p-test of a polynomial, the following criterion of the irreducibility of a polynomial is useful and it is studied in a high school mathematics. We call it a lemma but sometimes it is called **the Rational Root Theorem.**

Lemma 22.1.1. (Rational Root Theorem)
Suppose $f(x) = a_n x^n + \cdots + a_0 \in \mathbb{Z}[x]$ *and* $\frac{p}{q} \in \mathbb{Q}$ *is a zero of* $f(x)$ *with* $(p, q) = 1$ *for* $p, q \in \mathbb{Z}$. *Then* $p|a_0$ *and* $q|a_n$.

Solution. $\frac{p}{q}$ is a zero of $f(x)$, $a_n \frac{p^n}{q^n} + \cdots + a_1 \frac{p}{q} + a_0 = 0$ and this makes $a_n p^n + a_{n-1} p^{n-1} q + \cdots + a_1 p q^{n-1} + a_0 q^n = 0$. Then we can derive that $a_n p^n = -a_{n-1} p^{n-1} q - \cdots - a_1 p q^{n-1} - a_0 q^n$ and this implies that p divides $a_0 q^n$. Since

[2]What's the meaning of $\sqrt{i}$?
[3]The existence of a zero in an extension field is need to be proved.
[4]Such an $f(x)$ must be irreducible if $deg(f(x)) = 2$ or 3.

$(p,q)=1$, we have $p|a_0$. We can also have $q|a_n$ by changing the role of p and q. Thus the lemma is proved. □

Example 22.1.4. $\frac{-3}{2}$ *is a zero of the polynomial* $2x^2-5x-12$, *and* $-3|12$ *and* $2|2$.

By Lemma 22.1.1, we have the following property of zeros of a primitive polynomial.[5]

Proposition 22.1.1. *Let* $f(x)\in\mathbb{Z}[x]$ *be a monic polynomial.*[6] *Then any rational zero of* $f(x)$ *is an integer.*

Proof. If $\frac{p}{q}\in\mathbb{Q}$ with $(p,q)=1$ is a zero of a monic polynomial $f(x)\in\mathbb{Z}[x]$, then since the leading coefficient of $f(x)$ is 1, $q|1$ by Lemma 22.1.1 This implies that $q=\pm1$ and this makes $\frac{p}{q}$ an integer. □

Next theorem proves the existence of an extension field that include a zero of a given polynomial.

Theorem 22.1.1. *For* $f(x)\in\mathbb{F}$, *there exists an extension field* $\mathbb{E}$ *that contains a zero of* $f(x)$. *If* $f(x)$ *is irreducible over* $\mathbb{F}$, *then* $\mathbb{E}$ *can be the quotient field* $\mathbb{F}[x]/<f(x)>$.

Proof. If $f(x)$ is irreducible over $\mathbb{F}$, then $\mathbb{F}[x]/<f(x)>$ is a field. If $f(x)$ is not irreducible over $\mathbb{F}$, then $f(x)=g(x)h(x)$ where $g(x)$ is an irreducible factor of $f(x)$ over $\mathbb{F}$. This implies that $\mathbb{E}=\mathbb{F}[x]/<g(x)>$ is a field. Suppose

[5] There is no common factor of all its coefficients of a primitive polynomial.

[6] A polynomial with the leading coefficient 1.

$g(x) = a_n x^n + \cdots + a_0$ where $a_n, \cdots, a_0 \in \mathbb{F}$. Then $g(x) \in \mathbb{F}[x] \subset \mathbb{E}[x]$ and we can derive that

$$\begin{aligned} f(x+ < g(x) >) &= g(x+ < g(x) >)h(x+ < g(x) >) \\ &= \{a_n(x+ < g(x) >)^n + \cdots + a_0\}h(x+ < g(x) >) \\ &= \{a_n x^n + \cdots + a_0 + < g(x) >\}h(x+ < g(x) >) \\ &= < g(x) > h(x+ < g(x) >) =< g(x) > \end{aligned}$$

This implies that $x+ < g(x) >$ a zero of $f(x)$ over $\mathbb{E}$. If $f(x)$ is irreducible over $\mathbb{F}$, then it is easy to prove that $x+ < f(x) >$ is a zero of the polynomial $f(x)$ over the field $\mathbb{F}/< f(x) >$ and this completes the proof. □

Example 22.1.5. *Let* $f(x) = 3x^3 + 2x^2 - 4x + 6$. *Then* $f(x)$ *is irreducible over* $\mathbb{Q}$ *by Eisenstein Criterion and this makes* $\mathbb{Q}[x]/ < f(x) >$ *a field. Note that* $x+ < f(x) > \in \mathbb{Q}[x]/ < f(x) >$ *is a zero of* $f(x)$ *by Theorem 22.1.1. In fact,*

$$\begin{aligned} f(x+ < f(x) > &= 3(x+ < f(x) >)^3 + 2(x+ < f(x) >)^2 - 4(x+ < f(x) >) + 6 \\ &= 3(x^3 + < f(x) >) + 2(x^2 + < f(x) >) - 4x - 4 < f(x) > +6 \\ &= (3x^3 + 2x^2 - 4x + 6) + < f(x) > \\ &= < f(x) > \end{aligned}$$

and $< f(x) >$ *is the additive identity of* $\mathbb{Q}[x]/ < f(x) >$.

Example 22.1.6. $x^2 + 1$ *is irreducible over* $\mathbb{R}$ *since* i *and* $-i$ *are zeros and they are not in* $\mathbb{R}$. *This implies that* $\mathbb{R}[x]/< x^2 + 1 >$ *is a field and* $x+ < x^2 + 1 >$ *is a zero of* $x^2 + 1$ *in the field* $\mathbb{R}[x]/< x^2 + 1 >$.

Let's define a map θ from $\mathbb{R}[x]/< x^2+1 >$ to $\mathbb{C}$ as $\theta(x+ < x^2+1 >) = i$. Then $\theta(a+bx+ < x^2+1 >) = a+bi$ and this θ is a field isomorphism. So this confirms that i is a zero of x^2+1 and the following theorem generalize this property.

Lemma 22.1.2. *Let $\mathbb{F}$ be a field and $f(x)$ be an irreducible polynomial over $\mathbb{F}$. Then $\mathbb{F}$ is isomorphic to a subfield of the field $\mathbb{F}[x]/< f(x) >$.*

Proof. Since $f(x)$ is irreducible, by Theorem 17.3.1, $\mathbb{F}[x]/< f(x) >$ is a field. For $a \in \mathbb{F}$, let us define a map $\theta : \mathbb{F} \to \mathbb{F}[x]/< f(x) >$ as $\theta(a) = a+ < f(x) >$. Then θ is a field homomorphism. Suppose $a \in ker\theta$. Then $\theta(a) = a+ < f(x) >=< f(x) >$ and this implies that $a \in< f(x) >$. Since the degree of $f(x)$ is greater than one, $< f(x) >$ doesn't include a constant except 0. This means that $a = 0$ so θ is one to one. Therefore, $\mathbb{F}$ is isomorphic to a subfield $\theta(\mathbb{F})$ of $\mathbb{F}[x]/< f(x) >$. □

22.2 Zero of a Polynomial

Still our main ingredient to extend a field is an irreducible polynomial and its zero. We will look at several properties of a zero of a polynomial in this section.

Theorem 22.2.1. *a is a zero of $f(x) \in \mathbb{F}[x]$ if and only if $(x-a)|f(x)$.*

Proof. The proof of the theorem is straightforward by the Remainder Theorem. □

Definition 22.2.1. *Let a be a zero of $f(x) \in \mathbb{F}[x]$. If $(x-a)^m | f(x)$, but $(x-a)^{m+1} \not| f(x)$, then m is called the* **multiplicity** *of $f(x)$ of a. If $m = 1$, a is called a* **simple zero**.

Lemma 22.2.1. *$f(x)$ has a multiple zero a if and only if a is a common zero of $f(x)$ and $f'(x)$ where $f'(x)$ is the derivative of $f(x)$ with respect to x.*

Proof. If $f(x)$ has a zero with multiplicity $m \geq 2$, then $f(x) = (x-a)^m f_1(x)$ where a is not a zero of $f_1(x)$. Then $f'(x) = (x-a)^{m-1}(mf_1(x) + (x-a)f_1'(x))$ and since $m - 1$ is greater than zero, a is a zero of $f'(x)$. Now, assume that $f(x) = (x-a)^m f_1(x)$ has a simple zero, i.e., $m = 1$ with $f_1(a) \neq 0$. Then, $f'(a) = f_1(a) + (a-a)f_1'(a) = f_1(a) \neq 0$. This implies that a is not a common zero of $f(x)$ and $f'(x)$. Thus we have proven the lemma. □

Lemma 22.2.2. *For $f(x) \in \mathbb{F}[x]$, $f(x)$ has only a simple zero if and only if $gcd(f(x), f'(x))$ is a non-zero element of $\mathbb{F}$.*

Proof. Suppose $f(x)$ has only simple zeros. Then $f(x)$ can be written as

$$f(x) = c(x-a_1)\cdots(x-a_n)$$

over some extension field of $\mathbb{F}$ where $a_1, \cdots, a_n$ are all distinct numbers and $0 \neq c \in \mathbb{F}$, and $f'(x) = \sum_{1 \leq i \leq n} c(x-a_1)\cdots\widehat{(x-a_i)}(x-a_{i+1})\cdots(x-a_n)$ where $\widehat{(x-a_i)}$ means that the factor $(x-a_i)$ is missing. This implies that $a_1, \cdots, a_n$ cannot be a zero of $f'(x)$. This implies that $f(x)$ and $f'(x)$ are relatively prime. Conversely, suppose $f(x)$ and $f'(x)$ are relative prime. If $f(x)$ has a multiple zero a, then by Lemma 22.2.1, $f(x)$ and $f'(x)$ has a common factor $x-a$, a contradiction. Therefore $f(x)$ has only a simple zero and it proves the lemma. □

Lemma 22.2.3. *Let $f(x)$ be irreducible over a field $\mathbb{F}$ and a be a zero of $f(x)$ where a is an element of an extension field $\mathbb{E}$ of $\mathbb{F}$. If $g(x) \in \mathbb{F}[x]$ and $g(a) = 0$, then $f(x)$ divides $g(x)$.*

Proof. If $f(x)$ does not divide $g(x)$, then since $f(x)$ is irreducible, there are $s(x)$ and $t(x) \in \mathbb{F}[x]$ such that $f(x)s(x)+g(x)t(x) = 1$. Since a is a zero of $f(x)$ and $g(x)$, $x - a$ is a factor of $f(x) \in \mathbb{E}[x]$ and it is also a factor of $g(x) \in \mathbb{E}[x]$. This implies that $x - a$ divides 1, a contradiction. Therefore $f(x)$ divides $g(x)$. □

Proposition 22.2.1. *For an irreducible polynomial $f(x)$ over a field $\mathbb{F}$, $f(x)$ has a multiple zero in some extension field if and only if $f'(x) = 0$.*

Proof. Let us assume that $f(x)$ has a multiple zero a in some extension field. Then a is a zero of $f'(x)$ by Lemma 22.2.1 and this implies that $x - a$ is a common factor of $f(x)$ and $f'(x)$. Let's assume that $f'(x) \neq 0$. Since $f'(x) \in \mathbb{F}[x]$ and a is a zero of $f'(x)$, $f(x)$ divides $f'(x)$ by Lemma 22.2.3. This implies that $deg(f(x)) > deg(f'(x))$, a contradiction. Thus $f'(x) = 0$. Conversely let us assume that $f'(x) = 0$. Then since $f(x)$ and $f'(x)$ are not relatively prime, $f(x)$ has a multiple zero by Lemma 22.2.2 and this completes a proof of the proposition. □

Corollary 22.2.1. *If $f(x)$ is an irreducible polynomial over a field $\mathbb{F}$ of characteristic zero, then it has a simple zero.*

Proof. Without loss of generality, we can assume that the degree of $f(x)$ is greater than zero. Then $f'(x)$ is not zero and $f(x)$ has a simple zero by Proposition 22.2.1. □

Theorem 22.2.2. *For any polynomial $f(x)$ over a field $\mathbb{F}$ with $deg(f(x)) = n$, then $f(x)$ has at most n zeros over an extension field of $\mathbb{F}$.*

Proof. We prove the theorem by induction on the degree n of $f(x)$. If $n = 1$, then $f(x) = ax + b$ where $a, b \in \mathbb{F}$ and $a \neq 0$. Then $\frac{-b}{a} \in \mathbb{F}$ is the zero of $f(x)$ so the theorem is true for $n = 1$. Let's assume that the theorem holds for a polynomial whose degree is $n - 1$ or less than n. If $f(x)$ is reducible, then $f(x) = f_1(x)f_2(x)$ where $1 \leq n_1 = deg(f_1(x)), n_2 = deg(f_2(x)) < n$. By induction, $f(x)$ has at most $n_1 + n_2 \leq n$ zeros over an extension field of $\mathbb{F}$. So we have thus proven the theorem. If $f(x)$ is irreducible, there is an extension field $\mathbb{E}$ of $\mathbb{F}$ such that $a \in \mathbb{E}$ is a zero of the polynomial by Theorem 22.1.1. This implies that $f(x) = (x - a)f_3(x)$. Since the degree of $f_3(x)$ is $n - 1$, $f_3(x)$ has at most $n - 1$ zeros by induction. This implies that $f(x)$ has at most n zeros over an extension field of $\mathbb{F}$ and this proves the theorem. □

22.3 Splitting Field and Field Extensions

If we look at the proof of Theorem 22.2.2, we see that there exists an extension field that include a zero of a given polynomial. Therefore, we can say that there exists some extension fields that include all the zeros of a polynomial. We'd like to look at the smallest such a field.

Definition 22.3.1. *Suppose* $f(x) \in \mathbb{F}[x]$. *The smallest extension field* $\mathbb{E}$ *of* $\mathbb{F}$ *is called a* **splitting field** *of* $f(x)$ *if* $f(x)$ *is a product of linear polynomials of* $\mathbb{E}[x]$. *In this case, we say that* $f(x)$ **splits** *in* $\mathbb{E}$.[7]

If $\mathbb{E}$ is the splitting field of $f(x)$, then $\mathbb{E}$ includes all the zeros of $f(x)$. Let's take a look at several splitting fields.

Example 22.3.1. $\mathbb{Q}$ *is the splitting field of* x^2-3x+2 *since* 1 *and* 2 *are elements of* $\mathbb{Q}$.[8]

Example 22.3.2. $\mathbb{Q}(i)$ *is the splitting field of* x^4-1 *since* $x^4-1=(x+1)(x-1)(x+i)(x-i)$.

Example 22.3.3. *The splitting field of* $(x^2-2)(x^2+1)$ *is* $\mathbb{Q}(\sqrt{2}, i)$ *since* $(x^2-2)(x^2+1)=(x-\sqrt{2})(x+\sqrt{2})(x+i)(x-i)$.

Example 22.3.4. *Find an extension field of* $\mathbb{Z}_3$ *that has a zero of* $x^4+x^3+2x^2+1 \in \mathbb{Z}_3[x]$. *Identify all elements of the extension field and find the number of elements of the field.*

Solution. By $x^4+x^3+2x^2+1=(x+1)(x^3+2x+1)$, $x=-1$ is a root of $x^4+x^3+2x^2+1$. Since $f(\bar{0}) \neq \bar{0}, f(\bar{1}) \neq \bar{0}$, and $f(\bar{2}) \neq \bar{0}$, x^3+2x+1 is irreducible over $\mathbb{Z}_3$. So we have that $\mathbb{Z}_3[x]/< x^3+2x+1 >$ is a field and it is an extension field of $\mathbb{Z}_3$. Note that $x+ < x^3+2x+1 >$ a zero of x^3+2x+1 and $\mathbb{Z}_3[x]/< x^3+2x+1 > = \{ax^2+bx+c+ < x^3+2x+1 > | a,b,c \in \mathbb{Z}_3\}$. So the field $\mathbb{Z}_3[x]/< x^3+2x+1 >$ has 27 elements. □

[7]Even though $f(x)$ splits in $\mathbb{E}$, $\mathbb{E}$ might not be the splitting field of $f(x)$. Why?

[8]Why $\mathbb{Q}$ is the smallest one?

Definition 22.3.2. *Let* $\mathbb{F}$ *be a subfield of a field* $\mathbb{E}$ *and* $a \in \mathbb{E}$*. Then* $\mathbb{F}(a)$ *is the intersection of all subfields of* $\mathbb{E}$ *that contains* $\mathbb{F}$ *and* a*. Inductively,* $\mathbb{F}(a_1, \cdots, a_n)$ *is the intersection of all subfield of* $\mathbb{E}$ *that contains* $\mathbb{F}$ *and* $a_1, \cdots, a_n$*.* $\mathbb{F}(a)$ *is called a* **simple extension** *of* $\mathbb{F}$ *and* a *is called* **a primitive element**.

By Definition 22.3.2, we can say that $\mathbb{F}(a)$ is the smallest field that contains $\mathbb{F}$ and a. Inductively, $\mathbb{F}(a_1, \cdots, a_n)$ is the smallest field that contains $\mathbb{F}$ and $a_1, \cdots, a_n$.

Example 22.3.5. *If* $a \in \mathbb{F}$*, then* $\mathbb{F}(a) = \mathbb{F}$.

Example 22.3.6. $\mathbb{Q}(i) = \{a + bi | a, b \in \mathbb{Q}\}$ *and* $\mathbb{Q}(\sqrt{2}) = \{a + b\sqrt{2} | a, b \in \mathbb{Q}\}$.

Theorem 22.3.1. (Existence of Splitting Fields)
For a nonconstant polynomial $f(x) \in \mathbb{F}[x]$*, there exists a splitting field* $\mathbb{E}$ *of* $f(x)$ *over* $\mathbb{F}$.

Proof. This proof is similar to the one for Theorem 22.2.2. Let $f(x)$ be a nonconstant polynomial $f(x) \in \mathbb{F}[x]$. We prove the theorem by induction on $n = deg(f(x))$. If n is 1, then $f(x) = ax + b$ for $a, b \in \mathbb{F}$. So the zero of the polynomial is in the field $\mathbb{F}$ and this makes $\mathbb{E} = \mathbb{F}$. Assume that the theorem is true for all polynomials of degree less than n. If $f(x)$ is reducible, then by induction there is nothing to prove. Suppose that $f(x)$ is irreducible. Then there is an extension field $\mathbb{E}$ of $\mathbb{F}$ such that a_1 is a zero of $f(x)$ and $a_1 \in \mathbb{E}$ by Theorem 22.1.1. This means that $f(x) = (x - a_1)f_1(x)$ and $f_1(x)$ splits in the field $\mathbb{F}(a_2, \cdots, a_n)$ where $a_2, \cdots, a_n$ are zeros of $f_1(x)$ by induction. This implies that $f(x)$ splits in $\mathbb{F}(a_1, \cdots, a_n)$ and this proves the theorem. □

Using a zero of a polynomial, we can construct a simple field extension but the other way is not always true. That is, even though $\mathbb{F}(a)$ is a simple field extension of $\mathbb{F}$, a is not necessarily a zero of a polynomial over $\mathbb{F}$. The following definition separates these two cases.

Definition 22.3.3. *Let $\mathbb{E}$ be an extension field of a field $\mathbb{F}$. A non-zero element $a \in \mathbb{E}$ is called* **algebraic** *over $\mathbb{F}$ if a is a zero of a non-zero polynomial $f(x) \in \mathbb{F}[x]$; otherwise it is called* **transcendental**.

Example 22.3.7. *Since $\sqrt{2}$ is a zero of $x^2 - 2 \in \mathbb{Q}$, $\sqrt{2}$ is algebraic over $\mathbb{Q}$.*

Example 22.3.8. *$\sqrt{3} + \sqrt{5}$ is algebraic over $\mathbb{Q}$.*

Solution. Let $x = \sqrt{3} + \sqrt{5}$. Then, $(x - \sqrt{3})^2 = 5$ and this implies that $x^2 - 2x\sqrt{3} + 3 = 5$ or $(-2x\sqrt{3})^2 = (2 - x^2)^2$. This makes $12x^2 = 4 - 4x^2 + x^4$ or $x^4 - 16x^2 + 4 = 0$ and $\sqrt{3} + \sqrt{5}$ is a zero of this polynomial $x^4 - 16x^2 + 4$ over $\mathbb{Q}$. □

Recall that $\mathbb{Q}(\sqrt{3}, \sqrt{5}) = \mathbb{Q}(\sqrt{3} + \sqrt{5})$ so we can conclude that $\mathbb{Q}(\sqrt{3}, \sqrt{5})$ is a simple algebraic extension of $\mathbb{Q}$ by Example 22.3.8. The following are famous examples of transcendental numbers, π and e.

Example 22.3.9. *The number π and the number e are transcendental over $\mathbb{Q}$.*[9] *Since π is a zero of $x - \pi \in \mathbb{R}$, it is algebraic over $\mathbb{R}$.*

[9] It was proved first that e is transcendental and π was proved later using the fact of e. Actual proofs are omitted here.

22.4 Uniqueness of Splitting Fields

We turn our attention back to the splitting field of a polynomial and we want to prove that it is unique up to isomorphism.

Theorem 22.4.1. *Let* $\mathbb{F}$ *and* $\mathbb{F}_1$ *be isomorphic fields. Then the polynomial rings* $\mathbb{F}[x]$ *and* $\mathbb{F}_1[t]$ *are also isomorphic.*

Proof. Let θ be a field isomorphism from $\mathbb{F}$ to $\mathbb{F}_1$. Let us define the map $\overline{\theta}$ from $\mathbb{F}[x]$ to $\mathbb{F}_1[t]$ as follows:

$$\overline{\theta}(a_n x^n + \cdots + a_0) = \theta(a_n)t^n + \cdots + \theta(a_0) \tag{22.1}$$

for $a \in \mathbb{F}$, $\overline{\theta}(a) = \theta(a)$ and for $f(x) = a_n x^n + \cdots + a_0 \in \mathbb{F}[x]$. Then $\overline{\theta}$ is a ring homomorphism (why?). Also, $\overline{\theta}$ is onto and one to one (why?). Therefore, $\overline{\theta}$ is an isomorphism. □

Thanks to Theorem 22.4.1, we can define the following map f^θ and this map has an important role on field theory.

$$f^\theta(x) = \overline{\theta}(a_n x^n + \cdots + a_0) = \theta(a_n)t^n + \cdots + \theta(a_0)$$

where $f(x) = a_n x^n + \cdots + a_0$ and θ is a given field isomorphism.

Theorem 22.4.2. *Let* $f(x) \in \mathbb{F}[x]$ *be irreducible over a field* $\mathbb{F}$ *and let* a *be a zero of* $f(x)$ *where* a *is an element of some extension field of* $\mathbb{F}$*. Then there is an field isomorphism* θ_1 *from* $\mathbb{F}[x]/< f(x) >$ *to* $\mathbb{F}(a)$ *such that* $\theta_1(x+ < f(x) >) = a$ *for* $x+ < f(x) > \in \mathbb{F}[x]/< f(x) >$.

Proof. For any $g(x) \in \mathbb{F}[x]$, let us define a map θ from $\mathbb{F}[x]$ to $\mathbb{F}(a)$ as follows:

$$\theta(g(x)) = g(a) \tag{22.2}$$

for $g(x) \in \mathbb{F}[x]$. Then obviously θ is a ring homomorphism.[10] Since $f(a) = 0$, $< f(x) >$ is a subset of $ker(\theta)$ and since $f(x)$ is irreducible, $ker(\theta)$ is a subset of $< f(x) >$. So, since $ker(\theta) \neq \mathbb{F}[x]$ and $< f(x) >$ is maximal, we derive that $ker(\theta) =< f(x) >$ and $\mathbb{F}[x]/< f(x) >$ is a field. Then, θ induces the field homomorphism θ_1 from $\mathbb{F}[x]/< f(x) >$ to $\mathbb{F}(a)$ such that $\theta_1(x+ < f(x) >) = a$. Since $\mathbb{F}[x]/< f(x) >$ and $\mathbb{F}(a)$ are fields, θ_1 is onto and it implies that $\mathbb{F}[x]/< f(x) >$ and $\mathbb{F}(a)$ are isomorphic. $\square$

Corollary 22.4.1. *If a and b are zeros of an irreducible polynomial $f(x) \in \mathbb{F}[x]$ over $\mathbb{F}$, then $\mathbb{F}(a) \cong \mathbb{F}(b)$.*

Proof. It can be derived from $\mathbb{F}(a) \cong \mathbb{F}[x]/< f(x) >$ and $\mathbb{F}(b) \cong \mathbb{F}[x]/< f(x) >$ by Theorem 22.1.1. $\square$

Note that $\mathbb{F}(a) \ncong \mathbb{F}(b)$ in general for arbitrary a and b. In particular, Corollary 22.4.1 is not true if a or b are zeros of a reducible polynomial.

Corollary 22.4.2. *Let θ be a field isomorphism from $\mathbb{F}$ to $\mathbb{F}_1$ and $\overline{\theta}$ is a ring isomorphism from $\mathbb{F}[x]$ to $\mathbb{F}_1[t]$ as (22.1). Then if $p(x)$ is irreducible over $\mathbb{F}$, then $\overline{\theta}(p(x))$ is irreducible over $\mathbb{F}_1$ and vice versa. Furthermore the fields $\mathbb{F}[x]/< p(x) >$ and $\mathbb{F}_1[t]/< \overline{\theta}(p(x)) >$ are isomorphic.*

Proof. Let us assume that $\overline{\theta}(p(x))$ is reducible. Then $\overline{\theta}(p(x)) = g(t)h(t)$ for some $g(t), h(t) \in \mathbb{F}_1[t]$ with $1 \leq deg(g(t)), deg(h(t)) \leq deg(\overline{\theta}(p(x)))$ and we can have the following commutative diagram:

[10] It is called **the evaluation at** a.

Figure 22.4.1.

$$\begin{array}{ccc} \mathbb{F}[x] & \xrightarrow{\overline{\theta}} & \mathbb{F}_1[t] \\ \upharpoonleft & & \upharpoonleft \\ \mathbb{F} & \xrightarrow{\theta} & \mathbb{F}_1 \end{array}$$

where $\mathbb{F} \rightharpoonup \mathbb{F}[x]$ (resp. $\mathbb{F}_1 \rightharpoonup \mathbb{F}_1[t]$) means that $\mathbb{F}$ is a subring of $\mathbb{F}[x]$. This diagram makes $(\overline{\theta})^{-1}(g(t))(\overline{\theta})^{-1}(h(t)) = p(x)$, which implies that $p(x)$ is reducible over $\mathbb{F}$, a contradiction. Thus it is irreducible. Symmetrically, we can see that that $p(x)$ is irreducible over $\mathbb{F}$ if $\overline{\theta}(p(x))$ is irreducible over $\mathbb{F}_1$. This implies that $\mathbb{F}[x]/< p(x) >$ and $\mathbb{F}_1[t]/< \overline{\theta}(p(x)) >$ are fields. If we define $\psi : \mathbb{F}[x]/< p(x) > \to \mathbb{F}_1[t]/< \overline{\theta}(p(x)) >$ such that ψ is the restriction of $\overline{\theta}$ to $\mathbb{F}[x]/< p(x) >$, then obviously ψ is a field isomorphism and this proves the corollary. □

Theorem 22.4.3. *Let $p(x)$ be an irreducible polynomial over a field $\mathbb{F}$ and let a be a zero of $p(x)$. If θ be a field isomorphism from $\mathbb{F}$ to $\mathbb{F}_1$ and b be a zero of $\overline{\theta}(p(x))$, then there is a field isomorphism ψ from $\mathbb{F}(a)$ to $\mathbb{F}_1(b)$ such that $\overline{\theta}(a) = b$ where $\overline{\theta}$ is the map in the Corollary 22.4.2 and the restriction $\overline{\theta}|_{\mathbb{F}}$ of $\overline{\theta}$ to $\mathbb{F}$ is the map θ.*

Proof. Let θ be an isomorphism from $\mathbb{F}$ to a field $\mathbb{F}_1$. By Corollary 22.4.2, there is a field isomorphism $\overline{\theta}$ from $\mathbb{F}[x]/< p(x) >$ to $\mathbb{F}_1[x]/< \overline{\theta}(p(x)) >$ such that the restriction of $\overline{\theta}$ to $\mathbb{F}$ is θ where $\overline{\theta}$ is the map in the proof of Corollary 22.4.2. By Corollary 22.4.1 and Corollary 22.4.2, we have the following commutative diagram:

Figure 22.4.2.

$$\begin{array}{ccccccc} \mathbb{F}(a) & \overset{\theta_1^{-1}}{\rightarrow} & \mathbb{F}[x]/< p(x) > & \overset{\overline{\theta}}{\rightarrow} & \mathbb{F}_1[x]/< \overline{\theta}(p(x)) > & \overset{\theta_2}{\rightarrow} & \mathbb{F}_1(b) \\ & \searrow & \uparrow & & \uparrow & \nearrow & \\ & & \mathbb{F} & \overset{\theta}{\rightarrow} & \mathbb{F}_1 & & \end{array}$$

where $\theta_1 : \mathbb{F}[x]/< p(x) > \rightarrow \mathbb{F}(a)$ and $\theta_2 : \mathbb{F}_1[x]/< \overline{\theta}(p(x)) > \rightarrow \mathbb{F}_1(b)$ are appropriate maps in Theorem 22.4.2 with appropriate embeddings. Then $\theta_2 \circ \overline{\theta} \circ \theta_1^{-1}$ is the required map from $\mathbb{F}(a)$ to $\mathbb{F}_2(b)$ and this proves the theorem. □

Example 22.4.1. *$\sqrt{7}$ and $-\sqrt{7}$, the zeros of $p(x) = x^2 - 7$ are not elements of $\mathbb{Q}$. So $p(x)$ is irreducible over $\mathbb{Q}$. If we define θ from $\mathbb{Q}[\sqrt{7}]$ to itself by $\theta(a) = a$ for $a \in \mathbb{Q}$ and $\theta(\sqrt{7}) = -\sqrt{7}$, then*

$$\theta(a + b\sqrt{7}) = a - b\sqrt{7}$$

for all $a, b \in \mathbb{Q}$. Obviously, θ preserves the additivity and multiplicity. Also, it is easy to see that θ is a bijection and this makes θ a field automorphism.

Corollary 22.4.3. (Uniqueness of Splitting Fields)
Let $f(x)$ be a polynomial over a field $\mathbb{F}$. Then the splitting field of $f(x)$ is unique up to a field isomorphism.

Proof. If we put $\mathbb{F} = \mathbb{F}_1$ in Theorem 22.4.3, then the proof is straightforward. □

Now it is safe to say "the" splitting field of a polynomial by Corollary 22.4.3.

Example 22.4.2. *Find the splitting field $\mathbb{E}$ of $p(x) = x^6 + x^5 + x^4 + x^3 + x^2 + x + 1$ and field automorphisms of $\mathbb{E}$.*

Solution. $p(x)$ is irreducible over $\mathbb{Q}$ since $\dfrac{x^7-1}{x-1}=p(x)$. Note that all zeros of $p(x)$ are $\cos\dfrac{2k\pi}{7}+i\sin\dfrac{2k\pi}{7}$, $0\le k\le 6$. If we let $\omega=\cos\dfrac{2\pi}{7}+i\sin\dfrac{2\pi}{7}$, then all zeros of $p(x)$ are ω^i, $i=1,\cdots,6$ and this makes $\mathbb{Q}(\omega)=\mathbb{E}$. Now let's find field automorphisms of this splitting field. Let's choose i from 1 through 6 and define θ from $\mathbb{Q}(\omega)$ to itself by $\theta(a)=a$ and $\theta(\omega)=\omega^i$ for $a\in\mathbb{Q}$. Then $\theta(\omega^s\omega^t)=\theta(\omega^{s+t})=\omega^{i(s+t)}=\omega^{is}\omega^{it}=\theta(\omega^s)\theta(\omega^t)$ and this θ can be extended to a ring homomorphism on $\mathbb{Q}(\omega)$. Since it is easy to see that θ is a bijection, it is a field automorphism.

Chapter 22 Exercises.

1. (a) Prove that $\mathbb{Q}[x]/< x^2-7 >$ is a field.

   (b) Find a zero of x^2-7 in the field $\mathbb{Q}[x]/< x^2-7 >$.

   (c) Find all the elements of $< x^2-7 >$.

   (d) Find all the elements of $\mathbb{Q}[x]/ < x^2-7 >$.

   (e) Find the inverse of $x+3+ < x^2-7 >$ in $\mathbb{Q}[x]/ < x^2-7 >$.

2. Find a zero of $8x^3+4x^2-6x+1$.

3. (a) Prove that $\mathbb{Q}[x]/< x^3-3x^2+6x-15 >$ is a field.

   (b) Show that $x+< x^3-3x^2+6x-15 >$ is a zero of $x^3-3x^2+6x-15$.

4. (a) Prove that $5x^5-2x^3+6x^2-4x+10$ is irreducible over $\mathbb{Q}$.

   (b) Find an extension field $\mathbb{E}$ of $\mathbb{Q}$ such that $5x^5-2x^3+6x^2-4x+10$ has a zero in $\mathbb{E}$.

5. (a) Prove that $\mathbb{Z}[x]/< x^2+4 >$ is a field.

(b) Find the inverse of $x+3+ < x^2+4 >$ in $\mathbb{Z}[x]/< x^2+4 >$.

(c) Find the inverse of $x+3+ < x^2+5 >$ in $\mathbb{Z}[x]/< x^2+5 >$.

6. (a) Solve $x^8 = 1$ in the field $\mathbb{C}$.

   (b) Solve $x^8 = 3$ in $\mathbb{C}$.

   (c) Solve $x^n = 1$ in $\mathbb{C}$ where $2 \leq n \in \mathbb{N}$.

   (d) Solve $x^n = a$ in $\mathbb{C}$ where $2 \leq n \in \mathbb{N}$ and $1 \leq a \in \mathbb{N}$.

   (e) Find the splitting field of $x^8 - 1$ over $\mathbb{Q}$.

   (f) Find the splitting field of $x^6 - 1$ over $\mathbb{Q}$.

   (g) Find the splitting field of $x^4 - 9$ over $\mathbb{Q}$.

7. Find the splitting field of $x^3 - 6x^2 + 10x - 4 \in \mathbb{Z}[x]$.

8. Let's consider $\overline{\theta}$ of (22.1).

   (a) Prove that $\overline{\theta}$ is a ring homomorphism.

   (b) Prove that $\overline{\theta}$ is one to one.

   (c) Prove that $\overline{\theta}$ is onto. Therefore it's an isomorphism.

9. (a) Find an irreducible polynomial $f(x)$ over $\mathbb{Z}_5$.

   (b) Show that $f(x)$ has a zero in the field $\mathbb{Z}_5[x]/< f(x) >$.

10. (a) Find an irreducible polynomial $f(x)$ over $\mathbb{Z}_7$.

    (b) Show that $f(x)$ has a zero in the field $\mathbb{Z}_7[x]/< f(x) >$.

    (c) Find the cardinality of the field $\mathbb{Z}_7[x]/< f(x) >$, i.e., the number of the elements of $\mathbb{Z}_7[x]/< f(x) >$.

11. (a) Construct a field with 25 elements.

(b) Construct a field with 27 elements.

12. Let $\mathbb{F}_1$ and $\mathbb{F}_2$ be extension fields of $\mathbb{F}$ and subfields of $\mathbb{E}$, i.e.

$$\mathbb{F} \subseteq \mathbb{F}_1, \mathbb{F}_2 \subseteq \mathbb{E}$$

(a) Prove that $\mathbb{F}_1 \cap \mathbb{F}_2$ is a field.

(b) Prove that $\mathbb{F}_1 \cap \mathbb{F}_2$ is an extension field of $\mathbb{F}$.

(c) If $\mathbb{F}_i$, $i \in J$, is a subfield of $\mathbb{F}$, then prove that $\cap_{i \in J} \mathbb{F}_i$ is a subfield of $\mathbb{F}$. Here, J is an index set.

(d) Prove that there exists a smallest field $\mathbb{F}_1\mathbb{F}_2$, which contains $\mathbb{F}_1$ and $\mathbb{F}_2$. Note that we can construct the following diagram.

Figure 22.4.3.

$$\begin{array}{ccc} & \mathbb{F}_1\mathbb{F}_2 & \\ \nearrow & & \nwarrow \\ \mathbb{F}_1 & & \mathbb{F}_2 \\ \nwarrow & & \nearrow \\ & \mathbb{F}_1 \cap \mathbb{F}_2 & \end{array}$$

13. (a) Prove that $\mathbb{Q}(\sqrt{2}, \sqrt{3}) = \mathbb{Q}(\sqrt{2} + \sqrt{3})$.

(b) Prove that $\mathbb{Q}(\sqrt{2}, \sqrt{3}) = \mathbb{Q}(\sqrt{2} + 2\sqrt{3})$.

(c) For $0 \neq n \in \mathbb{Z}$, prove that $\mathbb{Q}(\sqrt{2}, \sqrt{3}) = \mathbb{Q}(\sqrt{2} + n\sqrt{3})$.

14. (a) Prove that $\mathbb{Q}(\sqrt{2} + \sqrt{3}) = \mathbb{Q}(\sqrt{5 + 2\sqrt{6}})$. (Hint: Make $5 + 2\sqrt{6} = (\sqrt{a} + \sqrt{b})^2$).

(b) Prove that $\mathbb{Q}(\sqrt{5} + \sqrt{7}) = \mathbb{Q}(\sqrt{12 + 2\sqrt{35}})$.

15. Find an irreducible polynomial $f(x)$ in $\mathbb{Z}[x]$ over $\mathbb{Q}$ such that $\sqrt[3]{3}+\sqrt[3]{2}$ is a zero of $f(x)$.

Chapter 23

Algebraic Extensions

For a given field, we defined a field extension of the field and introduced an algebraic extension in Chapter 22. For example, $\sqrt{2}$ is not an element of $\mathbb{Q}$, but it is a zero of $x^2 - 2 \in \mathbb{Q}[x]$. This implies that $\mathbb{Q}(\sqrt{2})$ is an algebraic extension of $\mathbb{Q}$ and every element of $\mathbb{Q}(\sqrt{2})$ is algebraic over $\mathbb{Q}$ (why?). The properties of algebraic extensions are very important part of the Galois theory and these are also important tools to investigate solutions of a polynomial equation. To see these, we will first look at a field extension as a vector space and make a connection with algebraic extensions.

23.1 Finite Extensions

Definition 23.1.1. *Let $\mathbb{E}$ be an extension field of a field $\mathbb{F}$. Then $\mathbb{E}$ is a vector space over $\mathbb{F}$.*[1] *If $\mathbb{E}$ is a finite dimensional vector space over $\mathbb{F}$, then the dimension of $\mathbb{E}$ over $\mathbb{F}$ is denoted by $[\mathbb{E} : \mathbb{F}]$ and $\mathbb{E}$ is called a* **finite extension field**

[1] Recall that an extension field is always a vector space over a subfield but a subfield is not necessarily a vector space over an extension field unless they are the same.

of $\mathbb{F}$. *If* $[\mathbb{E} : \mathbb{F}] = \infty$, *then this extension is called* **an infinite field extension.**

Sometime, $[\mathbb{E} : \mathbb{F}]$ is called **the degree of the field extension**.

Example 23.1.1. *Since* $\mathbb{Q}(\sqrt{2})$ *is an extension field of* $\mathbb{Q}$, $\mathbb{Q}(\sqrt{2})$ *is a vector space over* $\mathbb{Q}$ *and* $[\mathbb{Q}(\sqrt{2}) : \mathbb{Q}] = 2$. *For,* $\mathbb{Q}(\sqrt{2})$ *is spanned by* 1 *and* $\sqrt{2}$ *over* $\mathbb{Q}$ *(why?).*

Note that $\mathbb{Q}(\sqrt{2}) = \left\{ \dfrac{a + b\sqrt{2}}{c + d\sqrt{2}} \middle| a, b, c, d \in \mathbb{Q}, c + d\sqrt{2} \neq 0 \right\}$.

Theorem 23.1.1. *For given fields* $\mathbb{K} \supset \mathbb{E} \supset \mathbb{F}$, *if* $[\mathbb{K} : \mathbb{E}] = m$ *and* $[\mathbb{E} : \mathbb{F}] = n$, *then* $[\mathbb{K} : \mathbb{F}] = mn$.

Proof. Let $\{a_1, \cdots, a_n\}$ be a basis of $\mathbb{E}$ over $\mathbb{F}$ and $\{b_1, \cdots, b_m\}$ be a basis of $\mathbb{K}$ over $\mathbb{E}$. For any $k \in \mathbb{K}$, there are $\alpha_1, \cdots, \alpha_m \in \mathbb{E}$ such that $k = \alpha_1 b_1 + \cdots +$ $+\alpha_m b_m$. We also have that

$$\begin{aligned} \alpha_1 &= \beta_{11} a_1 + \cdots + \beta_{1n} a_n, \\ &\vdots \\ \alpha_m &= \beta_{m1} a_1 + \cdots + \beta_{mn} a_n \end{aligned}$$

with appropriate coefficients. This implies that

$$\begin{aligned} k = (\beta_{11} a_1 + \cdots + \beta_{1n} a_n) b_1 \\ + \cdots + \\ (\beta_{m1} a_1 + \cdots + \beta_{mn} a_n) b_m. \end{aligned}$$

So k is the linear sum of $a_1 b_1, \cdots, a_n b_m$, i.e., $\{a_1 b_1, \cdots, a_n b_m\}$ is a spanning set of $\mathbb{K}$ over $\mathbb{F}$. Let us assume that $0 = (\beta_{11} a_1 + \cdots + \beta_{1n} a_n) b_1 + \cdots + (\beta_{m1} a_1 +$

$\cdots + \beta_{mn}a_n)b_m = 0$. This implies that $\beta_{11}a_1 + \cdots + \beta_{1n}a_n = 0, \cdots, \beta_{m1}a_1 + \cdots + \beta_{mn}a_n = 0$. Since $a_1, \cdots, a_n$ is linearly independent, we have that $\beta_{11} = 0, \cdots, \beta_{mn} = 0$. This implies that $\{a_1b_1, \cdots, a_nb_m\}$ is linearly independent. So it is a basis of $\mathbb{K}$ over $\mathbb{F}$ and we derive that $[\mathbb{K} : \mathbb{F}] = dim_{\mathbb{F}}\mathbb{K} = mn$. □

Example 23.1.2. *Compute* $[\mathbb{Q}(\sqrt[7]{3}) : \mathbb{Q}]$.

Solution. Since $(\sqrt[7]{3})^7 - 3 = 3 - 3 = 0$, $\sqrt[7]{3}$ is a zero of $x^7 - 3$. Then, $x^7 - 3$ is irreducible by Eisenstein criterion using $p = 3$. Since $\{1, \sqrt[7]{3}, \sqrt[7]{3^2}, \cdots, \sqrt[7]{3^6}\}$ is a basis of $\mathbb{Q}(\sqrt[7]{3})$ over $\mathbb{Q}$ (why?), we derive that $[\mathbb{Q}(\sqrt[7]{3}) : \mathbb{Q}] = 7$. □

Corollary 23.1.1. *For given fields* $\mathbb{K} \supset \mathbb{E} \supset \mathbb{F}$, $[\mathbb{E} : \mathbb{F}]$ *is a divisor of* $[\mathbb{K} : \mathbb{F}]$.

Proof. This is an immediate consequence of Theorem 23.1.1. □

Example 23.1.3. *Show that* $[\mathbb{Q}(\sqrt{2}, \sqrt{7}) : \mathbb{Q}] = 4$.

Solution. Since $x^2 - 7$ is irreducible over $\mathbb{Q}$ and $\sqrt{7}$ is a zero of the polynomial, we have that $[\mathbb{Q}(\sqrt{7}) : \mathbb{Q}] = 2$. Since $\sqrt{2}$ is not an element of $\mathbb{Q}(\sqrt{7})$ (why?) but $\sqrt{2}$ is a zero of $x^2 - 2$ that is irreducible over $\mathbb{Q}(\sqrt{7})$, we have that $[\mathbb{Q}(\sqrt{2}, \sqrt{7}) : \mathbb{Q}(\sqrt{7})] = 2$. So $[\mathbb{Q}(\sqrt{2}, \sqrt{7}) : \mathbb{Q}] = [\mathbb{Q}(\sqrt{2}, \sqrt{7}) : \mathbb{Q}(\sqrt{7})][\mathbb{Q}(\sqrt{7}) : \mathbb{Q}] = 4$. □

Example 23.1.4. *Find the splitting field* $\mathbb{K}$ *of the polynomial* $x^4 - 4 \in \mathbb{Q}[x]$ *and find* $[\mathbb{K} : \mathbb{Q}]$.

Solution. Since $x^4 - x^2 - 2 = (x - \sqrt{2})(x + \sqrt{2})(x - i)(x + i)$, we can say that $\mathbb{Q}(\sqrt{2}, i)$ is the splitting field of $x^4 - x^2 - 2$. Then we can construct the following figure:

Figure 23.1.1.

$$\begin{array}{ccc} & \mathbb{Q}(\sqrt{2}, i) & \\ \nearrow & & \nwarrow \\ \mathbb{Q}(\sqrt{2}) & & \mathbb{Q}(i) \\ \nwarrow & & \nearrow \\ & \mathbb{Q} & \end{array}$$

where $\mathbb{Q} \to \mathbb{Q}(i)$ means that $\mathbb{Q}$ is a subfield of $\mathbb{Q}(i)$. This implies that $[\mathbb{K} : \mathbb{Q}] = 4$. □

Example 23.1.5. *Compute* $[\mathbb{E} : \mathbb{Q}]$ *where* $\mathbb{E}$ *is the splitting field of of* $x^4 + 2$.

Solution. Since $-2 = 2(cos(\pi + 2k\pi) + isin(\pi + 2k\pi))$, all the zeros of $x^4 + 2$ are $\sqrt[4]{2}(\cos\frac{\pi}{4} + i\sin\frac{\pi}{4})$, $\sqrt[4]{2}(\cos\frac{3\pi}{4} + i\sin\frac{3\pi}{4})$, $\sqrt[4]{2}(\cos\frac{5\pi}{4} + i\sin\frac{5\pi}{4})$, and $\sqrt[4]{2}(\cos\frac{7\pi}{4} + i\sin\frac{7\pi}{4})$. So

$$\begin{aligned} \mathbb{E} &= \mathbb{Q}\left(\sqrt[4]{2}\left(\pm\frac{\sqrt{2}}{2} + \pm i\frac{\sqrt{2}}{2}\right)\right) \\ &= \mathbb{Q}(\sqrt[4]{2^3}, i\sqrt[4]{2^3}). \end{aligned}$$

This means that any element of $\mathbb{E}$ can be written as $c_1 + c_2\sqrt[4]{2^3} + i\sqrt[4]{2^3} + \sqrt{2}$ and this implies that $\{1, \sqrt[4]{2^3}, i\sqrt[4]{2^3}, \sqrt{2}\}$ is a basis of $\mathbb{E}$ over $\mathbb{Q}$. Therefore, $[\mathbb{E} : \mathbb{Q}] = 4$. □

Theorem 23.1.2. *a is algebraic over a field $\mathbb{F}$ if and only if $[\mathbb{F}(a) : \mathbb{F}]$ is finite.*

Proof. Without loss of generality, let's assume that $a \neq 0$ since it is obvious if $a = 0$. Let us assume that $[\mathbb{F}(a) : \mathbb{F}] = n < \infty$. This implies that $1, a, a^2, \cdots, a^n$

is linearly dependent. So there exist $\alpha_0, \cdots, \alpha_n \in \mathbb{F}$ such that $\alpha_0 + \alpha_1 a + \cdots + \alpha_n a^n = 0$ and at least two of $\alpha_0, \cdots, \alpha_n$ are not zeros. This means that a is a zero of the non-zero polynomial $f(x) = \alpha_0 + \alpha_1 x + \cdots + \alpha_n x^n \in \mathbb{F}[x]$. This implies that a is algebraic over $\mathbb{F}$. Conversely let us assume that a is algebraic over $\mathbb{F}$. This implies that a is a zero of a polynomial of $\mathbb{F}[x]$. Let $p(x)$ be the monic polynomial of $\mathbb{F}[x]$ with the smallest degree such that $p(a) = 0$. Because of the minimality of the degree, it is irreducible (why?). Let us define a map θ from $\mathbb{F}[x]$ to $\mathbb{F}(a)$ as for any $f(x) \in \mathbb{F}[x]$, $\theta(f(x)) = f(a)$. Then θ is a ring homomorphism. $ker(\theta) = \{g(x) \in \mathbb{F}[x] | g(a) = 0\}$. This implies that $p(x)|g(x)$. Since $p(x)$ is irreducible, the ideal $< p(x) >$ of $\mathbb{F}[x]$ is maximal. This implies that $\mathbb{F}[x]/< p(x) >$ is a field. So the map θ induces a map $\overline{\theta}$ from $\mathbb{F}[x]/< p(x) >$ to $\mathbb{F}(a)$ as $\overline{\theta}(x+ < p(x) >) = a$. This implies that $\overline{\theta}(f(x)+ < p(x) >) = f(a) = 0$. Since $a \in im(\overline{\theta})$ and $\mathbb{F} \subset im(\overline{\theta})$, we have that $im(\overline{\theta}) = \mathbb{F}(a)$. We have that $\mathbb{F}[x]/< p(x) > \cong \mathbb{F}(a)$. Since the dimension of $\mathbb{F}[x]/< p(x) >$ over $\mathbb{F}$ is n, the dimension $\mathbb{F}(a)$ over $\mathbb{F}$ is n. Thus we have proven the theorem. □

Theorem 23.1.2 says that a finite extension is a simple algebraic extension and vice versa.

23.2 Minimal Polynomials

$p(x)$ in Theorem 23.1.2 has a as its zero and makes $\mathbb{F}[x]/< p(x) >$ a field because it is irreducible. This kind of $p(x)$ provides much more clear picture for field extensions by an algebraic element. To make such a polynomial unique, we have the following definition.

Definition 23.2.1. *The irreducible monic polynomial $p(x)$ over a field $\mathbb{F}$ is called the* **minimal polynomial** *of a over $\mathbb{F}$ if a is a zero $p(x)$ and $deg(p(x))$ is minimal.*

Example 23.2.1. *Find the minimal polynomial of $\sqrt{2}+\sqrt{3}$ over $\mathbb{Q}$.*

Solution. Let $\lambda = \sqrt{2}+\sqrt{3}$. Then $\lambda - \sqrt{2} = \sqrt{3}$ and if we square both sides of the equality, then we have that $\lambda^2 - 2\sqrt{2}\lambda + 2 = 3$. So we have that $\lambda^2 - 1 = 2\sqrt{2}\lambda$. If we square both sides of the equality, then we have that $\lambda^4 - 2\lambda^2 + 1 = 8\lambda^2$. This implies that λ is a zero of $x^4 - 10x^2 + 1 \in \mathbb{Q}[x]$ and this monic polynomial is irreducible over $\mathbb{Q}$ by mod 3 test. So it is the minimal polynomial of $\sqrt{2}+\sqrt{3}$ over $\mathbb{Q}$. □

Definition 23.2.2. *For a simple extension $\mathbb{E} = \mathbb{F}(a)$ of $\mathbb{F}$, a is called a* **primitive element** *of $\mathbb{E}$.*

Note that a primitive element can be algebraic or transcendental.

Example 23.2.2. *Since $\mathbb{Q}(\sqrt{2},\sqrt{7}) = \mathbb{Q}(\sqrt{2}+\sqrt{7})$, $\sqrt{2}+\sqrt{7}$ is a primitive element of $\mathbb{Q}(\sqrt{2},\sqrt{7})$ and it is algebraic over $\mathbb{Q}$.*

Example 23.2.3. *π is a primitive element of $\mathbb{Q}(\pi)$ over $\mathbb{Q}$ but it is not algebraic over $\mathbb{Q}$.*

From the definition of the minimal polynomial, we can state the following properties.

Corollary 23.2.1. *a is algebraic over $\mathbb{F}$ if and only if there exists the minimal polynomial of a in $\mathbb{F}[x]$.*

The following theorem explains how the "degree" of a polynomial and the "degree" of a field extension is related. We know that a polynomial has at most n zeros if n is the degree of the polynomial. When we count the number of zeros, we could count the same zero multiple times. For example, $f(x) = (x-1)^3(x-2)$ has 4 zeros but only 2 distinct zeros.

Theorem 23.2.1. *For any polynomial $f(x)$ over a field $\mathbb{F}$, if $deg(f(x)) = n$, then there is an extension field $\mathbb{E}$ of the field $\mathbb{F}$ of degree at most $n!$ in which $f(x)$ has all the n zeros of $f(x)$.*

Proof. We prove the theorem by induction on n of the polynomial $f(x)$. If n is 1, then it is obvious. Let's assume that the theorem holds for any polynomial whose degree is less than n. By Theorem 22.1.1, there exists an extension field $\mathbb{F}(a)$ such that a is a zero of $f(x)$. This implies that $[\mathbb{F}(a) : \mathbb{F}] \leq n$. We also have that $f(x) = (x-a)f_1(x) \in \mathbb{F}(a)[x]$. Let $a_1, \cdots, a_{n-1}$ be zeros of $f_1(x)$ in a certain extension field of $\mathbb{F}$. By induction, we have that $[\mathbb{F}(a, a_1, \cdots, a_{n-1}) : \mathbb{F}(a)] \leq (n-1)!$. This implies that $[\mathbb{F}(a, a_1, \cdots, a_{n-1}) : \mathbb{F}] \leq n!$ and this proves the theorem. □

Theorem 23.2.1 is for the splitting field of an arbitrary polynomial and the following corollary is for a simple extension by a zero of an irreducible polynomial.

Corollary 23.2.2. *If a is a zero of an irreducible polynomial with its degree n over a field $\mathbb{F}$, then $[\mathbb{F}(a) : \mathbb{F}] = n$.*

Corollary 23.2.2 can be stated as follows

Corollary 23.2.3. *If the minimal polynomial for a in $\mathbb{F}[x]$ has degree n, then* $[\mathbb{F}(a) : \mathbb{F}] = n$.

Definition 23.2.3. *For an extension field $\mathbb{E}$ of a field $\mathbb{F}$, $\mathbb{E}$ is called* **an algebraic extension** *if every element of $\mathbb{E}$ is algebraic over $\mathbb{F}$.*

Example 23.2.4. $\mathbb{Q}(\sqrt{2})$ *is an algebraic extension of $\mathbb{Q}$.*

If $\mathbb{E}$ is a finite extension field of a field $\mathbb{F}$, then $\mathbb{E}$ is algebraic over $\mathbb{F}$ but an infinite extension field can be algebraic.

Example 23.2.5. $[\mathbb{Q}(\sqrt{2}, \sqrt{3}, \cdots) : \mathbb{Q}] = \infty$, *but* $\mathbb{Q}(\sqrt{2}, \sqrt{3}, \cdots)$ *is algebraic over $\mathbb{Q}$.*

Theorem 23.2.2. *Let $\mathbb{K}$ be an algebraic extension of $\mathbb{E}$. If $\mathbb{E}$ is an algebraic extension of $\mathbb{F}$, then $\mathbb{K}$ is an algebraic extension of $\mathbb{F}$.*

Proof. For any a of $\mathbb{K}$, we need to prove that a is a zero of a polynomial over $\mathbb{F}$, and it is enough if we can show $[\mathbb{F}(a) : \mathbb{F}]$ is finite. Since a is algebraic over $\mathbb{E}$, there is a monic polynomial $x^m + a_{m-1}x^{m-1} + \cdots + a_0$ where $a_{m-1}, \cdots, a_0 \in \mathbb{E}$. We have that $[\mathbb{F}(a) : \mathbb{F}(a_{m-1}, \cdots, a_0)] = m$.

$$[\mathbb{F}(a) : \mathbb{F}] = [\mathbb{F}(a) : \mathbb{F}(a_{m-1}, \cdots, a_0)][\mathbb{F}(a_{m-1}, \cdots, a_0) : \mathbb{F}(a_{m-2}, \cdots, a_0)] \cdots$$
$$[\mathbb{F}(a_2, a_1, a_0) : \mathbb{F}(a_1, a_0)][\mathbb{F}(a_1, a_0) : \mathbb{F}(a_0)][\mathbb{F}(a_0) : \mathbb{F}]$$

Since a_0 is algebraic over $\mathbb{F}$, we have that $[\mathbb{F}(a_0) : \mathbb{F}]$ is finite. Since a_1 is algebraic over $\mathbb{F}$, it is algebraic over $\mathbb{F}(a_0)$. Inductively we can prove that a_{m-1} is algebraic over $\mathbb{F}(a_{m-2}, \cdots, a_0)]$. This implies that $\mathbb{F}(a)$ is algebraic over $\mathbb{F}$. So $\mathbb{K} = \mathbb{F}(a, a_{m-1}, \cdots, a_0)$ is an algebraic extension of $\mathbb{F}$. □

Theorem 23.2.3. *If a and b are algebraic over $\mathbb{F}$, then $a \pm b, ab$, and $\frac{a}{b}$, $b \neq 0$, are algebraic over $\mathbb{F}$ so the set of all algebraic elements over $\mathbb{F}$ is an extension field of $\mathbb{F}$.*

Proof. Let a and b be algebraic over $\mathbb{F}$. Then, $[\mathbb{F}(a) : \mathbb{F}] = m < \infty$ and $[\mathbb{F}(b) : \mathbb{F}] = n < \infty$ so $[\mathbb{F}(a,b) : \mathbb{F}(a)][\mathbb{F}(a) : \mathbb{F}] \leq mn < \infty$. This implies that $\mathbb{F}(a,b)$ is algebraic over $\mathbb{F}$ and since $a \pm b, ab$, and $\frac{a}{b}$, $b \neq 0$ are elements of $\mathbb{F}(a,b)$, they are algebraic over $\mathbb{F}$. Therefore, the set of all algebraic elements over $\mathbb{F}$ is an extension field of $\mathbb{F}$. □

An algebraic extension can be finite or infinite. A simple extension by a primitive element can be algebraic or transcendental. The following theorem connects these two concepts and it is called the **Primitive Element Theorem**.

Theorem 23.2.4. (Steinitz) *Let a and b be algebraic over a field $\mathbb{F}$ of characteristic zero. Then there is an element c of $\mathbb{F}(a,b)$ such that $\mathbb{F}(a,b) = \mathbb{F}(c)$.*

Proof. Let $f(x)$ and $g(x)$ be irreducible polynomials of a and b over $\mathbb{F}$ respectively. Let $a_1, \cdots, a_m$ and $b_1, \cdots, b_n$ be zeros of $f(x)$ and $g(x)$ respectively in an extension field $\mathbb{E}$ of $\mathbb{F}$. By Theorem 22.1.1, we can assume that $a_i \neq a_j, 1 \leq i \neq j \leq m$, and $b_i \neq b_j, 1 \leq i \neq j \leq n$. If m or n is one, then either $\mathbb{F}(a,b) = \mathbb{F}(b)$ or $\mathbb{F}(a,b) = \mathbb{F}(a)$ holds respectively. So there is nothing to prove. Let us assume that $m, n > 1$. Let us put $a_1 = a$ and $b_1 = b$. Since $\mathbb{F}$ has infinitely many elements, we can take $\lambda \in \mathbb{F}$ so that

$$\lambda \neq \frac{a_i - a}{b - b_j}$$

where $1 \leq i \leq m$ and $1 < j \leq n$. Let us put $c = a + b\lambda$. Let us show that $\mathbb{F}(c) = \mathbb{F}(a, b)$. Since $a + b\lambda \in \mathbb{F}(a, b)$, we have that $\mathbb{F}(c) \subset \mathbb{F}(a, b)$. Let us show that $\mathbb{F}(a, b)$ is a subset of $\mathbb{F}(c)$. Let us put $h(x) = f(c - \lambda x)$. This implies that $h(x) \in \mathbb{F}(c)[x]$ and $h(b) = f(c - \lambda b) = f(a) = 0$. We have that $h(x)$ and $g(x)$ have a common factor $x - b$. The irreducible polynomial of b over $\mathbb{F}(c)$ is a factor of $g(x)$. If $b_j \neq b$ is another zero of $g(x)$, then $h(b_j) = f(c - \lambda b_j) \neq 0$. Since $g(x)$ has a simple zero, we have that $(x - b)^2$ does not divide $g(x)$. This implies that $x - b$ is the greatest common divisor of $h(x)$ and $g(x)$ over some extension field of $\mathbb{F}(c)$. Since an irreducible polynomial of b over $\mathbb{F}(c)$ divides $h(x)$ and $g(x)$, $x - b$ is an irreducible polynomial of b over $\mathbb{F}(c)$. This implies that $x - b \in \mathbb{F}(c)[x]$, i.e., $b \in \mathbb{F}(c)$. Similarly we can prove that $a \in \mathbb{F}(c)$. So $\mathbb{F}(a, b) \subset \mathbb{F}(c)$ and we conclude that $\mathbb{F}(a, b) = \mathbb{F}(c)$. □

Let's recall that there exists a primitive element in a simple extension. (Definition 23.2.2)

Corollary 23.2.4. *Every finite extension field $\mathbb{E}$ of a field $\mathbb{F}$ of characteristic zero is a simple extension.*

Proof. Let $\mathbb{E}$ be the field extension $\mathbb{F}(a_1, \cdots, a_n)$ of $\mathbb{F}$ where $a_1, \cdots, a_n \notin \mathbb{F}$. Let us prove the corollary by induction on n. If $n = 1$, then the extension field is simple, so there is nothing to prove. Let us assume that the corollary is true for $n = k - 1$. Let us assume that $k = n$. We have that $\mathbb{F}(a_1, \cdots, a_n) = \mathbb{F}(a_1, \cdots, a_{n-1})(a_n) = \mathbb{F}(d_1)(a_n)$ for $d_1 \in \mathbb{E}$. So by induction, we have that

$\mathbb{F}(a_1, \cdots, a_n) = \mathbb{F}(d)(a_n) = \mathbb{F}(d_1, a_n) = \mathbb{F}(d_2)$ for $d_2 \in \mathbb{E}$. Thus we have proven the corollary. $\square$

By Example 23.2.5, every algebraic extension field of a field $\mathbb{F}$ is not necessarily a simple extension. If $\mathbb{F}$ is a perfect field, then by Theorem 22.3.1, every irreducible polynomial over $\mathbb{F}$ can be decomposed as the product of linear factors of multiplicity ones over a splitting field of the polynomial. If $\mathbb{F}$ is not a perfect field, we can not expect the product of linear factors of multiplicity ones. But we have the following results for a field which is not necessarily a perfect field. It does not say that the multiplicities are ones but it does say that all the multiplicities are the same.

Theorem 23.2.5. *Let $f(x)$ be an irreducible polynomial over a field $\mathbb{F}$ and $\mathbb{E}$ be a splitting field of $f(x)$. Suppose a and b are zeros of $f(x)$. Then if the multiplicity of a is m, so is the multiplicity of b.*

Proof. Since a has the multiplicity m, we have that $f(x) = (x-a)^m g(x)$ in the splitting field $\mathbb{E}$ of $f(x)$. Let n be the multiplicity of b. By Theorem 22.4.3, there is an isomorphism $\overline{\theta}$ from $\mathbb{E}[x]$ to itself such that the restriction of $\overline{\theta}$ to $\mathbb{F}$ is the identity map and $\overline{\theta}(a) = b$. This implies that $\overline{\theta}((x-a)^m g(x)) = (x-b)^m \overline{g}(x)$. This implies that $m \leq n$. By changing the role of a and b, we have that $n \leq m$. This implies that $m = n$ and it proves the theorem. $\square$

Corollary 23.2.5. *If $f(x)$ be an irreducible polynomial over a field $\mathbb{F}$, then $f(x) = (x-a_1)^m \cdots (x-a_n)^m$ in its splitting field of $f(x)$.*

Proof. The proof is straightforward by Theorem 23.2.5. $\square$

23.3 Fundamental Theorem of Algebra

Definition 23.3.1. *Let* $\mathbb{E}$ *be an extension field of* $\mathbb{F}$ *and*

$$\overline{\mathbb{F}} = \{a \in \mathbb{E} | a \text{ is algebraic over } \mathbb{F}\}$$

Then $\overline{\mathbb{F}}$ *is a subfield of* $\mathbb{E}$ *and it is called* **algebraic closure of $\mathbb{F}$ in $\mathbb{E}$**

Note that $\overline{\mathbb{F}}$ is the set of all possible algebraic elements over $\mathbb{F}$.

Definition 23.3.2. *A field* $\mathbb{F}$ *is called* **algebraically closed** *if* $\overline{\mathbb{F}} = \mathbb{F}$.

The following theorem is one important example for algebraically closed field and it is called (classical) **Fundamental Theorem of Algebra**, proved by **Gauss**.

Theorem 23.3.1. (Gauss) $\mathbb{C}$ *is algebraically closed.*

Its proof is beyond the scope of this text but there have been so many different proofs for the theorem.

Chapter 23 Exercises.

1. Compute the following degrees.

   (a) $[\mathbb{Q}(\sqrt{7}) : \mathbb{Q}]$

   (b) $[\mathbb{Q}(i) : \mathbb{Q}]$

   (c) $[\mathbb{Q}(\sqrt{7}, \sqrt{11}) : \mathbb{Q}]$

(d) $[\mathbb{Q}(\sqrt{2}, i) : \mathbb{Q}]$

2. (a) Prove that $\mathbb{Q}(\sqrt[3]{5})$ is a vector space over $\mathbb{Q}$.

   (b) Find $[\mathbb{Q}(\sqrt[3]{5}) : \mathbb{Q}]$.

   (c) Prove that $\{1, \sqrt[3]{5}, \sqrt[3]{5^2}\}$ is a basis of the vector space $\mathbb{Q}(\sqrt[3]{5})$ over $\mathbb{Q}$.

   (d) Find a basis of the vector space $\mathbb{Q}(\sqrt[3]{5})$ over $\mathbb{Q}$ other than $\{1, \sqrt[3]{5}, \sqrt[3]{5^2}\}$.

3. Find the splitting field $\mathbb{E}$ of $x^4 - 9 \in \mathbb{Q}[x]$ and compute $[\mathbb{E} : \mathbb{Q}]$.

4. Find the splitting field $\mathbb{E}$ of $x^3 - 9x^2 - 4x + 36 \in \mathbb{Q}[x]$ and compute $[\mathbb{E} : \mathbb{Q}]$.

5. Find the minimal polynomials of the following over $\mathbb{Q}$.

   (a) $2 + \sqrt{3}$

   (b) $\sqrt{2} + \sqrt{7}$

   (c) $2 - i$

   (d) $\sqrt{5 - \sqrt[3]{2}}$

   (e) $\sqrt[3]{3} + \sqrt[3]{2}$

6. (a) Prove that $\mathbb{Z}_3[x]/< x^2 + 1 >$ is a field.

   (b) Find the inverse of $x + 3 + < x^2 + 2 >$ in $\mathbb{Z}_3[x]/< x^2 + 1 >$.

   (c) Prove that $\mathbb{C}$ is isomorphic to $\mathbb{R}[x]/< x^2 + 1 >$.

7. (a) Find a degree three irreducible polynomial $f(x)$ over $\mathbb{Z}_5$ and show why it is irreducible.

   (b) Show that $\mathbb{Z}_5[x]/< f(x) >$ is a field.

   (c) Find the cardinality of $\mathbb{Z}_5[x]/< f(x) >$.

   (d) Find a zero of $f(x)$ in $\mathbb{Z}_5[x]/< f(x) >$.

8. (a) Prove that $5x^3 + 6x^2 + 9x - 6$ is irreducible over $\mathbb{Q}$.

   (b) Prove that $\mathbb{Q}[x]/< 5x^3 + 6x^2 + 9x - 6 >$ is a field and identify all its elements.

   (c) Find the unity and the zero element of $\mathbb{Q}[x]/< 5x^3 + 6x^2 + 9x - 6 >$.

   (d) Find a zero of $5x^3 + 6x^2 + 9x - 6$ in $\mathbb{Q}[x]/< 5x^3 + 6x^2 + 9x - 6 >$.

9. Prove the following.

   (a) $\mathbb{Q}(\sqrt{2}, \sqrt{3}) = \mathbb{Q}(\sqrt{2} + \sqrt{3})$

   (b) $\mathbb{Q}(\sqrt{2}, \sqrt{3}) = \mathbb{Q}(\sqrt{2} + 3\sqrt{3})$

   (c) Find $[\mathbb{Q}(\sqrt{2}, \sqrt{3}) : \mathbb{Q}]$

   (d) Find the minimal polynomial of $\sqrt{2} + \sqrt{3}$ over $\mathbb{Q}$

10. (a) Find the minimal polynomial of $\sqrt{3}$ over $\mathbb{Q}$. Prove your claim.

    (b) Find the minimal polynomial of $\sqrt[3]{3}$ over $\mathbb{Q}$. Prove your claim.

11. (a) Find the dimension of the vector space of $\mathbb{Q}(\sqrt{2})$ over $\mathbb{Q}$.

    (b) Find the dimension of the vector space of $\mathbb{Q}(\sqrt{2}, \sqrt{11})$ over $\mathbb{Q}(\sqrt{11})$.

    (c) Find the dimension of the vector space of $\mathbb{Q}(\sqrt{2}, \sqrt{11})$ over $\mathbb{Q}$.

    (d) Prove that $\mathbb{Q}(\sqrt{2}, \sqrt{11}) = \mathbb{Q}(\sqrt{2} + \sqrt{11})$.

    (e) Find the minimal polynomial of $\sqrt{2} + \sqrt{11}$ over $\mathbb{Q}$.

12. (a) Find all the zeros of $x^4 - 3$ in its splitting field $\mathbb{E}$

    (b) Find the splitting field $\mathbb{E}$ of $x^4 - 3$.

    (c) Find $[\mathbb{E} : \mathbb{Q}]$.

13. Let $\mathbb{K}$ be a finite extension field of a field $\mathbb{F}$. If $\mathbb{E}$ is a subfield of $\mathbb{K}$ such that $\mathbb{F} \subset \mathbb{E}$, prove that $\mathbb{E}$ is an algebraic extension of $\mathbb{F}$.

14. (a) If a is a transcendental number over a field $\mathbb{Q}$, then prove that $\mathbb{Q}[a]$ and $\mathbb{Q}[x]$ are isomorphic as rings.

    (b) If a is a transcendental number over a field $\mathbb{Q}$, then prove that $\mathbb{Q}(x)$ and $\mathbb{Q}(x)$ are isomorphic as rings where $\mathbb{Q}(x) = \{\frac{f(x)}{g(x)} | f(x), g(x) \in \mathbb{Q}[x], g(x) \neq 0\}$.

15. (a) Find a field $\mathbb{F}$ with 7^3 elements.

    (b) If $\mathbb{F}_1$ is another field with 7^3 elements, then is $\mathbb{F}$ isomorphic to $\mathbb{F}_1$? Why?

    (c) Find a field with p^3 elements where p is prime.

    (d) Find a field with p^k elements where p is prime and $k \in \mathbb{N}$.

16. Find a basis of the following extension field.

    (a) $\mathbb{Q}(\sqrt{7})$ over $\mathbb{Q}$.

    (b) $\mathbb{Q}(\sqrt[3]{7})$ over $\mathbb{Q}$.

    (c) $\mathbb{C}(\sqrt{7})$ over $\mathbb{C}$.

    (d) $\mathbb{Q}(\sqrt[4]{3})$ over $\mathbb{Q}$.

17. (a) Prove that $x^2 - 5$ is irreducible over $\mathbb{Q}(\sqrt[3]{3})$.

    (b) Prove that $x^2 - 5$ is irreducible over $\mathbb{Q}(\sqrt[3]{5})$.

18. Let $\mathbb{E}$ be an extension field of a field $\mathbb{F}$. For $a \in \mathbb{E}$, if a is algebraic over $\mathbb{F}$, then prove that $[\mathbb{F}(a) : \mathbb{F}]$ is the degree of the minimal polynomial of a over $\mathbb{F}$.

19. Let $\mathbb{E}$ be an extension field of a field $\mathbb{F}$. Prove that the algebraic closure of $\mathbb{F}$ in $\mathbb{E}$ is unique.

20. (a) Find an extension field of $\mathbb{E}$ of $\mathbb{Q}$ with $[\mathbb{E} : \mathbb{Q}] = 10$.

    (b) Find an extension field of $\mathbb{E}$ of $\mathbb{Q}$ with $[\mathbb{E} : \mathbb{Q}] = 7$.

    (c) Find an extension field of $\mathbb{E}$ of $\mathbb{Q}$ with $[\mathbb{E} : \mathbb{Q}] = 14$.

21. (a) Find all the zeros of $x^7 - 1$.

    (b) If $\omega^7 = 1$ and $\omega \neq 1$, then compute $[\mathbb{Q}(\omega) : \mathbb{Q}]$.

22. Let $\mathbb{F}(a)$ be an extension field of $\mathbb{F}$ such that $a \notin \mathbb{F}$. Let $p_1(x)$ and $p_2(x)$ be two irreducible polynomials such that a is a zero of them. Prove that $p_1(x) = cp_2(x)$ for some non-zero $c \in \mathbb{F}$.

23. Let $\mathbb{E}$ be an extension field of the field $\mathbb{F}$, and let $p_1(x)$ and $p_2(x)$ be two minimal polynomials of $a \in \mathbb{E}$ over $\mathbb{F}$. Prove that $p_1(x) = p_2(x)$.

Chapter 24

Geometric Constructions

The ancient Greeks had only two tools to measure their lands for charging a tax: an unmarked straight edge and a compass. Since then, there have been many geometric constructions using those two tools. A well-known example is the bisection of an arbitrary angle and a not-so-well-known exam is the construction of a regular 17-gon made by Gauss. In this chapter, we will find several constructible examples. We will also prove that the trisection of an arbitrary angle is not possible using only an unmarked straight edge and a compass.[1] First, we define the word 'constructible' geometrically.

24.1 Constructible Constructions

Definition 24.1.1. *A real number* r *is called* **constructible** *if we can construct a line segment with length* r *using only an unmarked straight edge, a compass, and a unit length* 1 *finite number of times.*

[1]The trisection of some particular angles, like 90° is possible.

Similarly, we say that a point $(a, b) \in \mathbb{R}^2$ is **constructible** if a and b are constructible by Definition 24.1.1. Now we look at several constructible examples without proofs.

Lemma 24.1.1. *1. All natural numbers are constructible.*

2. All integers are constructible.

3. If a and b are constructible, then so are $a \pm b$, ab, and $\frac{a}{b}$ for $b \neq 0$.

4. All rational numbers are constructible.

5. $\mathbb{Q}^2$ is a set of constructible points.[2]

6. a is constructible if and only if $\sqrt{a}$ is constructible.

7. θ is constructible if and only if $\cos\theta$ or $\sin\theta$ is constructible.

Lemma 24.1.2. *The set of all constructible numbers is a subfield of $\mathbb{R}$.*

Let $\mathbb{F}$ be a set of constructible real numbers. Then $\mathbb{F}^2 = \{(x, y) | x, y \in \mathbb{F}\}$ is a set of constructible points in $\mathbb{R}^2$. If we have two different points of $\mathbb{F}^2$, then we can construct a line by a ruler and its equation can be written as

$$c_1 x + c_2 y + c_3 = 0$$

where $c_1, c_2, c_3 \in \mathbb{F}$. Similarly if we have a constructible center and a constructible radius, then we can construct a circle by a compass and its equation can be written as

$$x^2 + y^2 + c_4 x + c_5 y + c_6 = 0$$

[2]It is not THE set of all constructible points.

where $c_4, c_5, c_6 \in \mathbb{F}$. Note that not all points on this line or the circle are constructible but only identifiable points are called constructible.

Other than constructible number in $\mathbb{F}$, we only have three possible ways to make new constructible numbers (or points) and those are the intersections of

1. two lines,

2. two circles, or

3. one line and one circle.

Now, let $\mathbb{F} = \mathbb{Q}$. From 1, we only construct rational numbers (why?) so there is no new constructible point. But using 2 or 3, we can construct $\sqrt{b_1}$ where $b_1 \in \mathbb{Q}$. Note that $[\mathbb{F}(\sqrt{b_1}) : \mathbb{F}] = 2$ or 1. If $\sqrt{b_2}$ is another constructible number, we have that $[\mathbb{F}(\sqrt{b_1}, \sqrt{b_2}) : \mathbb{F}(\sqrt{b_1})] = 2$ or 1, and this implies that $[\mathbb{F}(\sqrt{b_1}, \sqrt{b_2}) : \mathbb{F}] = 2^n$ for $n = 0, 1, 2$. If we repeat this process, we can derive the following theorem.

Theorem 24.1.1. *$a \in \mathbb{R}$ is constructible if and only if there exist finitely many real numbers $a_1, \cdots, a_n$ such that*

1. *$a \in \mathbb{Q}(a_1, \cdots, a_n)$*

2. *$[\mathbb{Q}(a_1) : \mathbb{Q}] = 2$ or 1*

3. *$[\mathbb{Q}(a_1, \cdots, a_n) : \mathbb{Q}(a_1, \cdots, a_{n-1})] = 2$ or 1.*

Corollary 24.1.1. *A real number a is constructible if and only if a is an element of a subfield $\mathbb{F}$ of $\mathbb{R}$ with $[\mathbb{F} : \mathbb{Q}] = 2^m$ for some $m \in \mathbb{N}$.*

Corollary 24.1.2. *Suppose $a \in \mathbb{R}$ is a zero of an irreducible polynomial over $\mathbb{Q}$. If the degree of the polynomial does not divide 2^m for $m \in \mathbb{N}$, then a is not constructible.*

Proof. Let us assume that a is constructible. Then there exists a subfield $\mathbb{F}$ of $\mathbb{R}$ such that $a \in \mathbb{F}$ and $[\mathbb{F} : \mathbb{Q}] = 2^m$ for $m \in \mathbb{N}$ by Corollary 24.1.1. Since $\mathbb{Q}(a)$ is an intermediate field between $\mathbb{F}$ and $\mathbb{Q}$, $[\mathbb{Q}(a) : \mathbb{Q}]$ divides $[\mathbb{F} : \mathbb{Q}]$. But this contradicts the assumption of the corollary. So a is not constructible. □

Because of the degree 2^m, this type of field extension is called **a two tower.**

24.2 Trisection

Now we are ready to prove that the trisection of an arbitrary angle is not possible if we use only an unmarked straight edge and a compass. Since we only need to show a counter-example, we use $60°$.

Proposition 24.2.1. *It is impossible to trisect* $60°$ *by use of only an unmarked straight edge and a compass.*

Proof. Assume that the trisection of $60°$ is possible. This implies that $\cos 20°$ is constructible. Let $x = \cos 20°$. Since $\cos(3\alpha) = 4\cos^3\alpha - 3\cos\alpha$, we have that $\cos 60° = 4\cos^3 20° - 3\cos 20°$ and this implies that $8x^3 - 6x - 1 = 0$. Note that $f(x) = 8x^3 - 6x - 1$ is irreducible by the *mod* 5-test. For, if $f(x)$ is reducible over $\mathbb{Z}$, then it is reducible in $\mathbb{Z}_5$ since $8 \not\equiv 0$ in $\mathbb{Z}_5$ and it is easy to check that $f(x)$ is irreducible over $\mathbb{Z}_5$. Also, note that $[\mathbb{Q}(\cos 20°) : \mathbb{Q}] = 3$. But this implies that 3 divides 2^m for some $m \in \mathbb{N}$ by Corollary 24.1.1, a contradiction.

Therefore the trisection of 60° is not possible using only an unmarked straight edge and a compass. $\square$

Proposition 24.2.2. *It is impossible to construct $\sqrt[3]{2}$ using only an unmarked straight edge and a compass. Furthermore it is impossible to duplicate a cube.*

Proof. Assume that the construction of $\sqrt[3]{2}$ is possible. Note that $\sqrt[3]{2}$ is a zero of $g(x) = x^3 - 2$ and $g(x)$ is irreducible by Eisenstein Criterion (or by *mod* 7-test). But this implies that $3|2^m$ for some $m \in \mathbb{N}$ by Corollary 24.1.1, a contradiction. So it is not possible and we can use this result to prove the 2nd part of the proposition. $\square$

Chapter 24 Exercises.

1. Construct the following number geometrically using only an unmarked straight edge and a compass.

   (a) 7

   (b) $\frac{9}{7}$

   (c) $\sqrt{2}$ (Hint: Use a square)

   (d) $\sqrt{3}$

   (e) $\sqrt{n}$, where n is a square free integer.

2. Prove that if a and b are constructible, then so are $a \pm b$, ab, and $\frac{a}{b}$ for $b \neq 0$.

3. Prove that every element of $\mathbb{Q}$ is constructible.

4. a is constructible if and only if $\sqrt{a}$ is constructible.

5. θ is constructible if and only if $\cos\theta$ or $\sin\theta$ is constructible.

6. If $\mathbb{F}$ is the set of all constructible real numbers, prove that $\mathbb{Q}$ is a subfield of $\mathbb{F}$.

7. If possible, construct the following numbers geometrically using only an unmarked straight edge and a compass.

   (a) $\sqrt[4]{2}$ (Hint: Use the Pythagorean Theorem)

   (b) $\sqrt[8]{2}$

   (c) $\sqrt[4]{3}$

   (d) $\sqrt[3]{3}$

   (e) $\sqrt[5]{2}$

   (f) π

   (g) e

8. If $a \in \mathbb{R}$ is constructible, then prove that a is an intersection of a line and a circle.

9. Prove that regular $3, 4, 10, 15$-gons are constructible.

10. Prove that it is impossible to construct the following.

    (a) a regular 9-gon

    (b) a regular 18-gon

    (c) a regular 36-gon

11. Prove that the following is constructible.

    (a) a regular pentagon

(b) a regular hexagon

(c) a regular 20-gon

(d) a regular 30-gon

12. Prove that a regular n-gon is constructible if and only if a regular $2n$ is constructible.

13. Prove that 72° can be trisected.

14. Explain why it is not constructible to twice the volume of a cube to a cube.

15. Explain why it is not constructible to equal the area of a circle to a square.

16. Explain why a circle can not be cubbed.

Chapter 25

Separable and Inseparable Extensions

Now we are approaching a pinnacle of an undergraduate abstract algebra, Galois theory. There are two important concepts when we study Galois theory in a finite degree field extension. One of them is a separable extension[1] and it is based on the simplicities of zeros of an irreducible polynomial. Recall that a zero can be a simple zero or a multiple zero.

25.1 Perfect Fields and Multiplicity of a Zero

We need to define two things first to understand a separable extension.

Definition 25.1.1. *A field $\mathbb{F}$ is called* **perfect** *if $\mathbb{F}$ is a field of characteristic zero or $\mathbb{F}^p = \mathbb{F}$ where $\mathbb{F}^p = \{x^p | x \in \mathbb{F}\}$ for a prime p.*

[1]The other one is a normal extension.

For example, all infinite fields with characteristic zero are perfect. $\mathbb{Q}, \mathbb{R}$, and $\mathbb{C}$ are all perfect fields. Also, all finite fields are perfect since their characteristic are primes. An example of "imperfect" fields is $\mathbb{F}(x)$, the field of rational functions when $\mathbb{F}$ is a finite field. The following proposition proves one of these claims.

Proposition 25.1.1. *If $\mathbb{F}$ is a field of characteristic p, $\mathbb{F} = \mathbb{F}^p$.*

Proof. Let's define a map $\theta : \mathbb{F} \longrightarrow \mathbb{F}^p$ as $\theta(x) = x^p$ for $x \in \mathbb{F}$. Then $\theta(x+y) = (x+y)^p$ for any $x, y \in \mathbb{F}$ and $(x+y)^p = x^p + C(p,1)x^{p-1}y + \cdots + C(p,p-1)xy^{p-1} + y^p = x^p + y^p = \theta(x) + \theta(y)$ since the characteristic of $\mathbb{F}$ is p. We also have that $\theta(xy) = (xy)^p = x^p y^p = \theta(x)\theta(y)$. This implies that θ is a homomorphism. Since $\mathbb{F}$ is a field, θ is an injection. Also, θ is a surjection since $\theta(x) = x^p$ for $x^p \in \mathbb{F}^p$. So θ is an isomorphism and this proves the proposition. □

Lemma 25.1.1. *Let $f(x) \in \mathbb{F}[x]$. If $p(x)$ is an irreducible polynomial over $\mathbb{F}$, and $f(x)$ and $p(x)$ has a common factor $h(x)$ with $deg(h(x)) \geq 1$, then $p(x)$ is a factor of $f(x)$.*

Proof. There is an extension field $\mathbb{E}$ of $\mathbb{F}$ such that $h(a) = 0$ for $a \in \mathbb{E}$. This implies that $p(x)$ is an irreducible polynomial over $\mathbb{F}$. We have that $f(x) = p(x)q(x) + r(x)$ such that $r(x) = 0$ or $deg(r(x)) < deg(p(x))$ for $q(x)$ and $r(x)$ over some extension field of $\mathbb{F}$. If $r(x)$ is not zero, then since $deg(r(x)) < deg(p(x))$, $r(x)$ is the irreducible polynomial of a. This contradicts the fact that $p(x)$ is the irreducible polynomial of a, i.e., $r(x) = 0$. This implies that $p(x)$ divides $f(x)$. □

Definition 25.1.2. *For* $f(x) = a_n x^n + \cdots + a_k x^k + \cdots + a_0$ *over a field* $\mathbb{F}$, **the derivative** $D : \mathbb{F}[x] \to \mathbb{F}[x]$ *is defined as follows:*

$$D\big(f(x)\big) = na_n x^{n-1} + \cdots + ka_k x^{k-1} + \cdots + a_1$$

The following properties of the derivative are simple and easy to prove but they are useful. They are called the sum or difference rule and the product rule in Calculus.

Proposition 25.1.2. *For any* $f(x), g(x) \in \mathbb{F}[x]$,

1. $D\big(f(x) \pm g(x)\big) = D\big(f(x)\big) \pm D\big(g(x)\big)$
2. $D\big(f(x)g(x)\big) = D\big(f(x)\big)g(x) + f(x)D\big(g(x)\big)$

We denote $D\big(f(x)\big)$ as $f'(x)$. Note that for an irreducible polynomial $f(x)$ over a field $\mathbb{F}$, $f(x)$ has a zero of multiplicity $m > 1$ if and only if $f(x)$ and $f'(x)$ has a common factor in $\mathbb{F}$ with a positive degree.

25.2 Separable Polynomials and Extensions

Definition 25.2.1. *Let* $f(x) \in \mathbb{F}[x]$ *and* $f(x) = cp_1(x) \cdots p_n(x)$ *be the decomposition of* $f(x)$ *into irreducibles over* $\mathbb{F}$ *where* $c \in \mathbb{F}$. $f(x)$ *is called* **separable** *if each* $p_i(x)$ *has a simple zero over its splitting field.*

This means that a separable polynomial has no multiple zeros so the zeros of a separable polynomial are all distinct. It also means that the degree of a separable polynomial is the number of distinct zeros of the polynomial.

Example 25.2.1. $(x^3 - 1)(x^2 + 5) \in \mathbb{Q}[x]$ *is separable. Also,* $x^2 - 1 \in \mathbb{Q}[x]$ *is separable.*

Example 25.2.2. $x^4 + \overline{1}$ *is not separable over* $\mathbb{Z}_2$ *since* $x^4 + \overline{1} = (x + \overline{1})^4$ *over* $\mathbb{Z}_2$*. It is also reducible over* $\mathbb{Z}_2$*.*

Example 25.2.3. $x^6 + \overline{2}x^3 + \overline{1}$ *is not separable over* $\mathbb{Z}_3$ *since* $x^6 + \overline{2}x^3 + \overline{1} = (x + \overline{1})^6$ *over* $\mathbb{Z}_3$*. It is also reducible over* $\mathbb{Z}_3$*.*

Note that the polynomials in Example 25.2.2 and 25.2.3 are not separable but their irreducible factors are separable since they are linear. More examples are $x^3 + \overline{2} = (x + \overline{2})^3$ and $x^6 + x^3 + \overline{2} = (x^2 + x + \overline{2})^3$ over $\mathbb{Z}_3$.

We can extend Definition 25.2.1 as following.

Definition 25.2.2. *Let* $\mathbb{E}$ *be an extension field of a field* $\mathbb{F}$*.* $a \in \mathbb{E}$ *is called* **separable** *over* $\mathbb{F}$ *if either it is transcendental over* $\mathbb{F}$ *or its minimal polynomial over* $\mathbb{F}$ *is separable.* $\mathbb{E}$ *is called* **a separable extension** *of* $\mathbb{F}$ *if every element of* $\mathbb{E}$ *is separable over* $\mathbb{F}$*.*

Example 25.2.4. $\mathbb{Q}(\sqrt{2}, \sqrt{7})$ *is a separable extension of* $\mathbb{Q}$*.*

Proposition 25.2.1. *Any algebraic extension of a perfect field is separable.*

Proof. Let $\mathbb{F}$ be a perfect field and $\mathbb{E}$ be an algebraic extension field of $\mathbb{F}$. Suppose that $a \in \mathbb{E}$. Then a is a zero of an irreducible polynomial $p(x) = \sum_{i=0}^{n} a_i x^i \in \mathbb{F}[x]$ with $deg(p(x)) \geq 1$ since this field extension is algebraic. Let's assume that a is a multiple zero. Then $deg(p(x)) \geq 2$ and $p'(a) = \sum_{i=0}^{n} ia_i a^{i-1} = 0$. If char $\mathbb{F} = 0$, then $a_i = 0$ for all a_i, a contradiction since $deg(p(x)) \geq 2$. If char $\mathbb{F} = p$, we have that either $a_i = 0$ or $p|i$, i.e., $i = jp$ for some $j \in \mathbb{N}$. We have that $p(x) = \sum_{i=0}^{n} a_i x^i = \sum_j a_j x^{jp} = g(x^p)$ for some $g(x) \in \mathbb{F}[x]$. Since $\mathbb{F}$ is perfect, $\mathbb{F} = \mathbb{F}^p$ and this implies that $a_j = b_j^p$ for some $b_j \in \mathbb{F}$. Then we derive that $p(x) = \sum_j b_j^p x^{jp} = g(x^p) = g(x)^p$ and this contradicts the fact that $p(x)$ is irreducible. So a has a simple zero. This makes $p(x)$ separable and thus we have proved the proposition. $\square$

Chapter 25 Exercises.

1. (a) Find an irreducible polynomial $f(x)$ over $\mathbb{Q}$ with $\sqrt{2} + \sqrt{7}$ as its zero.

   (b) Is the polynomial $f(x)$ separable?

2. (a) Find all zeros of $x^3 - 2$ over its splitting field.

   (b) Find the splitting field $\mathbb{E}$ of $x^3 - 2 \in \mathbb{Q}[x]$.

   (c) Is $\mathbb{C}$ the splitting field of $x^3 - 2 \in \mathbb{Q}[x]$?

   (d) Is $\mathbb{E}$ an separable extension of $\mathbb{Q}$?

3. Find all zeros over $\mathbb{C}$ for the following polynomials and prove that they are separable

   (a) $x^5 - 1$

   (b) $x^6 - 2$

   (c) $(x^2 + 4x + 2)(x^2 - 6x - 7)$

   (d) $x^7 - 7$

4. (a) Prove that $\left(\mathbb{Z}_3\right)^3 = \mathbb{Z}_3$.

   (b) Prove that $\left(\mathbb{Z}_5\right)^5 = \mathbb{Z}_5$.

5. Let $f(x) = 2x^{3\cdot 3} + x^{3\cdot 2} - x^3 + 2 \in \mathbb{Z}_3[x]$.

   (a) Find a polynomial $g(x) \in \mathbb{Z}_3[x]$ such that $f(x) = g(x^3)$.

   (b) Prove that $f(x) = (g(x))^3$.

6. (a) Find all the zeros of $x^5 - 3 \in \mathbb{Q}[x]$ using De Moivres' Theorem.

   (b) Find the splitting field $\mathbb{F}$ of $x^5 - 3$ over $\mathbb{Q}$.

   (c) Find $[\mathbb{F} : \mathbb{Q}]$.

   (d) Is this extension separable?

7. (a) Prove that $x^3 + 2x + 1$ is irreducible over $\mathbb{Z}_5$.

   (b) Find all the elements of the ideal $< x^3 + 2x + 1 >$ of $\mathbb{Z}_5[x]$.

   (c) Prove that $\mathbb{Z}_5[x]/< x^3 + 2x + 1 >$ is a field and find all the elements of the field $\mathbb{Z}_5[x]/< x^3 + 2x + 1 >$.

   (d) Find the order of $\mathbb{Z}_5[x]/< x^3 + 2x + 1 >$.

   (e) Is $\mathbb{Z}_5[x]/< x^3 + 2x + 1 >$ a separable extension of $\mathbb{Z}_5$?

8. Show that each polynomial is separable over $\mathbb{Q}$, i.e., show that it has a simple zero or not over its splitting field.

   (a) $x^2 + 6x + 5$

   (b) $x^3 + 6x^2 + 3x + 2$

   (c) $x^4 - 2x^2 + 2$

   (d) $x^5 - 6x^3 + 8x - 2$

9. Prove that a finite extension of a perfect field is separable.

10. Let $\mathbb{E}$ be an algebraic extension of a field $\mathbb{F}$ and $\mathbb{K}$ be the set of all separable elements of $\mathbb{E}$ over $\mathbb{F}$. Prove that $\mathbb{K}$ is a subfield of $\mathbb{E}$.

11. Show that the following is reducible. Are their irreducible factors separable?

    (a) $x^4 + x^2 + \overline{1} \in \mathbb{Z}_2[x]$.

    (b) $x^9 + \overline{2}x^6 + x^3 + \overline{2} \in \mathbb{Z}_3[x]$.

    (c) $x^{4p} + x^{2p} + x^p + \overline{1} \in \mathbb{Z}_p[x]$.

Chapter 26

Galois Theory

The quadratic formula of a quadratic equation provides the complete solutions of the equation. Since there are such formulas of a cubic equation and a fourth-degree equation, many people had tried to find such formulas for an arbitrary polynomial equation with the degree greater than or equal to five. But Abel and Galois proved that there do not exist such formulas. In particular, Galois derived a necessary and sufficient condition for a solvable (algebraically) polynomial equation using Galois Theory. The core part of Galois theory is that there is a one to one correspondence between the subgroups of the Galois group and the subfields in a given field extension. Recall that if a polynomial $f(x) \in \mathbb{F}[x]$ has distinct zeros, the splitting field of $f(x)$ over $\mathbb{F}$ is an algebraic and separable extension. We will also see that it is "normal".

26.1 Normal Extensions

Definition 26.1.1. *Let $\mathbb{E}$ be an extension field of a field $\mathbb{F}$ and σ be an automorphism of $\mathbb{E}$. We say that σ* **fixes** *$\mathbb{F}$ if $\sigma(a) = a$ for all $a \in \mathbb{F}$ and call*

σ **an $\mathbb{F}$-automorphism.** *σ is called* **an $\mathbb{F}$-homomorphism** *if σ is a field homomorphism from $\mathbb{K}$ to $\mathbb{E}$ for field extensions $\mathbb{F}$ and $\mathbb{K}$ of $\mathbb{E}$ and if it fixes $\mathbb{F}$.*

Note that **an $\mathbb{F}$-monomorphism** and **an $\mathbb{F}$-isomorphism** can be defined in similar ways.

Example 26.1.1. *$\mathbb{C}$ is an extension field of $\mathbb{R}$. If we define a map σ from $\mathbb{C}$ to itself as follows:*

$$\sigma(a+bi) = a - bi$$

for $a + bi \in \mathbb{C}$, then σ is an $\mathbb{R}$-automorphism (the complex conjugation map). For, $\sigma(a) = a$ for any $a \in \mathbb{R}$, $\sigma((a+bi)(a_1+b_1i)) = \sigma(a+bi)\sigma(a_1+b_1i)$, and σ is one to one and onto (why?).

Example 26.1.2. *Using a similar argument, the following map σ from $\mathbb{Q}(\sqrt{7})$ to itself is a $\mathbb{Q}$-automorphism:*

$$\sigma(a+b\sqrt{7}) = a - b\sqrt{7}$$

for $a+b\sqrt{7} \in \mathbb{Q}(\sqrt{7})$. In fact, it is easy to prove that the only $\mathbb{Q}$-automorphisms from $\mathbb{Q}(\sqrt{7})$ to itself are the identity map ι on $\mathbb{Q}$ and this conjugation map σ.

Definition 26.1.2. *Let $\mathbb{E}$ be an extension field of a field $\mathbb{F}$. We say that $\mathbb{E}$* **is normal over** $\mathbb{F}$ *if $\mathbb{E}$ contains all zeros of an irreducible polynomial $f(x)$ in $\mathbb{F}[x]$ whenever $\mathbb{E}$ contains one zero of $f(x)$.*

Note that if $\mathbb{E}$ is normal over $\mathbb{F}$ and $\mathbb{E}$ includes one zero of an irreducible polynomial $f(x)$ in $\mathbb{F}[x]$, then $f(x)$ splits in $\mathbb{E}$ since $f(x)$ is factored into linear factors in $\mathbb{E}[x]$.

Example 26.1.3. *Let ω be a primitive n-th zero of unity, i.e., $\omega^n = 1$ and $\omega \neq 1$. Show that $\mathbb{Q}(\omega)$ is a normal extension over $\mathbb{Q}$.*

Solution. $\omega^2, \cdots, \omega^{n-1}$ are all the zeros of $x^n - 1$ because ω is a zero of $x^n - 1 \in \mathbb{Q}[x]$ where $n > 1$. Since $\omega, \cdots, \omega^{n-1} \in \mathbb{Q}(\omega)$, the field $\mathbb{Q}(\omega)$ is the splitting field of $x^n - 1$. Thus $\mathbb{Q}(\omega)$ is normal over $\mathbb{Q}$. □

A normal extension is not always finite. The following is a good example

Example 26.1.4. $\mathbb{Q}(\sqrt{2}, \sqrt[3]{2}, \sqrt[4]{2}, \cdots)$ *is an normal extension of* $\mathbb{Q}$ *and this extension is not finite.*

26.2 Galois Extensions

Definition 26.2.1. *A field extension $\mathbb{E}$ of a field $\mathbb{F}$ is called* **a Galois extension** *if this extension is finite, normal, and separable.*

For example, the extension field $\mathbb{Q}(\omega)$ in Example 26.1.3 is a Galois extension of $\mathbb{Q}$ but the extension field $\mathbb{Q}(\sqrt{2}, \sqrt[3]{2}, \sqrt[4]{2}, \cdots)$ in Example 26.1.4 is not a Galois extension of $\mathbb{Q}$. In this sense, a normal extension is sometimes called **a quasi-Galois extension**.

Definition 26.2.2. *Let $\mathbb{E}$ be an extension field of a field $\mathbb{F}$. Then the set of all $\mathbb{F}$-automorphisms of $\mathbb{E}$ is called* **the Galois group of $\mathbb{E}$ over $\mathbb{F}$** *and denoted by $Gal(\mathbb{E}/\mathbb{F})$.*

Note that $Gal(\mathbb{E}/\mathbb{F}) = \{\sigma \in Aut(\mathbb{E}) | \ \sigma(a) = a \text{ for all } a \in \mathbb{F}\}$ and it is a group under composition of functions.

Definition 26.2.3. *Let $\mathbb{E}$ be an extension field of a field $\mathbb{F}$ and H be a subgroup of the Galois group $Gal(\mathbb{E}/\mathbb{F})$. Then we call the following set* **the fixed field of $\mathbb{E}$ under the group H:**

$$\{a \in \mathbb{E}|\ \theta(a) = a \text{ for all } \theta \in H\} \tag{26.1}$$

It is denoted by $\mathbb{E}_H$.

It is easy to prove that $\mathbb{E}_H$ is a field satisfying $\mathbb{F} \subseteq \mathbb{E}_H \subseteq \mathbb{E}$. Note that $\mathbb{E}_{Gal(\mathbb{E}/\mathbb{F})} = \mathbb{F}$ and $\mathbb{E}_{\{e\}} = \mathbb{E}$ so when H becomes a larger subgroup of $Gal(\mathbb{E}/\mathbb{F})$, $\mathbb{E}_H$ becomes a smaller intermediate field in between $\mathbb{F}$ and $\mathbb{E}$.

Example 26.2.1. *Prove that $Gal(\mathbb{C}/\mathbb{R}) \cong \mathbb{Z}_2$.*

Solution. Suppose $\sigma \in Gal(\mathbb{C}/\mathbb{R})$. Then $\sigma(i^2) = \sigma(-1) = -1$ (why?). This implies that $\sigma(i)\sigma(i) = -1$. So $\sigma(i) = i$ or $\sigma(i) = -i$. If $\sigma(i) = i$, σ is the identity map on $\mathbb{C}$ since $\sigma(a + bi) = a + bi$ for any $a + bi \in \mathbb{C}$. If $\sigma(i) = -i$, then σ is the complex conjugation map since $\sigma(a + bi) = a - bi$. Note that this σ^2 is the identity map. So $Gal(\mathbb{C}/\mathbb{R})$ has two elements whose order are 1 and 2 respectively. Therefore it is isomorphic to $\mathbb{Z}_2$. $\square$

Note that $\mathbb{C}_{Gal(\mathbb{C}/\mathbb{R})} = \mathbb{R}$.

Example 26.2.2. *Prove that the Galois group $Gal(\mathbb{Q}(\sqrt[3]{3}))/\mathbb{Q}) \cong \{e\}$ where $\{e\}$ is the trivial group with only one element.*

Solution. Every element of $\mathbb{Q}(\sqrt[3]{3})$ has the form $a_0 + a_1\sqrt[3]{3} + a_2\sqrt[3]{9}$ where $a_0, a_1, a_2 \in \mathbb{Q}$. Suppose $\sigma \in Gal(\mathbb{Q}(\sqrt[3]{3})/\mathbb{Q})$. Then $\sigma(a_0 + a_1\sqrt[3]{3} + a_2\sqrt[3]{9}) = a_0 + a_1\sigma(\sqrt[3]{3}) + a_2\sigma(\sqrt[3]{3}))^2$. Since $\sigma(\sqrt[3]{3}))^3 = \sigma((\sqrt[3]{3})^3) = \sigma(3) = 3$, $\sigma(\sqrt[3]{3}))$ is a zero of $x^3 - 3 = 0$. But $\sqrt[3]{3}$ is the only zero of $x^3 - 3$ in $\mathbb{Q}(\sqrt[3]{3})$. This implies that $\sigma(\sqrt[3]{3}) = \sqrt[3]{3}$. So σ is the identity map of $\mathbb{Q}(\sqrt[3]{3})$. Therefore $G(\mathbb{Q}\sqrt[3]{3})/\mathbb{Q})$ has only one element. $\square$

Lemma 26.2.1. *Let $\mathbb{E}$ be an extension field of a field $\mathbb{F}$, $f(x) \in \mathbb{F}[x]$, and $\sigma \in Gal(\mathbb{E}/\mathbb{F})$. Then if a is a zero of $f(x)$, $\sigma(a)$ is a zero of $f(x)$.*

Proof. Suppose a is a zero of $f(x) = a_n x^n + \cdots + a_1 x + a_0 \in \mathbb{F}[x]$, i.e., $a_n a^n + \cdots + a_1 a + a_0 = 0$. This implies that $\sigma(a_n a^n + \cdots + a_1 a + a_0) = \sigma(0) = 0$. Also, $\sigma(a_n a^n + \cdots + a_1 a + a_0) = a_n(\sigma(a))^n + \cdots + a_1\sigma(a) + a_0 = 0$ since σ preserves the multiplication. This implies that $\sigma(a)$ is also a zero of $f(x)$. $\square$

Corollary 26.2.1. *Let a be algebraic over a field $\mathbb{F}$ and $f(x)$ be its minimal polynomial over $\mathbb{F}$ where $deg(f(x)) = k$. Let $\mathbb{E}$ be an extension field of $\mathbb{F}$ that contains the splitting field of $f(x)$. Then the number of all $\mathbb{F}$-homomorphisms from $\mathbb{F}(a)$ to $\mathbb{E}$ is equal to the number r of distinct zeros of $f(x)$. Note that $r \leq k = [\mathbb{F}(a) : \mathbb{F}]$. In particular, $r = k$ if and only if $f(x)$ is separable.*

Proof. Since $f(x)$ is the minimal polynomial of a over $\mathbb{F}$, it is irreducible over $\mathbb{F}$ and $[\mathbb{F}(a) : \mathbb{F}] = deg(f(x)) = k$. Let $a_1, \cdots, a_r \in \mathbb{E}$ be distinct zeros of $f(x)$ where $r \leq k$. By Lemma 26.2.1, there is an $\mathbb{F}$-automorphism $\sigma_i : \mathbb{F}(a) \to \mathbb{F}(a_i)$, $1 \leq i \leq r$. Thus the number of all $\mathbb{F}$-homomorphisms is at least r. Now, suppose $\sigma : \mathbb{F}(a) \to \mathbb{E}$ is an $\mathbb{F}$-homomorphism. Since $f^\sigma(x) = f(x)$, using Lemma 26.2.1 again, we can conclude that $\sigma(a) = a_i$ for $1 \leq i \leq r$, which implies that $\sigma = \sigma_i$. Thus the number of all $\mathbb{F}$-homomorphisms is exactly r. Note that $f(x)$ is separable if and only if $f(x)$ has exactly $r = k$ zeros. $\square$

Theorem 26.2.1. *Let* $\mathbb{F}$ *be a field and* $\mathbb{E}$ *be the splitting field of* $f(x)$ *with* $deg(f(x)) = n$*. Then if* $f(x)$ *have* n *distinct zeros,* $Gal(\mathbb{E}/\mathbb{F})$ *is isomorphic to a subgroup of the symmetric group* $\mathfrak{S}_n$ *on* $\{1, \cdots, n\}$.

Proof. Let $X = \{a_1, \cdots, a_n\}$ be the set of all zeros of $f(x)$ in $\mathbb{E}$. For $\sigma \in Gal(\mathbb{E}/\mathbb{F})$, $\sigma(a_i) = a_j$, $1 \leq i, j \leq n$. This implies that $\sigma(X) = X$ and the restriction $\sigma|_X$ of σ to the set X is a permutation. So if we define a map $\theta : Gal(\mathbb{E}/\mathbb{F}) \to \mathfrak{S}_n$ as $\theta(\sigma) = \sigma|_X$, then θ is a bijection, which implies that $Gal(\mathbb{E}/\mathbb{F})$ is isomorphic to a subgroup of $\mathfrak{S}_n$. □

Example 26.2.3. *Find the splitting field* $\mathbb{E}$ *of* $f(x) = (x^2 - 5)(x^2 - 11)$ *and show that* $Gal(\mathbb{E}/\mathbb{Q})$ *is actually isomorphic to a subgroup of* $\mathfrak{S}_4$.

Solution. Since all zeros of $f(x)$ are $\pm\sqrt{5}$ and $\pm\sqrt{11}$, $f(x)$ is separable and $\mathbb{Q}(\sqrt{5}, \sqrt{11})$ is the splitting field $\mathbb{E}$ of $f(x)$. Also, by Corollary 26.2.1, $Gal(\mathbb{Q}(\sqrt{5}, \sqrt{11})/\mathbb{Q})$ has 4 elements. Note that $\sigma_0, \sigma_1, \sigma_2, \sigma_3$ are all different $\mathbb{Q}$-automorphisms of $\mathbb{Q}(\sqrt{5}, \sqrt{11})$ where σ_0 is the identity map, $\sigma_1(\sqrt{5}) = -\sqrt{5}$, $\sigma_2(\sqrt{11}) = -\sqrt{11}$, and $\sigma_3(\sqrt{5}) = -\sqrt{5}$ and $\sigma_3(\sqrt{11}) = -\sqrt{11}$ fixing everything else respectively. So $Gal(\mathbb{Q}(\sqrt{5}, \sqrt{11})/\mathbb{Q}) = \{\sigma_i | 0 \leq i \leq 3\}$ that is isomorphic to $\mathbb{Z}_2 \times \mathbb{Z}_2$. Obviously, $\mathbb{Z}_2 \times \mathbb{Z}_2$ is isomorphic to a subgroup of $\mathfrak{S}_4$. □

Lemma 26.2.2. *Let* θ *be an isomorphism from a field* $\mathbb{F}$ *to a field* $\mathbb{F}_1$*. Let* $f(x) \in \mathbb{F}[x]$ *be a non constant polynomial. If* $\mathbb{E}$ *and* $\mathbb{E}_1$ *be splitting fields of* $f(x)$ *and* $f^{\theta}(x)$ *respectively, then there is an isomorphism* $\overline{\theta}$ *from* $\mathbb{E}$ *to* $\mathbb{E}_1$ *such that* $\overline{\theta}|_{\mathbb{F}} = \theta$ *as in Corollary 22.4.2. The number of all such isomorphisms is less than or equal to* $[\mathbb{E} : \mathbb{F}]$.

Proof. Let us prove the theorem by induction on $[\mathbb{E} : \mathbb{F}] = n$. If $n = 1$, then $\mathbb{E} = \mathbb{F}$ and $\mathbb{E}_1 = \mathbb{F}_1$. This implies that $\theta = \overline{\theta}$. Now let us assume that the lemma holds for $[\mathbb{E} : \mathbb{F}] < n$. Let $g(x)$ be an irreducible factor of $f(x)$, and a and a_1 be zeros of $g(x)$ and $g^\theta(x)$. By Theorem 22.4.3, there is an $\mathbb{F}$-isomorphism $\overline{\theta} : \mathbb{F}(a) \to \mathbb{F}_1(a_1)$ such that $\overline{\theta}|_\mathbb{F} = \theta$ and $\overline{\theta}(a) = a_1$. Note that $f(x) = (x-a)h(x)$ for some $h(x) \in \mathbb{F}(a)[x]$. Since $\mathbb{E}$ and $\mathbb{E}_1$ are splitting fields of $h(x)$ and $h^\theta(x)$ respectively. By induction and Corollary 26.2.1, there is an isomorphism from $\mathbb{E}$ to $\mathbb{E}_1$. Since the number of different zeros of $f(x)$ is less than or equal to n, the remaining proof of the lemma is straightforward. □

Corollary 26.2.2. *For a given polynomial $f(x)$ over a field $\mathbb{F}$, the splitting field of $f(x)$ is unique up to $\mathbb{F}$-isomorphism.*

Proof. If we take the identity map from $\mathbb{F}$ to itself, then by Lemma 26.2.2, the splitting field of $f(x)$ is unique up to $\mathbb{F}$-isomorphism. □

Proposition 26.2.1. *For a finite field extension $\mathbb{E}$ of a field $\mathbb{F}$. The following statements are equivalent.*

1. *$\mathbb{E}$ is a normal extension of $\mathbb{F}$.*
2. *$\mathbb{E}$ is a splitting field of a polynomial in $\mathbb{F}[x]$.*
3. *Any $\mathbb{F}$-monomorphism θ from $\mathbb{E}$ to itself is an automorphism.*

Proof.

1. $\to$ 2. : Since $[\mathbb{E} : \mathbb{F}]$ is finite, we have that $\mathbb{E} = \mathbb{F}[a_1, \cdots, a_n]$ for $a_1, \cdots, a_n \in \mathbb{E}$. Let $f_i(x)$ be the minimal polynomial of a_i, $1 \leq i \leq n$, over $\mathbb{F}$ and $f(x) = \prod_{i=1}^n f_i(x)$. Since $\mathbb{E}$ is a normal extension of $\mathbb{F}$ and $a_i \in \mathbb{E}$ is a zero of $f(x) \in \mathbb{F}[x]$, $\mathbb{E}$ includes all zeros of $f(x)$, i.e., all zeros of $f(x)$ are in $\mathbb{E} = \mathbb{F}[a_1, \cdots, a_n]$. Thus all linear factors of $f(x)$ are in $\mathbb{E}[x]$, which means that $\mathbb{E}$ is a splitting field of $f(x)$.

2. $\to$ 3. : Let $\mathbb{E} = \mathbb{F}[a_1, \cdots, a_n]$ be a splitting field of $f(x) \in \mathbb{F}[x]$ where $a_1, \cdots, a_n$ are all zeros of $f(x)$. If $\theta : \mathbb{E} \to \mathbb{E}$ be an $\mathbb{F}$-monomorphism, then $\overline{\theta}(f(x)) = f^{\theta}(x)$ and each $\theta(a_i)$ is a zero of $f(x)$ where $\overline{\theta} : \mathbb{E}[x] \to \mathbb{E}[x]$. Since θ is one to one, it simply permutes all the zeros of $f(x)$. This implies that $\theta(\mathbb{E}) = \mathbb{E}$ since $\mathbb{E}$ is generated by $\mathbb{F}$ and all the zeros of $f(x)$. Thus 2 implies 3.

3. $\to$ 1. : Let $f(x)$ be a irreducible polynomial and a be a zero of $f(x)$ where $a \in \mathbb{E}$ and $\theta(a) = b$. For a zero of $f(x)$, let $\mathbb{E}_1$ be the splitting field of $f(x)$ over $\mathbb{F}$. This implies that $b \in \mathbb{E}_1$. By Theorem 22.4.3, there exists an $\mathbb{F}$-isomorphism $\theta : \mathbb{F}(a) \to \mathbb{F}(b)$ such that $\theta(a) = b$ and this θ can be extended to an $\mathbb{F}$-isomorphism $\overline{\theta} : \mathbb{E} \to \mathbb{E}_1$. Let $\overline{\theta}|_{\mathbb{F}} = \theta$ where $\overline{\theta}|_{\mathbb{F}}$ is the restriction of $\overline{\theta}$ to $\mathbb{F}$. This implies that $b = \theta(a) \in \mathbb{E}$. Since $\mathbb{E}$ is the splitting field of $f(x)$ and $\overline{\theta}$ is an automorphism by 3, we have that $\mathbb{E} = \mathbb{E}_1$. Therefore we have proven the proposition. $\square$

Lemma 26.2.3. *If $\mathbb{E}$ is a Galois extension of a field $\mathbb{F}$, then $Gal(\mathbb{E}/\mathbb{F})$ is a finite group of order $[\mathbb{E} : \mathbb{F}]$ and $\mathbb{F}$ is its fixed field.*

Proof. Since $[\mathbb{E} : \mathbb{F}] < \infty$, by Theorem 23.2.4, $\mathbb{E}$ is a simple extension, i.e., $\mathbb{E} = \mathbb{F}(a)$ for $a \in \mathbb{E}$. Let $p(x)$ be the minimal polynomial of a and $deg(p(x)) = n$. Then $n = [\mathbb{E} : \mathbb{F}]$. Since $\mathbb{E}$ be a Galois extension of the field $\mathbb{F}$, $p(x)$ has n distinct zeros in $\mathbb{E}$. By Theorem 22.4.3, there are exactly n distinct $\mathbb{F}$-automorphisms of $\mathbb{E}$ i.e., $|Gal(\mathbb{E}/\mathbb{F})| = n$. Obviously $\mathbb{F}$ is a subfield of the fixed field of $Gal(\mathbb{E}/\mathbb{F})$. If b is in the fixed field of $Gal(\mathbb{E}/\mathbb{F})$ but $b \notin \mathbb{F}$, then we can find b' of $\mathbb{E}$ such that $b' \neq b$ and b' is a zero of $p(x)$. By Theorem 22.4.3, there exists an $\mathbb{F}$-isomorphism $\theta : \mathbb{F}(b) \to \mathbb{F}(b')$ with $\theta(b) = b'$. So by Theorem 22.4.3, this can be extended to an $\mathbb{F}$-automorphism $\sigma : \mathbb{E} \to \mathbb{E}$. Since $\sigma \in Gal(\mathbb{E}/\mathbb{F})$ and $\sigma(b) = b' \neq b$, we have a contradiction. Therefore $\mathbb{F}$ is the fixed field of $Gal(\mathbb{E}/\mathbb{F})$. $\square$

Lemma 26.2.4. *If $\mathbb{E}$ is an extension field of a field $\mathbb{F}$ and H is a finite subgroup of $Gal(\mathbb{E}/\mathbb{F})$, then $\mathbb{E}$ is a Galois extension of $\mathbb{E}_H$ and $Gal(\mathbb{E}/\mathbb{E}_H) = H$.*

Proof. Let $a \in \mathbb{E}$ and $H = \{\theta_1, \cdots, \theta_n\}$ where $\theta_1, \cdots, \theta_n$ are distinct elements. Then $H(a) = \{\theta_1(a), \cdots, \theta_m(a)\}$ for $1 \leq i \neq j \leq m \leq n$ with $\theta_i(a) \neq \theta_j(a)$. Let's pick $\theta \in H$ where θ is one to one. If $\theta(\theta_i(a)) = \theta(\theta_j(a))$, then $\theta_i(a) = \theta_j(a)$. This implies that $i = j$, so $\{\theta\theta_1(a), \cdots, \theta\theta_m(a)\} = \{\theta_1(a), \cdots, \theta_m(a)\}$. Let

$$f(x) = \prod_{i=1}^{m}(x - \theta_i(a)).$$

Since $f(x) = f^{\theta}(x)$ for any $\theta \in H$, we have that $f(x) \in \mathbb{E}_H[x]$. Since one of θ_i's must be the identity map, a is a zero of $f(x)$. So if $g(x)$ is the minimal polynomial of a over $\mathbb{E}_H$, then we have that $g(\theta_i(a)) = \theta_i(g(a)) = 0$, $1 \leq i \leq m$. So $deg(g(x)) \geq deg((f(x))$. Since $g(x)$ is the minimal polynomial of a over $\mathbb{E}_H$, we have that $g(x)$ and $f(x)$ are the same. This implies that a is algebraic and separable over $\mathbb{E}_H$. We also have that $[\mathbb{E}_H(a) : \mathbb{E}_H] = m \leq n = |H|$. Now, let's pick $a \in \mathbb{E}$ so that $[\mathbb{E}_H(a) : \mathbb{E}_H]$ is maximal. Then we need to show that $\mathbb{E} = \mathbb{E}_H(a)$. To show this, let us assume that $\mathbb{E} \neq \mathbb{E}_H(a)$. Then there must be $b \in \mathbb{E}$ such that $b \notin \mathbb{E}_H$. By Corollary 26.2.1, we have that $[\mathbb{E}_H(a, b) : \mathbb{E}_H] > [\mathbb{E}_H(a) : \mathbb{E}_H]$. Since $\mathbb{E}_H(a, b)$ is a simple extension, this contradicts the maximality of $[\mathbb{E}_H(a) : \mathbb{E}_H]$. So $\mathbb{E} = \mathbb{E}_H(a)$ and $\mathbb{E}$ is a Galois extension of $\mathbb{E}_H$. For the second part of the lemma, we also have that H is a subgroup of $Gal(\mathbb{E}/\mathbb{E}_H)$. By Lemma 26.2.3, we have that $Gal(\mathbb{E}/\mathbb{E}_H) = [\mathbb{E} : \mathbb{E}_H] \leq |H|$. So

$H = Gal(\mathbb{E}/\mathbb{E}_H)$. Therefore we have proven the lemma. □

26.3 Fundamental Theorem of Galois Theory

The following is the main theorem of this chapter.

Theorem 26.3.1. (Fundamental Theorem of Galois Theory)
If $\mathbb{E}$ is a Galois extension of a field $\mathbb{F}$, then $Gal(\mathbb{E}/\mathbb{F})$ is a finite group such that $|Gal(\mathbb{E}/\mathbb{F})| = [\mathbb{E} : \mathbb{F}]$ and there is a one to one correspondence between a subfield $\mathbb{K}$ of $\mathbb{E}$ containing $\mathbb{F}$ and a subgroup H of $Gal(\mathbb{E}/\mathbb{F})$ such that $H = Gal(\mathbb{E}/\mathbb{K})$ for each $\mathbb{K}$ and $\mathbb{K} = \mathbb{E}_H$ for each H. Furthermore $\mathbb{K}$ is the fixed field $\mathbb{E}_{Gal(\mathbb{E}/\mathbb{K})}$ of $Gal(\mathbb{E}/\mathbb{K})$.

Proof. Suppose that $\mathbb{E}$ is a Galois extension of $\mathbb{F}$ and $\mathbb{K}$ is an intermediate field between $\mathbb{E}$ and $\mathbb{F}$, i.e., $\mathbb{F} \subseteq \mathbb{K} \subseteq \mathbb{E}$. Then $Gal(\mathbb{E}/\mathbb{K})$ is the set of all automorphisms of $\mathbb{E}$ that fix each element in $\mathbb{K}$. And this automorphisms obviously fix elements in $\mathbb{F}$ because $\mathbb{F} \subseteq \mathbb{K}$. This means that $Gal(\mathbb{E}/\mathbb{K})$ is a subset of $Gal(\mathbb{E}/\mathbb{F})$. So it is a subgroup under the function composition and $\mathbb{E}_{Gal(\mathbb{E}/\mathbb{F})} = \mathbb{F}$. Now, suppose that H is a subgroup of $Gal(\mathbb{E}/\mathbb{F})$. Then there exists an intermediate field $\mathbb{E}_H$ between $\mathbb{E}$ and $\mathbb{F}$ such that $Gal(\mathbb{E}/\mathbb{E}_H) = H$ holds by Lemma 26.2.4. So we have proven that there is a one to one correspondence between $\mathbb{K}$ and H. □

Note that we have that the composite map given by $\mathbb{K} \to Gal(\mathbb{E}/\mathbb{K})$ and $H \to \mathbb{E}_H$ is the identity map by Theorem 26.3.1.

Corollary 26.3.1. *Let $\mathbb{E}$ be a Galois extension of a field $\mathbb{F}$, and $\mathbb{K}$ be a subfield of $\mathbb{E}$ and containing $\mathbb{F}$. Then the following hold.*

1. *$\mathbb{E}$ is a finite separable extension of $\mathbb{K}$.*

2. *$\mathbb{K}$ is a normal extension of $\mathbb{F}$ if and only if $Gal(\mathbb{E}/\mathbb{K})$ is a normal subgroup of $Gal(\mathbb{E}/\mathbb{F})$.*

3. *If $\mathbb{K}$ is a Galois extension of $\mathbb{F}$, then $Gal(\mathbb{K}/\mathbb{F})$ is isomorphic to the group $Gal(\mathbb{E}/\mathbb{F})/Gal(\mathbb{E}/\mathbb{K})$.*

Proof.

1. Since $[\mathbb{E} : \mathbb{F}] < \infty$ and $\mathbb{E} \supseteq \mathbb{K} \supseteq \mathbb{F}$, $[\mathbb{E} : \mathbb{K}] < \infty$. Also, since $\mathbb{E}/\mathbb{F}$ is separable, so is $\mathbb{E}/\mathbb{K}$.

2. For the proof of 2, first we will show that $\mathbb{K}$ is a normal extension of $\mathbb{F}$ if and only if for any $\sigma \in Gal(\mathbb{E}/\mathbb{F})$, $\sigma(\mathbb{K}) \subset \mathbb{K}$ holds. Let us assume that $\sigma(\mathbb{K}) \subset \mathbb{K}$ holds. By Theorem 23.2.4, $\mathbb{K}$ is a simple extension, i.e., $\mathbb{K} = \mathbb{F}(a)$ for $a \in \mathbb{K}$. This implies that for any $\sigma \in Gal(\mathbb{E}/\mathbb{F})$, $\sigma(a) \in \mathbb{K}$. So $\mathbb{K}$ is a splitting field of the polynomial $p(x) = \prod_{\sigma \in Gal(\mathbb{E}/\mathbb{F})}(x - \sigma(a)) \in \mathbb{F}[x]$. So by Proposition 26.2.1, $\mathbb{K}$ is a normal extension of $\mathbb{F}$. Conversely, let us assume that $\mathbb{K}$ is a normal extension of $\mathbb{F}$. Thus $\mathbb{K} = \mathbb{F}(a)$ is a splitting field of the minimal polynomial $p(x)$ of a over $\mathbb{F}$. So for any $\sigma \in Gal(\mathbb{E}/\mathbb{F})$, $\sigma(a)$ is a zero of the polynomial $p(x)$ and this makes $\sigma(a) \in \mathbb{K}$. This implies that $\sigma(\mathbb{K}) \subset \mathbb{K}$. So we have proven the statement. Now let's prove 2. Suppose that $\mathbb{K}$ is a normal extension of $\mathbb{F}$. Then equivalently we can say that $\sigma(k) \in \mathbb{K}$ for any $k \in \mathbb{K}$ by the proved statement. If we pick $\mu \in Gal(\mathbb{E}/\mathbb{K})$, $\mu(\sigma(k)) = \sigma(k) \Leftrightarrow \sigma^{-1}(\mu(\sigma(k))) = k \Leftrightarrow \sigma^{-1}\mu\sigma \in Gal(\mathbb{E}/\mathbb{K}) \Leftrightarrow Gal(\mathbb{E}/\mathbb{K})$ is a normal subgroup of $Gal(\mathbb{E}/\mathbb{F})$. This completes the proof of 2.

3. Suppose that $\mathbb{K}$ is a Galois extension of $\mathbb{F}$. Let's define a map T from $Gal(\mathbb{E}/\mathbb{F})$ to $Gal(\mathbb{K}/\mathbb{F})$ as $T(\theta) = \theta|_{\mathbb{K}}$. Then by Lemma 26.2.4, T is a surjective homomorphism. Note that $Gal(\mathbb{E}/\mathbb{K})$ is the kernel of T. Therefore $Gal(\mathbb{K}/\mathbb{F})$ is isomorphic to $Gal(\mathbb{E}/\mathbb{F})/Gal(\mathbb{E}/\mathbb{K})$. $\square$

Example 26.3.1. *For the extension field $\mathbb{Q}(\sqrt{5}, \sqrt{7})$ of the field $\mathbb{Q}$, find $Gal(\mathbb{Q}(\sqrt{5}, \sqrt{7})/\mathbb{Q})$, $Gal(\mathbb{Q}(\sqrt{5}, \sqrt{7})/\mathbb{Q}(\sqrt{5}))$, $Gal(\mathbb{Q}(\sqrt{5}, \sqrt{7})/\mathbb{Q}(\sqrt{7}))$, and $Gal(\mathbb{Q}(\sqrt{5}, \sqrt{7})/\mathbb{Q}(\sqrt{5}, \sqrt{7}))$. Also show that $Gal(\mathbb{Q}(\sqrt{5})/\mathbb{Q})$ is isomorphic to $Gal(\mathbb{Q}(\sqrt{5}, \sqrt{7})/\mathbb{Q})/Gal(\mathbb{Q}(\sqrt{5}, \sqrt{7})/\mathbb{Q}(\sqrt{5}))$.*

Solution. Obviously, an identity map $\iota : \mathbb{Q}(\sqrt{5}, \sqrt{7}) \to \mathbb{Q}(\sqrt{5}, \sqrt{7})$ such that

$$\iota(\sqrt{5}) = \sqrt{5}, \quad \iota(\sqrt{7}) = \sqrt{7}, \text{ and } \iota(a) = a \text{ for } a \in \mathbb{Q}$$

is a $\mathbb{Q}$-automorphism of $\mathbb{Q}(\sqrt{5}, \sqrt{7})$. Similarly to ι, if we define a map α from $\mathbb{Q}(\sqrt{5}, \sqrt{7})$ to itself such that $\alpha(\sqrt{5}) = -\sqrt{5}$ and fixing everything else, then α is a $\mathbb{Q}$-automorphism of $\mathbb{Q}(\sqrt{5}, \sqrt{7})$ with order 2. Also, a map β from $\mathbb{Q}(\sqrt{5}, \sqrt{7})$ to itself with $\beta(\sqrt{5}) = \sqrt{5}$ and fixing everything else, then β is a $\mathbb{Q}$-automorphism of $\mathbb{Q}(\sqrt{5}, \sqrt{7})$ with order 2. Note that $\alpha\beta = \beta\alpha = \gamma$ is a $\mathbb{Q}$-automorphism of $\mathbb{Q}(\sqrt{5}, \sqrt{7})$ such that $\gamma(\sqrt{5}) = -\sqrt{5}$ and $\gamma(\sqrt{7}) = -\sqrt{7}$. Also note that $\gamma^2 = \iota$ and there is no other $\mathbb{Q}$-automorphism of $\mathbb{Q}(\sqrt{5}, \sqrt{7})$. So we have that $Gal(\mathbb{Q}(\sqrt{5}, \sqrt{7})/\mathbb{Q}) = \{\iota, \alpha, \beta, \gamma\}$ which is isomorphic to the Klein 4-group K_4. Now, it is easy to see that $Gal(\mathbb{Q}(\sqrt{5}, \sqrt{7})/\mathbb{Q}(\sqrt{5})) = \{\iota, \beta\}$, $Gal(\mathbb{Q}(\sqrt{5}, \sqrt{7})/\mathbb{Q}(\sqrt{7})) = \{\iota, \alpha\}$, and $Gal(\mathbb{Q}(\sqrt{5}, \sqrt{7})/\mathbb{Q}(\sqrt{5}, \sqrt{7})) = \{\iota\}$. $Gal(\mathbb{Q}(\sqrt{5}, \sqrt{7})/\mathbb{Q})/Gal(\mathbb{Q}(\sqrt{5}, \sqrt{7})/\mathbb{Q}(\sqrt{5})) = \{\iota, \alpha\}$. Note that $Gal(\mathbb{Q}(\sqrt{5})/\mathbb{Q}) = \{\iota, \alpha'\}$ where $\alpha'(\sqrt{5}) = -\sqrt{5}$ and $\alpha'(b) = b$ for any $b \in \mathbb{Q}$. Therefore $Gal(\mathbb{Q}(\sqrt{5})/\mathbb{Q}) \cong Gal(\mathbb{Q}(\sqrt{5}, \sqrt{7})/\mathbb{Q})/Gal(\mathbb{Q}(\sqrt{5}, \sqrt{7})/\mathbb{Q}(\sqrt{5}))$. $\square$

Definition 26.3.1. *A Galois extension* $\mathbb{E}$ *over a field* $\mathbb{F}$ *is called* **Abelian (or cyclic)** *if the Galois group* $Gal(\mathbb{E}/\mathbb{F})$ *is Abelian (or cyclic).*

Recall that G is called a solvable group, if $G^{(n)} = \{e\}$ for some positive integer n where $G^{(n)} = [G^{(n-1)}, G^{(n-1)}], G^{(1)} = G$ and every finite Abelian group G is solvable (see Chapter 9).

Theorem 26.3.2. *Let* $\mathbb{F}$ *be a field of characteristic zero and* a *be an element of* $\mathbb{E}$. *If* $\mathbb{E}$ *is the splitting field of* $x^n - a \in \mathbb{F}[x]$, *then the Galois group* $Gal(\mathbb{E}/\mathbb{F})$ *is solvable.*

Proof. Let ω be an n-th root of unity, i.e., $\omega^n = 1$ where $\omega \neq 1$. We have two cases: Case I: $\omega \in \mathbb{F}$ and Case II: $\omega \notin \mathbb{F}$.

Case I. Assume that $\omega \in \mathbb{F}$. Then by assumptions of the theorem, all the solutions of $x^n - a$ are $\sqrt[n]{a}$, $\sqrt[n]{a}\omega$, $\cdots$, $\sqrt[n]{a}\omega^{n-1}$ and they are all in $\mathbb{E}$. So $\mathbb{E} = \mathbb{F}(\sqrt[n]{a})$. If we can prove $Gal(\mathbb{F}(\sqrt[n]{a})/\mathbb{F})$ is Abelian, then the theorem is proved. For any $\theta, \psi \in Gal(\mathbb{F}(\sqrt[n]{a})/\mathbb{F})$, we have that $\theta(\sqrt[n]{a}) = \sqrt[n]{a}\omega^i$ and $\psi(\sqrt[n]{a}) = \sqrt[n]{a}\omega^j$ for some $i, j \in \mathbb{N}$. Then we have that

$$(\theta\psi)(\sqrt[n]{a}) = \theta(\psi(\sqrt[n]{a})) = \theta(\sqrt[n]{a}\omega^j) = \sqrt[n]{a}\omega^i\omega^j = \sqrt[n]{a}\omega^{i+j}.$$

On the other hand, we also have that

$$(\psi\theta)(\sqrt[n]{a}) = \psi(\theta(\sqrt[n]{a})) = \psi(\sqrt[n]{a}\omega^i) = \sqrt[n]{a}\omega^j\omega^i = \sqrt[n]{a}\omega^{i+j}.$$

This implies that $\psi\theta = \theta\psi$, i.e., $Gal(\mathbb{F}(\sqrt[n]{a})/\mathbb{F})$ is Abelian, which proves the theorem.

Case II. Assume that $\omega \notin \mathbb{F}$. Without loss of generality, we can assume that a

is not zero. Since $\sqrt[n]{a}$ and $\sqrt[n]{a}\omega^i$, $1 \leq i \leq n-1$, are all the zeros of $x^n - a$, the splitting field $\mathbb{E}$ of $x^n - a$ is $\mathbb{F}(\sqrt[n]{a}, \omega)$. Note that $\mathbb{F}(\omega)$ is a subfield of $\mathbb{E}$ and by Case I, $Gal(\mathbb{F}(\omega)/\mathbb{F})$ is Abelian. Since $\mathbb{E}$ is a splitting field of $x^n - a$ over $\mathbb{F}(\omega)$, $Gal(\mathbb{E}/\mathbb{F}(\omega))$ is also Abelian by Case I. So we have the following normal series:

$$\{e\} \subset Gal(\mathbb{E}/\mathbb{F}(\omega)) \subset Gal(\mathbb{E}/\mathbb{F}).$$

Then since $Gal(\mathbb{F}(\omega)/\mathbb{F})$ is isomorphic to $Gal(\mathbb{E}/\mathbb{F})/Gal(\mathbb{E}/\mathbb{F}(\omega))$, $Gal(\mathbb{E}/\mathbb{F})$ is solvable by Theorem **??**. Therefore we have proven the theorem. □

Example 26.3.2. *Let ω be a 7-th zero of $x^7 = 1$ such that $\omega \neq 1$. Since $\mathbb{Q}(\omega)$ is the splitting field of $x^7 - 1$ over $\mathbb{Q}$, by Case I of Theorem 26.3.2, $\mathbb{Q}(\omega)$ is an Abelian extension over $\mathbb{Q}$*

Chapter 26 Exercises.

1. (a) Find $Gal(\mathbb{Q}(\sqrt{3}, \sqrt{7})/\mathbb{Q}(\sqrt{7}))$.

   (b) Find $Gal(\mathbb{Q}(\sqrt{3}, \sqrt{7})/\mathbb{Q})$.

   (c) Find $Gal(\mathbb{Q}(\sqrt{p}, \sqrt{q})/\mathbb{Q}(\sqrt{p}))$ where p and q are distinct primes.

   (d) Find $Gal(\mathbb{Q}(\sqrt{p}, \sqrt{q})/\mathbb{Q})$ where p and q are distinct primes.

   (e) Find the Galois group $Gal(\mathbb{Q}\sqrt[3]{p})/\mathbb{Q})$ where p is a prime number.

2. Prove that the Galois group $Gal(\mathbb{C}/\mathbb{R})$ is solvable.

3. (a) Let ω be a 5-th zero of $x^5 = 2$. Then prove that $\mathbb{Q}(\omega)$ is normal over $\mathbb{Q}$.

(b) Let ω be a p-th zero of $x^p = 2$ where p is a prime. Prove that $\mathbb{Q}(\omega)$ is normal over $\mathbb{Q}$.

4. In Definition 26.2.3, prove that $\mathbb{E}_H$ is an intermediate field of $\mathbb{E}$ and $\mathbb{F}$ with $\mathbb{E} \supseteq \mathbb{E}_H \supseteq \mathbb{F}$.

5. (a) Show that $\mathbb{E} = \mathbb{Q}(e^{2\pi i/7})$ is a Galois extension of $\mathbb{Q}$.

   (b) Find $Gal(\mathbb{E}/\mathbb{Q})$.

   (c) Show that $Gal(\mathbb{E}/\mathbb{Q})$ is Abelian.

   (d) Find $\mathbb{E}_{Gal(\mathbb{E}/\mathbb{Q})}$.

6. (a) Let $\omega \neq 1$ be an 11-th zero of $x^{11} = 1$. Prove that $\mathbb{Q}(\omega)$ is an Abelian extension over $\mathbb{Q}$.

   (b) Let ω be a p-th zero of $x^p = 1$. Prove that $\mathbb{Q}(\omega)$ is an Abelian extension over $\mathbb{Q}$ if p is a prime number.

7. Let $\mathbb{Q}(\omega)$ be an extension field of $\mathbb{Q}$ where ω is a 17-th zero of $x^{17} = 1$ such that $\omega \neq 1$. If $\mathbb{E}$ is the splitting field of $x^{17} - a$ over $\mathbb{Q}(\omega)$, then

   (a) Prove that $\mathbb{E}$ is a normal extension of $\mathbb{Q}$.

   (b) Prove that $\mathbb{E}$ is a separable extension of $\mathbb{Q}$.

   (c) Prove that $\mathbb{E}$ is a Galois extension of $\mathbb{Q}$.

   (d) Prove that $Gal(\mathbb{E}, \mathbb{Q}(\omega))$ is Abelian.

   (e) Prove that $Gal(\mathbb{E}, \mathbb{Q})$ is solvable.

8. If $\sigma_1, \cdots, \sigma_n$ are distinct automorphisms of a field $\mathbb{F}$, then prove that there do not exist non-zeros $a_1, \cdots, a_n \in \mathbb{F}$ such that $a_1\sigma_1 + \cdots + a_n\sigma_n$ is the zero map from $\mathbb{F}$ to itself.

9. If $\mathbb{E}$ is a Galois extension of a field $\mathbb{F}$, then prove that there are only finitely many fields between $\mathbb{E}$ and $\mathbb{F}$.

10. Let $\mathbb{E}$ be the splitting field in $\mathbb{C}$ of x^4+1 over $\mathbb{Q}$. Find $Gal(\mathbb{E}/\mathbb{Q})$ and show that $[\mathbb{E}:\mathbb{Q}]=4$.

11. Let $\mathbb{E}$ be the splitting field in $\mathbb{C}$ of x^3-3 over $\mathbb{Q}$. Find $Gal(\mathbb{E}/\mathbb{Q})$ and show that $[\mathbb{E}:\mathbb{Q}]=6$.

12. (a) Show that $\mathbb{Q}(\sqrt[3]{5})$ is not a Galois extension of $\mathbb{Q}$.

    (b) Find a Galois extension $\mathbb{E}$ of $\mathbb{Q}$ such that $\mathbb{E}$ is an extension field of $\mathbb{Q}(\sqrt[3]{5})$.

13. (a) Show that $\mathbb{Q}(\sqrt[3]{p})$ is not a Galois extension of $\mathbb{Q}$ where p is a prime number.

    (b) Find a Galois extension $\mathbb{E}$ of $\mathbb{Q}$ such that $\mathbb{E}$ is an extension field of $\mathbb{Q}(\sqrt[3]{p})$.

14. Let ω be a 11-th zero of $x^{11}=1$ such that $\omega \neq 1$.

    (a) Prove that $\mathbb{Q}(\omega)$ is an Abelian extension over $\mathbb{Q}$.

    (b) Prove that $\mathbb{Q}(\omega)$ is a normal extension of $\mathbb{Q}$.

    (c) For a prime p, let ω be a p-th zero of $x^p=1$ such that $\omega \neq 1$. Then $\mathbb{Q}(\omega)$ is an Abelian extension over $\mathbb{Q}$.

15. In Definition 26.2.2, prove that $Gal(\mathbb{E}/\mathbb{F})$ is a group under the function composition.

Chapter 27

Appendix: Matrices and Their Determinants

27.1 Matrices

Example 27.1.1. *Let* $A = \begin{pmatrix} 2 & 3 \\ 5 & 4 \\ -2 & 4 \end{pmatrix}$.

1. A is a 3×2 matrix with six entries.

2. The matrix A has three rows $(2 \quad 3), (5 \quad 4)$ and $(-2 \quad 4)$

3. It has two columns $\begin{pmatrix} 2 \\ 5 \\ -2 \end{pmatrix}$ and $\begin{pmatrix} 3 \\ 4 \\ 4 \end{pmatrix}$.

4. The 2nd row and 1st column entry of A is 5 and the 1st row and 2nd column entry is 3.

For a given field $\mathbb{F}$, an $m \times n$ matrix over $\mathbb{F}$ has m rows and n columns such that the entries of A are elements of $\mathbb{F}$. We only consider $\mathbb{F} = \mathbb{R}$ in this chapter.

Let $A = (a_{ij})$ and $B = (b_{ij})$ be $m \times n$ matrices. Then the addition $A + B$ and the subtraction $A - B$ of A and B are defined as follows:

$$A \pm B = (a_{ij}) \pm (b_{ij}) = (a_{ij} \pm c_{ij}).$$

Example 27.1.2. *Let* $A = \begin{pmatrix} 3 & 3 & -1 \\ 4 & 1 & 2 \end{pmatrix}$ *and* $B = \begin{pmatrix} 1 & 4 & 2 \\ 3 & 2 & 4 \end{pmatrix}$. *Find* $A + B$ *and* $A - B$.

Solution.

$$\begin{aligned} A + B &= \begin{pmatrix} 3+1 & 3+4 & -1+2 \\ 4+3 & 1+2 & 2+4 \end{pmatrix} = \begin{pmatrix} 4 & 7 & 1 \\ 7 & 3 & 6 \end{pmatrix} \\ A - B &= \begin{pmatrix} 3-1 & 3-4 & -1-2 \\ 4-3 & 1-2 & 2-4 \end{pmatrix} = \begin{pmatrix} 2 & -1 & -3 \\ 1 & -1 & -2 \end{pmatrix}. \quad \square \end{aligned}$$

Let $A = (a_{ij})$ be an $m \times n$ matrix. For a given scalar $r \in \mathbb{R}$, we define the scalar multiplication rA is defined as follows:

$$cA = r(a_{ij}) = (ra_{ij}).$$

Example 27.1.3.

$$5A = \begin{pmatrix} 3 & 3 & -1 \\ 4 & 1 & 2 \end{pmatrix} = \begin{pmatrix} 5\cdot 3 & 5\cdot 3 & 5\cdot(-1) \\ 5\cdot 4 & 5\cdot 1 & 5\cdot 2 \end{pmatrix} = \begin{pmatrix} 15 & 15 & -5 \\ 20 & 5 & 10 \end{pmatrix}.$$

Let $A = (a_{ij})$ be an $m \times p$ matrix and $B = (b_{ij})$ be a $p \times m$ matrix. Then the matrix multiplication AB is the $m \times n$ matrix $C = (c_{ij})$ such that

$$c_{ij} = \sum_{k=1}^{p} a_{ik} b_{kj},$$

i.e.,

$$AB = \begin{pmatrix} a_{11} & \cdots & a_{1p} \\ & \cdots & \\ a_{i1} & \cdots & a_{ip} \\ & \cdots & \\ a_{m1} & \cdots & a_{mp} \end{pmatrix} \begin{pmatrix} b_{11} & \cdots & b_{1j} & \cdots & b_{1n} \\ & & \cdots & & \\ b_{p1} & \cdots & b_{pj} & \cdots & b_{pn} \end{pmatrix} = (c_{ij}) = C.$$

Note that the number of columns (=p) of the 1st matrix must be equal to the number of rows of the 2nd matrix to have a valid multiplication between matrices, and this is only way to have a well defined c_{ij}. Also the resultant is an $m \times n$ matrix where m is the number of rows of the 1st matrix and n is the number of columns of the 2nd matrix.

Example 27.1.4. *Let*

$$A = \begin{pmatrix} 1 & 2 \\ 1 & 3 \end{pmatrix} \quad \textit{and} \quad B = \begin{pmatrix} 1 & 3 & -1 \\ 4 & 1 & 2 \end{pmatrix}.$$

Compute $AB = C = (c_{ij})$.

Solution. Note that C is a 2×3 matrix since A is a 2×2 matrix and B is a 2×3 matrix. Each c_{ij} is computed as following

$$
\begin{aligned}
c_{11} &= \begin{pmatrix}1 & 2\end{pmatrix}\begin{pmatrix}1\\4\end{pmatrix} = 1\cdot 1 + 2\cdot 4 = 1 + 8 = 9\\
c_{12} &= \begin{pmatrix}1 & 2\end{pmatrix}\begin{pmatrix}3\\1\end{pmatrix} = 1\cdot 3 + 2\cdot 1 = 3 + 2 = 5\\
c_{13} &= \begin{pmatrix}1 & 2\end{pmatrix}\begin{pmatrix}-1\\2\end{pmatrix} = 1\cdot (-1) + 2\cdot 2 = -1 + 4 = 3\\
c_{21} &= \begin{pmatrix}1 & 3\end{pmatrix}\begin{pmatrix}1\\4\end{pmatrix} = 1\cdot 1 + 3\cdot 4 = 1 + 12 = 13\\
c_{22} &= \begin{pmatrix}1 & 3\end{pmatrix}\begin{pmatrix}3\\1\end{pmatrix} = 1\cdot 3 + 3\cdot 1 = 3 + 3 = 6\\
c_{23} &= \begin{pmatrix}1 & 3\end{pmatrix}\begin{pmatrix}-1\\2\end{pmatrix} = 1\cdot (-1) + 3\cdot 2 = -1 + 6 = 5
\end{aligned}
$$

This implies that $C = \begin{pmatrix}9 & 5 & 3\\13 & 6 & 5\end{pmatrix}$. □

The $n \times n$ matrix I_n is called **the identity matrix** if $a_{ii} = 1$ and $a_{ij} = 0$ for $i \neq j$ when $I_n = (a_{ij})$.

Example 27.1.5. $I_2 = \begin{pmatrix}1 & 0\\0 & 1\end{pmatrix}$ *and* $I_3 = \begin{pmatrix}1 & 0 & 0\\0 & 1 & 0\\0 & 0 & 1\end{pmatrix}$.

The identity matrix can be defined as $I_n = (\delta_{ij})$. This δ_{ij} is a function such that

$$\delta_{ij} = \begin{cases} 1, & i = j \\ 0, & i \neq j \end{cases}$$

and it is called **the Kronecker delta**.

Example 27.1.6. *Show that, $AI_n = I_nA = A$ for any $n \times n$ matrix A*

Solution. Let $A = (a_{ij})$ and $\begin{pmatrix} 0 \\ \vdots \\ 0 \\ 1 \\ 0 \\ \vdots \\ 0 \end{pmatrix}$ be a column matrix with 1 on the jth column and 0 elsewhere. Then

$$\begin{pmatrix} a_{i1} & \cdots & a_{ij} & \cdots & a_{in} \end{pmatrix} \begin{pmatrix} 0 \\ \vdots \\ 0 \\ 1 \\ 0 \\ \vdots \\ 0 \end{pmatrix} = a_{ij}$$

and this implies that $AI_n = A$. The $I_nA = A$ case is similar. □

The identity matrix is a special matrix and there are other special matrices.

Definition 27.1.1. *Let $A = (a_{ij})$ be an $m \times n$ matrix.*

1. *A is called* **a zero matrix** *is* $a_{ij} = 0$ *for all* i *and* j.

2. *A is called* **a diagonal matrix** *if* $m = n$ *and* $a_{ij} = 0$ *for* $i \neq j$. *Then A is denoted by* $diag\{a_{11}, \cdots a_{nn}\}$.

3. *A is called* **an upper triangular matrix** *if* $a_{ij} = 0$ *for* $i > j$.

4. *A is called* **a strictly upper triangular matrix** *if* $a_{ij} = 0$ *for* $i \geq j$.

5. *A is called* **a lower triangular matrix** *if* $a_{ij} = 0$ *for* $i > j$.

6. *A is called* **a strictly lower triangular matrix** *if* $a_{ij} = 0$ *for* $i \leq j$.

The following is an example of a strictly upper triangular matrix and it is useful in Linear Algebra.

Example 27.1.7. *Let* $N = \begin{pmatrix} 0 & 1 & 0 \\ 0 & 0 & 1 \\ 0 & 0 & 0 \end{pmatrix}$. *Then* $N^2 = \begin{pmatrix} 0 & 0 & 1 \\ 0 & 0 & 0 \\ 0 & 0 & 0 \end{pmatrix}$ *and* N^3 *is a zero matrix.*

We called this type of matrices **nilpotent**. For example, $\begin{pmatrix} 1 & 3 \\ 6 & 2 \end{pmatrix}$ is not nilpotent.

27.2 Determinants

The ij**-minor** of an $n \times n$ matrix $A = (a_{ij})$ is an $(n-1) \times (n-1)$ matrix such that the ith row and jth column of A are deleted. It is denoted by C_{ij}

Example 27.2.1. *The minor* C_{23} *of* $\begin{pmatrix} 1 & 3 & 3 \\ 2 & 4 & -3 \\ 6 & 2 & 5 \end{pmatrix}$ *is* $\begin{pmatrix} 1 & 3 \\ 6 & 2 \end{pmatrix}$.

For an $n \times n$ matrix $A = (a_{ij})$, **the determinant of** A is denoted by $|A|$ or $det(A)$, and defined by induction on n as follows:

$$det(A) = \sum_{j=1}^{n} (-1)^{i+j} a_{ij} |C_{ij}|$$

where C_{ij} is the ij-minor of A. This definition is called **the cofactor expansion** using the jth column of A. Similarly, we define the cofactor expansion using the ith row of A as

$$det(A) = \sum_{i=1}^{n} (-1)^{i+j} a_{ij} |C_{ij}|$$

Example 27.2.2. *Let* $A = \begin{pmatrix} 1 & 3 \\ 2 & 5 \end{pmatrix}$. *Then*

$$\begin{aligned} |A| &= (-1)^{1+1} \cdot 1 \cdot C_{11} + (-1)^{1+1} \cdot 3 \cdot C_{12} \\ &= 1 \cdot 5 - 3 \cdot 2 = 5 - 6 = -1 \end{aligned}$$

The following is useful properties of determinants.

Lemma 27.2.1. *Recall that* $M_n(\mathbb{R})$ *is the set of all* $n \times n$ *matrices with entries in* $\mathbb{R}$. *Let* $A, B \in M_n(\mathbb{R})$. *Then*

1. $|AB| = |A||B|$

2. *The determinants of a diagonal matrix, an upper triangular matrix, or a lower triangular matrix are the product of the diagonal entries for each matrix. For example* $|diag\{a_{11}, \cdots, a_{nn}\}| = a_{11} \cdots a_{nn}$

3. *The determinant of a strictly upper or lower triangular matrix is* 0.

4. $|I_n| = 1$

5. $|A^{-1}| = \dfrac{1}{|A|}$ *if A is invertible.*

6. $|cA| = c^n|A|$ *for* $c \in \mathbb{R}$

7. $|A| = 0$ *if A has a zero row or column.*

8. $|A| = 0$ *if A has a row (or a column) that is linearly dependant to other rows (or columns).*

9. *The determinant of a strictly upper or lower triangular matrix is* 0.

Note that $|A + B| \neq |A| + |B|$ in general. So we say that the determinant is not additive but multiplicative as a function.

Chapter 27 Exercises.

1. Let $A = \begin{pmatrix} 1 & 2 & -1 \\ 3 & 3 & -1 \\ 4 & 1 & 2 \end{pmatrix}$ and $B = \begin{pmatrix} -2 & 2 & 3 \\ 1 & 4 & 2 \\ 3 & 2 & 4 \end{pmatrix}$. Compute $A + B$, $A - B$, and $-2B$. Also confirm that $A + A = 2A$.

2. Compute AB where $A = \begin{pmatrix} 1 & 1 & 2 \\ 2 & 1 & 3 \end{pmatrix}$ and $B = \begin{pmatrix} 2 & 1 & -1 \\ 1 & 2 & -1 \\ 4 & 1 & 2 \end{pmatrix}$

3. Find the 33-minor C_{33} of $\begin{pmatrix} 2 & 1 & 3 \\ -1 & 4 & 3 \\ 4 & 2 & 3 \end{pmatrix}$.

4. Compute the following determinants

(a) $\begin{vmatrix} 1 & 3 \\ 1 & 2 \end{vmatrix}$

(b) $\begin{vmatrix} 2 & 1 & 3 \\ 1 & 2 & -1 \\ 2 & 1 & -2 \end{vmatrix}$

(c) $\begin{vmatrix} 2 & 1 & 3 & 1 \\ 1 & 2 & -1 & 3 \\ 2 & 1 & -2 & 0 \\ 0 & 1 & 1 & 2 \end{vmatrix}$

5. Let $A = \begin{pmatrix} 2 & 1 & 3 \\ 1 & 2 & -1 \\ 2 & 1 & -2 \end{pmatrix}$ and $B = \begin{pmatrix} 1 & 2 & -1 \\ 2 & 1 & 1 \\ 0 & 2 & 1 \end{pmatrix}$.

   (a) Compute AB.

   (b) Compute $|A|$ and $|B|$.

   (c) Compute $|AB|$ to confirm that $|AB| = |A||B|$.

6. Prove that $|A^{-1}| = \dfrac{1}{|A|}$ if A is invertible.

7. Prove that $|A| = 0$ if A has a zero row or column.

8. Prove that $|cA| = c^n|A|$ for $c \in \mathbb{R}$ where A is an $n \times n$ matrix.

Bibliography

[1] *Contemporary Abstract Algebra*, Joseph A. Gallian, Houghton Mifflin, 2006.

[2] *Notes on Sylow Theorems*, Denisa Goia, 8th MAA Wisc. Section, Marshfield, April, 5-6, 2013.

[3] *Topics in Algebra*, I. Herstein, 2nd Ed., 1975.

[4] *ABSTRACT ALGEBRA An Introduction*, Thomas. W. Hungerford, Brooks/Cole, 1977.

[5] *Linear ALGEBRA*, Ki-Bong Nam and Hong Goo Park, Kyungmoon-Sa, 2011.

[6] *Introductory Abstract Algebra*, Ki-Bong Nam, Xueqing Chen, Moon-Ok Wang, and Ki-Suk Lee, 1st, 2nd, 3rd editions, Kyung Moon Sa, 2011.

Index

■ Woo Jeon
University of Wisconsin-Green Bay
■ Xueqing Chen
University of Wisconsin-Whitewater
■ Moon-Ok Wang
Hanyang University, Korea
■ Ki-Suk Lee
Korea National University of Education
■ Ki-Bong Nam
University of Wisconsin-Whitewater

5th Edition

Introductory Abstract Algebra

ISBN 978-89-6105-937-4

KM KYUNG MOON SA
174, Wausan-ro, Mapo-gu, Seoul, Korea
Tel: (82) 2 332 2004
Fax: (82) 2 336 5193

Printed in Korea